Climate Change and Risk Management – Strategies, Analysis, and Adaptation

Edited by Antonio Di Pietro, José R. Martí and Vinay Kumar

Published in London, United Kingdom

Climate Change and Risk Management - Strategies, Analysis, and Adaptation
http://dx.doi.org/10.5772/intechopen.111028
Edited by Antonio Di Pietro, José R. Martí and Vinay Kumar

Contributors
Abbasali Dehghani Tafti, Adriano Camps, Alexander Buber, Amalia Kouskoura, Asghar Tavan, Badr-Eddine Boudriki Semlali, Carlos Molina, Carolina Velásquez, Consuelo Agnesi, Edgard Gonzales, Eleni Kalliontzi, Evangelos Katsaros, Gabriella Duca, Hyuk Park, Ioannis Bakouros, Jiayao Wang, Johanes A. Belle, José Luís Zêzere, Kam Tim Tse, Kasturi Singh, Kenny Gonzales, Ketevan Getiashvili, Mahmood Nekoie-Moghadam, Marcus Oxley, Mikhail Bolgov, Mireia Carvajal Librado, Monica Crişan, Nelson Mileu, Philippe Quevauviller, Samin Ansari Mahabadi, Sofia Karma, Stefano Zanut, Yu Chang

First published in London, United Kingdom, 2024 by IntechOpen
IntechOpen is the global imprint of INTECHOPEN LIMITED, registered in England and Wales, registration number: 11086078, 167-169 Great Portland Street, London, W1W 5PF, United Kingdom

British Library Cataloguing-in-Publication Data
A catalogue record for this book is available from the British Library

Additional hard and PDF copies can be obtained from orders@intechopen.com

Climate Change and Risk Management - Strategies, Analysis, and Adaptation
Edited by Antonio Di Pietro, José R. Martí and Vinay Kumar
p. cm.

This title is part of the Sustainable Development Book Series, Volume 23
Topic: Climate Change and Environmental Sustainability
Series Editor: Usha Iyer-Raniga
Topic Editor: Yixin Zhang

Print ISBN 978-0-85466-560-0
Online ISBN 978-0-85466-559-4
eBook (PDF) ISBN 978-0-85466-561-7
ISSN 2753-6580

For EU product safety concerns:
IN TECH d.o.o., Prolaz Marije Krucifikse Kozulić 3, 51000 Rijeka, Croatia,
info@intechopen.com or visit our website at intechopen.com.

IntechOpen Book Series

Sustainable Development

Volume 23

Aims and Scope of the Series

Transforming our World: the 2030 Agenda for Sustainable Development endorsed by United Nations and 193 Member States, came into effect on Jan 1, 2016, to guide decision making and actions to the year 2030 and beyond. Central to this Agenda are 17 Goals, 169 associated targets and over 230 indicators that are reviewed annually. The vision envisaged in the implementation of the SDGs is centered on the five Ps: People, Planet, Prosperity, Peace and Partnership. This call for renewed focused efforts ensure we have a safe and healthy planet for current and future generations.

This Series focuses on covering research and applied research involving the five Ps through the following topics:

1. Sustainable Economy and Fair Society that relates to SDG 1 on No Poverty, SDG 2 on Zero Hunger, SDG 8 on Decent Work and Economic Growth, SDG 10 on Reduced Inequalities, SDG 12 on Responsible Consumption and Production, and SDG 17 Partnership for the Goals

2. Health and Wellbeing focusing on SDG 3 on Good Health and Wellbeing and SDG 6 on Clean Water and Sanitation

3. Inclusivity and Social Equality involving SDG 4 on Quality Education, SDG 5 on Gender Equality, and SDG 16 on Peace, Justice and Strong Institutions

4. Climate Change and Environmental Sustainability comprising SDG 13 on Climate Action, SDG 14 on Life Below Water, and SDG 15 on Life on Land

5. Urban Planning and Environmental Management embracing SDG 7 on Affordable Clean Energy, SDG 9 on Industry, Innovation and Infrastructure, and SDG 11 on Sustainable Cities and Communities.

The series also seeks to support the use of cross cutting SDGs, as many of the goals listed above, targets and indicators are all interconnected to impact our lives and the decisions we make on a daily basis, making them impossible to tie to a single topic.

Meet the Series Editor

Usha Iyer-Raniga is a professor in the School of Property and Construction Management at RMIT University. Usha co-leads the One Planet Network's Sustainable Buildings and Construction Programme (SBC), a United Nations 10 Year Framework of Programmes on Sustainable Consumption and Production (UN 10FYP SCP) aligned with Sustainable Development Goal 12. The work also directly impacts SDG 11 on Sustainable Cities and Communities. She completed her undergraduate degree as an architect before obtaining her Masters degree from Canada and her Doctorate in Australia. Usha has been a keynote speaker as well as an invited speaker at national and international conferences, seminars and workshops. Her teaching experience includes teaching in Asian countries. She has advised Austrade, APEC, national, state and local governments. She serves as a reviewer and a member of the scientific committee for national and international refereed journals and refereed conferences. She is on the editorial board for refereed journals and has worked on Special Issues. Usha has served and continues to serve on the Boards of several not-for-profit organisations and she has also served as panel judge for a number of awards including the Premiers Sustainability Award in Victoria and the International Green Gown Awards. Usha has published over 100 publications, including research and consulting reports. Her publications cover a wide range of scientific and technical research publications that include edited books, book chapters, refereed journals, refereed conference papers and reports for local, state and federal government clients. She has also produced podcasts for various organisations and participated in media interviews. She has received state, national and international funding worth over USD $25 million. Usha has been awarded the Quarterly Franklin Membership by London Journals Press (UK). Her biography has been included in the Marquis Who's Who in the World® 2018, 2016 (33rd Edition), along with approximately 55,000 of the most accomplished men and women from around the world, including luminaries as U.N. Secretary-General Ban Ki-moon. In 2017, Usha was awarded the Marquis Who's Who Lifetime Achiever Award.

Meet the Volume Editors

Antonio Di Pietro earned his master's degree in Informatics Engineering from Sapienza University of Rome, Italy, in 2004 and his Ph.D. in the same field from Roma Tre University, Italy, in 2015. Since 2007, he has been working as a researcher at the Italian National Agency for New Technologies, Energy and Sustainable Economic Development (ENEA). His current research focuses on the application of Unmanned Aircraft Systems (UAS), as well as the modeling and simulation of critical infrastructures. Furthermore, he is involved in developing decision support systems that integrate natural hazard modeling to enhance infrastructure resilience. He has participated in numerous European and Italian national research projects and has served as an advisor for several evaluation studies commissioned by the European Union, specifically in the field of Critical Infrastructure Protection. He is the Operations Manager of ENEA UAS operations for the monitoring of critical infrastructures. In addition to his research, Dr. Di Pietro has advised numerous MSc students and has taught courses in software engineering, programming languages, databases, and distributed applications. He is the author of more than 50 publications.

José R. Martí has a Ph.D. in Electrical Engineering from the University of British Columbia (UBC), Vancouver, Canada. He is a professor at this institution, where he has made contributions to real-time solutions of large power networks and the modeling of interdependent critical infrastructures for resiliency and real-time response. Contributions include models for traveling wave propagation of electromagnetic waves, hydraulic waves, seismic waves, and wildfire spread. He is the lead investigator of the Complex Systems Integration (CSI) Laboratory at the Institute for Computing, Information and Cognitive Systems (ICICS) at UBC. Current projects include considering the effect of climate change in risk evaluation, real-time earthquake early warning preparation and response, and real-time wildfire spread and suppression.

Vinay Kumar has a Ph.D. in Meteorology/Physics from the University of Pune, India. He is Assistant Professor of Meteorology at the University of the Incarnate Word, San Antonio, TX, USA. His responsibilities include teaching undergraduate students, weather-climate research, and university services. Since obtaining his Ph.D., Dr. Kumar has been actively involved in weather and climate modeling. In the past three years he has taught courses on meteorology, climatology, geology, and astronomy. He was an associate editor of the journal of the American Meteorological Society and a five-time guest editor of MDPI journals. His areas of research and teaching include extreme events (e.g., floods), natural hazards (e.g., fog), monsoons, and hurricanes.

Contents

Preface

Climate change represents one of the most significant challenges of our time, impacting natural systems, human health, economies, and societies worldwide. The increasing frequency and intensity of extreme weather events, such as hurricanes, floods, and heatwaves, underscore the urgent need for comprehensive risk management strategies and adaptive measures. This edited volume, *Climate Change and Risk Management – Strategies, Analysis, and Adaptation*, brings together a diverse collection of research that addresses the multifaceted nature of climate change and proposes actionable solutions.

Section 1: Foundations and Risk Management Strategies

The book begins with Chapter 1, "Adapting to Climate Change through Risk Management." This chapter outlines key principles of risk management, including exposure, sensitivity, adaptive capacity, and vulnerability, providing a conceptual framework for understanding and mitigating the impacts of climate change. Following this, Chapter 2, "Perspective Chapter: Advancements in Disaster Risk Mitigation Strategies," delves into recent advancements in disaster risk mitigation, emphasizing macro regional development and policy implementation. Chapter 3, "Risk Management in Mass Gatherings," addresses the unique challenges of managing risks during mass gatherings, focusing on health hazards and risk assessment.

Section 2: Climate Data Analysis and Technological Tools

The second section presents advanced techniques and tools for analyzing climate data and assessing risks. Chapter 4, "Analysis of Climate Indices to Determine Global Climate Patterns: Techniques for Summarizing Complex Climate Data," introduces methods for summarizing complex climate data to identify global patterns, which is crucial in understanding phenomena such as the 2017 Atlantic hurricane season. Chapter 5, "Synthesis of Tropical Cyclones: Understanding, Modeling, and Adapting to Climate Change Impacts," focuses on modeling tropical cyclones and assessing their impacts in the context of climate change, including the increasing intensity of storms like Hurricane Maria. Chapter 6, "Current and Future Multirisk Analysis in Climate Change Scenarios with Riskcoast WebGIS," explores the use of WebGIS technology in multi-risk analysis for coastal areas, demonstrating its application in regions affected by rising sea levels. Chapter 7, "Potential Earthquake Proxies from Remote Sensing Data," examines remote sensing techniques for identifying earthquake proxies and assessing risks, showcasing how these tools can predict and mitigate the effects of climate-induced seismic activity.

Section 3: Case Studies and Scenario Analysis

The third section focused on real-world applications and case studies. Chapter 8, "Forecast of Catastrophic Floods Based on Hydrodynamic Modeling: The 2013 Flood on the Amur River Case Study," provides an in-depth analysis of catastrophic flood forecasting using hydrodynamic models, with a specific case study of the 2013 flood on the Amur River, highlighting the importance of accurate modeling in disaster preparedness. Chapter 9, "Perspective Chapter: The Coastal Migration of the Locations of Tropical Cyclone Rapid Intensification over the North Indian Ocean," investigates the patterns of tropical cyclone rapid intensification over the North Indian Ocean, underscoring the implications of a warming climate on these events and their devastating impacts on coastal communities.

Section 4: Vulnerability, Resilience, and Policy Engagement

The final section addresses the human and policy dimensions of climate change. Chapter 10, "People Vulnerability before, during and after a Disaster: A Dynamic Taxonomic Approach," explores the dynamic nature of human vulnerability and resilience before, during, and after disasters, using examples from recent events like the 2020 Australian bushfires. Chapter 11, "Science-Policy Interfacing and Community-Building for Disaster Risk Reduction in the Area of Natural Hazards," emphasizes the importance of bridging the gap between science and policy for effective disaster risk reduction, illustrating successful community-building efforts in earthquake-prone regions. Chapter 12, "Understanding and Integrating Systemic Risk (SR) into Disaster Risk Reduction (DRR) and Risk Informed Development (RID) for Long-Term Resilient Development," provides insights into integrating systemic risk into disaster risk reduction and sustainable development strategies, crucial for building long-term resilience against climate change impacts.

This volume serves as a valuable resource for researchers, practitioners, and policymakers engaged in the critical task of managing and adapting to the risks posed by climate change. By presenting both theoretical frameworks and practical case studies, it aims to enhance our collective understanding and capability to address one of the most pressing issues of our time.

Antonio Di Pietro,
Energy Technologies and Renewable Sources Department,
ENEA Casaccia Research Centre,
Rome, Italy

Josè R. Martì,
Department of Electrical and Computer Engineering,
The University of British Columbia,
Vancouver, Canada

Vinay Kumar
Assistant Professor of Meteorology,
Department of Atmospheric Science, Environment Science, and Physics,
University of the Incarnate Word,
San Antonio, USA

Section 1

Foundations and Risk Management Strategies

Chapter 1

Adapting to Climate Change through Risk Management

Samin Ansari Mahabadi

Abstract

Climate change, along with changes in hydrological variables, causes alterations in access to water resources, the intensification of extreme phenomena (such as droughts and floods), and economic, social, and environmental instability. Risk management emerges as an appropriate approach for increasing adaptation to climate change, characterized by its inherent flexibility and the reduction of uncertainties associated with climate change. This approach improves adaptive capacity through transformation and reversibility processes, ultimately reducing the system's exposure and vulnerability to risks. In this chapter, we delve into key concepts and components related to risk and adaptation, including resilience, exposure, sensitivity, adaptive capacity, vulnerability, and their connections and interactions. Subsequently, we elucidate the methodology for enhancing climate change adaptation through risk management, utilizing a variety of processes and tools. Furthermore, we provide an illustrative example of the application of the portfolio robust decision-making tool for climate change risk management in the integrated water resources system.

Keywords: risk management, exposure, sensitivity, adaptive capacity, vulnerability, adaptation to climate change

1. Introduction

The ongoing phenomenon of climate change, characterized by long-term alterations in average climatic conditions and fluctuations, has brought about significant transformations in the components of the hydrological cycle. This includes changes in precipitation patterns, snow melting, evaporation, and soil moisture dynamics, leading to a cascade of consequences, including heightened risks associated with extreme events, such as droughts and floods, and uncertainties in accessing water sources [1]. Therefore, one of the paramount challenges facing the twenty-first century is the imperative to mitigate the risks associated with climate change and maintain the sustainability of systems under the effects of this phenomenon. This challenge is particularly pronounced in sectors reliant on water resources, where the impact of climate change poses a significant threat.

In response to the inescapable impacts of climate change on susceptible systems and their sustainability, the imperative of addressing adaptation to climate change is raised. The concept of adaptation, originally applied to the study of biological systems in the 1970s [2], has since transcended disciplinary boundaries, finding applications in diverse scientific fields, from ecology to social sciences, psychology, and economics. Although the term is not novel, it has gained prominence in recent years as a primary response to the impacts of climate change.

Various definitions of climate change adaptation have been proposed, with the Intergovernmental Panel on Climate Change (IPCC) characterizing it as the adjustment of natural or human systems in response to occurred or expected stimuli or their effects. Adaptation aims to minimize damages and vulnerability or seize opportunities [3, 4]. The United Nations Framework Convention on Climate Change (UNFCCC) defines adaptation as practical measures to shield countries and societies from potential disruptions and damages caused by climate change effects. Furthermore, adaptation is viewed as a dynamic process during which strategies for adjustment, coping, and utilization of the consequences of climate events are reinforced, developed, and implemented [5]. In essence, adaptation represents the process or outcome that reduces harm or the risk of harm, while achieving benefits associated with climate variability and change [6].

The IPCC has delineated various forms of adaptation into three distinct categories: self-adaptation, independent adaptation, and planned adaptation. Self-adaptation refers to measures taken before the tangible effects of climate change become apparent. An example of self-adaptation is social learning in society to adapt to possible risks, such as flooding [7]. On the other hand, independent and planned adaptation strategies come into play after the observation of climate change effects. The disparity between independent and planned adaptation lies in the approach to responding to climate stimuli. Independent adaptation relies on individual characteristics, unfettered by public organizational intervention. It draws from people's perceptions, scientific information, and experiential knowledge, allowing for swift implementation [8]. In contrast, planned adaptation involves the formulation and implementation of policies that are adopted based on awareness and assessment of changing conditions, and the planning and execution of these measures (such as process, operational, and institutional measures) may necessitate a substantial amount of time [9]. It is noteworthy that the risk of climate change and the vulnerability of systems that lack adaptation measures are significantly higher compared to systems where adaptation is planned.

Numerous sources of uncertainty associated with climate change and its adaptation, encompassing factors such as the accuracy of climate change predictions, the impact of climate change on systems, the nature of adaptation strategies, the level of stress experienced by the system, and the efficacy of adaptation measures, underscore the appropriateness of adopting a risk management approach for climate change adaptation. In this regard, the significant impact of climate change on the water sector and challenges arising from climate-induced uncertainties in water supply and demand, along with their consequential economic, social, and environmental effects, have amplified the ineffectiveness of the traditional management approach (command and control). This issue has led to a growing inclination toward risk management for climate change adaptation within this sector.

This shift prompts crucial questions: Within the risk management approach, adaptation can be formed under which processes, and what mechanisms are employed at each process to reduce the effects and risks associated with climate change? The elucidation and design of these processes pave the way for effective risk management implementation, within diverse systems. Further questions extend to the factors influencing the reduction of vulnerability of an adaptive system to external threats like climate change. In other words, how is the connection and the impact pathway of the adaptive capacity on vulnerability? Considering the concepts of resilience and vulnerability, how is it possible to increase adaptive capacity and reduce risk through adaptation measures? This understanding not only clarifies the interplay of these concepts but also serves as a guiding framework for bolstering adaptive capacity within systems.

Additionally, the selection of adaptation measures prompts an essential question: what are the characteristics that define an appropriate method and model for evaluating the effects of climate change and selecting suitable adaptation strategies to reduce risk? These considerations are paramount in navigating the complex landscape of climate change impacts and ensuring the judicious choice and implementation of adaptation measures.

To address the above questions, various review studies in the fields of water resources management under climate change, adaptation concepts and processes, and assessment of adaptation strategies were reviewed [10–19]. While these studies are categorized as review papers, their scope has been confined to specific dimensions of adaptation, neglecting the crucial exploration of interrelationships among various dimensions of adaptation and risk. Therefore, they are unable to answer the questions raised in this field.

This chapter addresses this gap by delving into diverse dimensions of adaptation. It scrutinizes the risk management approach, the adaptation processes, and the intricate connections between adaptation, vulnerability, and resilience concepts for adaptation, especially in the water sector. Through an examination of conceptual advancements, along with an assessment of the pros and cons of different approaches and processes, the chapter presents a combination of the authors' findings to create an adaptation system to climate change. Therefore, it provides valuable knowledge and insights, enhancing our comprehension of adaptation and delineating its various facets. Consequently, it offers a model for augmenting adaptation to climate change through a risk management framework.

2. Concepts related to adaptation

2.1 Impact of vulnerability components changes on adaptation and risk

Effectively responding to the challenges posed by climate change necessitates a strategic understanding of the relationship between adaptation and vulnerability. The initial step in adapting to the conditions created by climate change involves identifying vulnerable points. Subsequently, solutions are adopted to enhance the system's capacity to mitigate these vulnerabilities. Therefore, vulnerability analysis is necessary to achieve adaptation.

Various definitions of vulnerability exist, one of which characterizes it as the extent of a natural system's or human society's inability to cope with the adverse effects of climate change, variability, and extreme events. This depends on the changes in climate, sensitivity, and adaptive capacity within the system or society [20]. Similarly, vulnerability is described as a condition influenced by physical, social, economic, and environmental factors or processes, intensifying society's sensitivity to hazards [21]. Vulnerability also pertains to the amount of damage caused by specific dangerous events. Climate vulnerability, therefore, signifies the degree of sensitivity or inability of a system to withstand the adverse effects of climate change, encompassing fluctuations and extreme events that depend not only on the sensitivity of a system but also on its adaptive capacity [22]. According to the IPCC definition [23], vulnerability is a function of three factors: exposure, sensitivity, and adaptive capacity. In the context of climate change, "level of exposure " is defined as the intensity of changes in climate

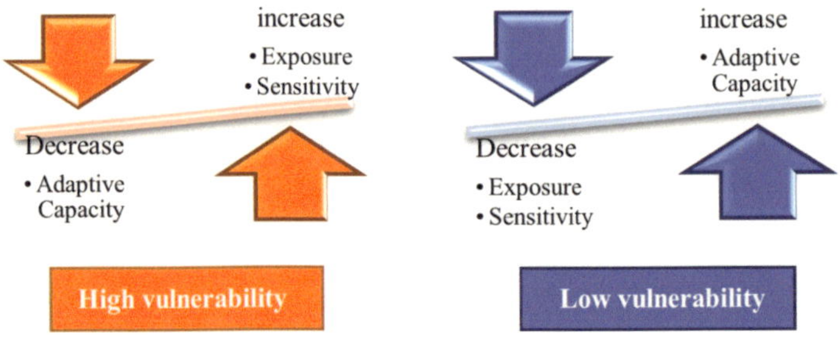

Figure 1.
Relationship between vulnerability components and their effects.

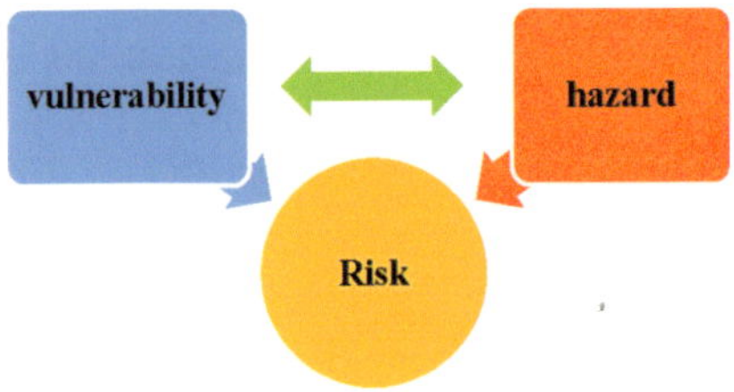

Figure 2.
The connection between vulnerability, hazard, and risk.

variables, while "sensitivity" denotes the extent of the effects of climate change on the system. "Adaptive capacity" reflects the system's ability to reduce potential damages and use opportunities. Consequently, increasing exposure and sensitivity components heightens vulnerability, whereas an increase in the system's adaptive capacity reduces vulnerability. **Figure 1** illustrates the relationship between vulnerability components and the impact of adaptive capacity on vulnerability.

Risk emerges from the interaction between hazards and vulnerability. Consequently, alterations in vulnerability components impact the level of risk. By mitigating vulnerability through adjustments in its constituent elements, such as enhancing adaptive capacity and reducing exposure, the risk is concurrently diminished. **Figure 2** illustrates the connection between vulnerability, hazard, and risk.

2.2 The relationship between adaptive capacity and vulnerability and resilience

The concept of vulnerability exhibits deficiencies in several key dimensions, with shortcomings including a disproportionate emphasis on evaluating system issues over its core functionality, more attention to the state of the system than its active processes and mechanisms, an oversight in neglecting to analyze the system's mission within the drawn landscape, and a disproportionate focus on risks or damaging components.

Simultaneously, prioritizing the system's survival takes precedence over scrutinizing its separate issues. Consequently, to assess the system's stability in fulfilling its expected function, introducing an additional concept becomes imperative, and this can be accomplished by incorporating the notion of "resilience." It is important to highlight that potential connections and related concepts exist between the frameworks of vulnerability and resilience, as elucidated by [24], and these can be regarded as mutually complementary.

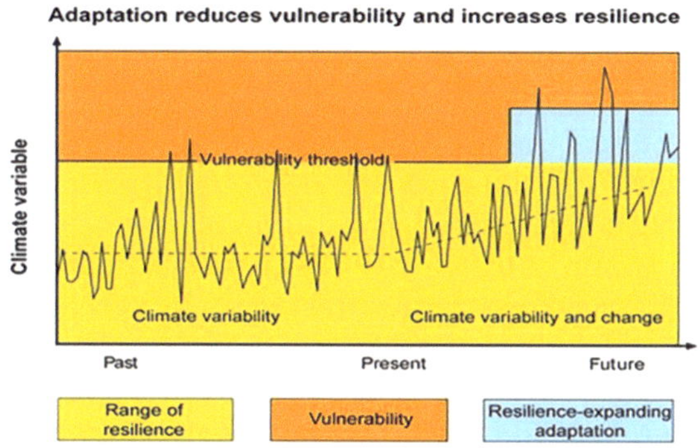

Figure 3.
Conceptual relationship between climate change, vulnerability, and adaptation (in the sense of resilience). Reference: Adapted from UKCIP [22].

The shift in perspective from the vulnerability-oriented focus on system factors to the resilience-oriented emphasis on processes and system relationships is evident in the literature on adaptation [25]. Until 2008, adaptation measures in the continents of Asia, Africa, and Europe predominantly centered on vulnerability, addressing processes such as ensuring water availability for pastures or enhancing the efficiency of irrigation systems. However, in line with the 2014 recommendations from the IPCC, there has been a discernible transition toward greater consideration of system functions in adaptation measures. This is reflected in initiatives like local development planning and the integrated management of water and soil resources [26].

The interrelationship and conceptual framework involving "vulnerability," "resilience," and "adaptive capacity" within the context of climate change impacts are elucidated in **Figure 3**. This visual representation illustrates the historical trajectory of the system, encountering climate risks within the bounds of natural fluctuations (climate changes). During these fluctuations, the system operates within the "resilience" range, sustaining stability under constant climatic conditions. However, the advent of climate change imposes new trends on the system, which is beyond the limit of its resilience and puts the system in a vulnerable area.

In reaction to these alterations, Adaptation measures strengthen the adaptive capacity of the system through two paths: modifying the flexibility range and reducing vulnerability. Ultimately, combining these pathways delineates a new boundary (blue part) that protects the system against emerging climate fluctuations—a boundary known as anti-climate.

Section 3 delves into a comprehensive explanation of the execution of these two pathways, elucidating the enhancement of adaptive capacity under the heading of adaptation processes.

3. Adaptation process

In the majority of research, the term "adaptation" has historically been construed as the idea of reverting to the past [27]. Within this conceptualization, adaptation represents a system's capacity to return to its initial state and a stable point of equilibrium following the disturbance. This static form of adaptation is often referred to as reversibility adaptation or incremental adaptation in certain studies, wherein the system's structure, performance, and feedback are maintained to reduce vulnerability [28].

Over time, this definition has evolved, moving away from a static interpretation to embracing a dynamic concept [29]. In the revised definition, adaptation does not aim to achieve the system's initial equilibrium but emphasizes sustaining the system, absorbing changes, and adapting to disturbance conditions through gradual transformation, ultimately reaching a new equilibrium. In other words, the adaptive capacity of the system in dealing with changes and disturbances increases by transferring the characteristics of the system from one state to another. This perspective, known as transformation adaptation, allows for the attainment of a new equilibrium point by altering the structure and purpose of the system, a concept aligned with the modification of the resilience range (Section 2). **Figure 4** illustrates a schematic representation of the system's initial equilibrium state and the newly created equilibrium state.

The viewpoint of transformation adaptation distinguishes between the two concepts of sustainability and stability; instead, it introduces criteria such as adaptability, flexibility, and dynamic instead of control, resistance, and static. In transformation adaptation, the concepts of diversification and modularization (transformation into smaller elements with independent functions) along with decentralization are proposed [30]. In this adaptive methodology, the trait of diversification enables the system to manifest different responses amid conditions of pressure and disturbance. An exemplification of this adaptive strategy is observed in the cultivation of crops with varying sensitivities to climate changes. Moreover, the features of modularization and decentralization, particularly when elements share similar functions, can reduce the risk temporally and spatially, preventing the entire system from being under pressure simultaneously and in the same location. An illustrative example is the cultivation of a desired crop in diverse locations to distribute and minimize potential risks. **Table 1** provides a comparative analysis of two adaptation processes.

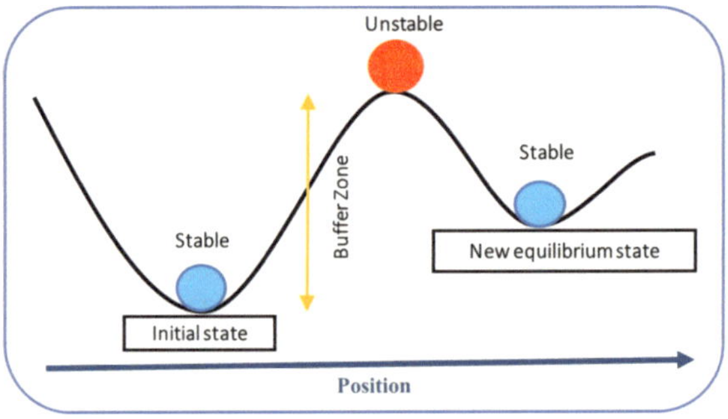

Figure 4.
System equilibrium in the initial state and the new state.

	Reversibility adaptation	Transformation adaptation
Objectives	Maintain the purpose and structure of the system	Change the purpose and structure of the system
	Return to the initial equilibrium state	Create a new equilibrium state
Attributes	Static equilibrium adaptation	Dynamic equilibrium adaptation
	Dynamic-resilience-adaptation flexibility	Static-fortification control-resistance

Table 1.
Objectives and characteristics of the two adaptation approaches.

3.1 Adaptation process in water resources

Recognizing the significance of adapting to climate change within the integrated water resources system, this section delves into two distinct paths of the adaptation process aimed at restoring balance to the water resources system. As depicted in **Figure 5**, factors such as climate change and excessive consumption contribute to a decline in available water. Given the pivotal role of water as an interface between various subsystems, this decline precipitates adverse effects on the economic, social, and environmental performance of the system, ultimately leading to its degradation.

In this situation, adaptation can be pursued through two distinct pathways. In the reversibility process, actions are undertaken to revert to the original state and the initial equilibrium point. Depending on the system's capacity to recover, these actions may either fully restore the system to its initial state (restoration), or partially reinstate the conditions preceding the degradation (rehabilitation), thereby enhancing system conditions [31, 32]. Conversely, in the transformation adaptation process, structural changes increase the system's resilience by creating a new equilibrium point. The adaptive method of transformation in water resource management with a protective approach can be applied by reallocating water resources and creating new equilibrium conditions for the system. So, by changing the structure of the system through diversifying economic activities and allocating water resources to them, the system's sensitivity and adaptive capacity will decrease and increase, respectively. Samples of various implemented measures through both reversibility and transformation processes at different temporal and spatial scales in countries are shown in **Table 2**. In addition, the comprehensive classification of these adaptation processes within the system is illustrated in **Figure 6**.

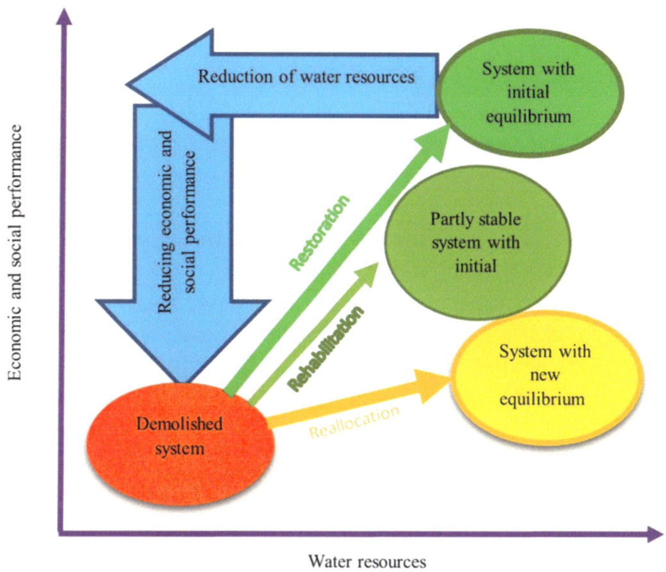

Figure 5.
The effectiveness of different approaches to system reconstruction and restoration. Source: modified Gurr et al. [31] for water resources.

Reversibility approach	Source	Transformational approach	Source
Water management and agricultural water-saving technologies in China and India	[33–36]	Changing the cultivation pattern (changing rice r to rice and cotton in China), (changing rice cultivation to apple in India), (changing rice to alfalfa and safflower in Kazakhstan), and (wheat to fruit trees in Morocco)	[37–41]
Building Local Irrigation Infrastructure in Chile	[42]		
Restoration of destroyed areas through the creation of protection zones in Ethiopia	[43–46]	Change in economic activities (allocation of agricultural land to fish farming in China), (allocation of rice cultivation land to aquaculture in Bangladesh), (change from animal husbandry to rainfed farming and beekeeping in Kenya), and (switches between cropping and transhumant livelihoods in Burkina Faso)	[47–52]
Dikes and inlet barriers for the Gulf Coast in the United States	[53]		

Table 2.
Adaptability measures through reversibility and transformation approaches.

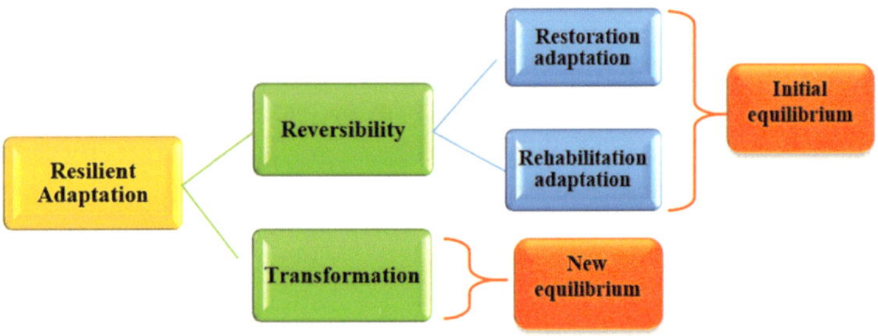

Figure 6.
Classification of adaptation methods. Source: Author.

3.2 Advantages and disadvantages of climate change adaptation processes in the water resources system

Adapting to climate change within the transformational process holds the potential for heightened success in ensuring the sustainability and preservation of the system. This success is attributed to the removal of flawed structures and the establishment of a more flexible structure that exhibits reduced sensitivity to the impacts of climate change. Such achievements gain particular significance within the integrated water resources system grappling with the challenges posed by climate change.

In the realm of water scarcity, the challenge extends beyond the repercussions of climate change alone; we consistently face mechanisms that perpetuate the escalating trend of consumption. However, adaptation strategies centered around reversibility cannot merely address the water shortage resulting from climate change. Consequently, a shift toward adaptation strategies is imperative, focusing on structural changes to control the escalating mechanisms of consumption within a transformation process. Furthermore, beyond its impact on the quantity of available water resources, climate change also induces alterations in the intensity and frequency of extreme events. In such circumstances, the transformation adaptation process proves to be more effective than the reversibility adaptation process due to the flexibility and heightened system carrying capacity.

It is essential to note that the prescription of transformation adaptation results in a significant shift in water consumption and may leave no alternative, but we should embrace this approach in some cases. However, implementing such a change requires a fundamental change in the economic and social structures that have evolved over many years in a given region, inevitably giving rise to path dependency.

Path dependence emerges as a formidable obstacle to the acceptance and success of this form of adaptation. Conversely, reversibility adaptation strategies refrain from instigating substantial changes to the region's structure. Since they do not deviate significantly from the path taken, they hold the potential for greater acceptance and success compared to transformation adaptation. Nevertheless, the development capacities and preparations created through reversibility adaptation can serve as facilitating factors in transitioning toward transformation adaptation [54].

The cultivation of awareness among stakeholders becomes paramount for the success of climate change adaptation processes [55]. The local capacities and flexibility for transformation adaptation can be created by local stakeholders through social learning processes, representing the best option from a community-oriented perspective for transformation adaptation measures. This form of adaptation is regarded as reversibility adaptation [56]. In this regard, reversible adaptation strategies can be effective in the short to medium term, and when aligned with training and capacity building, they lay the groundwork for transformation adaptation strategies [57].

Considering the merits and drawbacks of distinct adaptation processes, research endeavors have been undertaken to optimize their application across various countries. In New Zealand, in response to challenges faced by farmers arising from shifts in water availability, climate change, and land use alterations, localized research was conducted. Building on these findings, a comprehensive program for the regional adaptation path was formulated. This program recommended reversibility adaptation for the short-term and proposed transformation adaptation for long-term sustainability [58]. Subsequent, prerequisites for transformation adaptation were identified, including a shift in prevailing attitudes and public understanding [59] and legislative modifications [58]. Also, in Australia, to achieve sustainability, four adaptation pathways were defined based on farmers' experiences. Research outcomes indicated that sustainability could be attained by following several paths together, both individually and as a group [60].

4. Risk management approach for climate change adaptation

The risk management (RM) approach, initially introduced by Wideman in 1992, has undergone enhancements based on user experiences in diverse projects marked by uncertainties [61, 62]. The uncertainty in climate forecasts, stemming from the numerous global climate change scenarios and models, has prompted the adoption of a risk management approach in climate change adaptation efforts in recent years [63].

The essence of the risk management approach is based on the development of adaptation strategies for both current and anticipated risks. These strategies are evaluated using a wide range of possible future scenarios to create robust adaptation strategies and measures [64]. Rather than utilizing quantitative data of future climate scenarios solely for impact assessment, this approach employs such data to assess the robustness of proposed strategies, aligning with the principles of robust decision-making [65].

In the risk management approach, the risk is characterized as a function of the probability of repeating the consequences, directing attention toward the consequences of climatic events rather than solely focusing on the probability of those events occurring [64]. A notable advantage of the risk-based approach, as opposed to a top-down approach, lies in early stakeholder involvement. They facilitate the identification of potential adaptation strategies, which are subsequently subjected to analysis across a diverse array of scenarios.

Systems encounter uncertainties in predicting climate change, its effects, and changing conditions over time, leading to uncertainties regarding the potential impact of adaptation measures on the system.

Over the past two decades, numerous methodologies have been developed to address these uncertainties. In this regard, the risk management approach, owing to its inherent flexibility, can adeptly leverage an extensive array of these methodologies. Moreover, climate change impact assessment, adaptation, and vulnerability have common elements with risk management, including the need to manage uncertainty, communication between hazards and consequences, communication between technical experts and stakeholders, risk reduction by reducing hazards and the consequences of these hazards, and formal processes to link all these activities are [63].

Due to the high impact of climate change on water resources, the risk management approach has been extensively applied for climate change adaptation across various countries (refer to **Table 3**). Drawing from the experiences of scientists [66] in managing transboundary water resources, the utilization of this approach in climate change-affected resource management is viewed as a natural development. In this study, six key steps were suggested in the risk management process for transboundary water resources, encompassing risk assessment planning, risk identification, risk assessment, risk response plan, control plan, and continuous evaluation of the program.

Many governments worldwide, including New Zealand and Australia, have embraced risk management as a main approach to assess adaptation options to climate change [67]. For instance, the South Australian Government's Department of Environment, Water, and Natural Resources has established a process framework for risk management in water planning and management. This framework addresses natural resource risks, community values, and the effective performance of management practices. It incorporates key components such as communication and consultation, monitoring and evaluation, context building, risk assessment (including risk identification, analysis, and estimation), and risk reduction, ensuring the implementation of activities of communication and consultation, and monitoring and evaluation throughout all stages of the process [71].

Application	Source
Risk management for transboundary water conflict resolutions	[66]
A risk management approach to climate change adaptation (New Zealand and Australia)	[67]
Management of water resources planning by US agencies, including the US Army Corps of Engineers	[68]
Thames River Barrier, energy production in the Niger Basin, water management in Yemen, and flood risk management in a large southeast Asian metropolis	[69]
Risk management to reduce the effects of climate change in the Southern Mediterranean Countries	[70]

Table 3.
Examples of risk management applications in dealing with climate change in water resources.

Various methods underpin risk management and assessment, developed in diverse frameworks [69, 71, 72]. Among risk management and assessment methods, the flexible robustness methods provide a viable means of planning for changes in water supply and demand under evolving climate conditions. Methods of risk management have found widespread use in American agencies, including the United States Army Corps of Engineers (USACE), for efficient water resource management [68]. Examples such as the Thames River Dam, energy production in the Niger Basin, water management in Yemen, and flood risk management in a major Southeast Asian metropolis, introduced as the applications of the framework of risk management with robust decision-making tools for uncertainty management by World Resources Institute [69].

In the realm of adaptation assessment at both national and local levels, it is suggested to employ risk management approach. Within this context, evaluating adaptation strategies in alignment with key concepts such as vulnerability/resilience, sustainable development, and disaster risk, facilitates comprehension of prevailing policies, managerial practices within communities, existing programs and knowledge, and identifying risks [63].

5. Evaluation and decision-making tools for adaptation strategies

It is imperative to conduct a thorough evaluation of the impacts resulting from adaptation strategies and measures to gauge the efficacy of enhancing the system's adaptive capacity and resilience, while concurrently mitigating vulnerability and risk associated with climate change. The significance of scrutinizing the effects of adaptation strategies and measures has given rise to various methodological approaches and tools, which are broadly categorized into groups encompassing natural sciences, social sciences, overlapping natural and social sciences, and transdisciplinary [73].

Within the realm of natural sciences-based methods, assessments are carried out through hydrological modeling, hydrodynamic modeling, and integrated evaluation modeling. However, the limitations of hydrological and hydrodynamic modeling methods lie in their challenges of accessing data related to other factors influencing risks at the same resolution, coupled with a deficiency in integrated assessment capabilities. Integrated assessment modeling (IAM) emerges as a suitable option for large-scale (global) assessments, enabling the evaluation of human-environment interactions and the interplay between adaptation and mitigation [74, 75]. Nonetheless, the uncertainty in estimating the costs and benefits of adaptation measures constrains the broader application of these models.

In contrast, methods rooted in social sciences adopt a bottom-up approach and include risk assessment, participatory evaluation, and cost–benefit analysis. The risk assessment method possesses the advantage of determining risk reduction measures and conducting uncertainty analyses, thereby overcoming the limitations associated with IAM [76]. However, drawbacks of the risk assessment method include the mismatch between the time scale of adaptation to climate change and short-term community concerns, as well as the absence of high-resolution future climate information at the local scale and the associated risks.

Recent years have witnessed the development of participatory evaluation methods aimed at evaluating the impacts of climate change and its adaptation [77–79]. Notwithstanding the merits of these methods, they exhibit certain limitations, such as subjectivity in stakeholder identification, the selected assessment method, and the stakeholders' level of knowledge.

Overlapping methods try to solve the shortcomings of social and natural sciences-based methods by combining the elements of them. Environmental assessment models, hydroeconomics models, and water resource systems assessment modeling are examples of overlapping methods employed in this context.

Hydroeconomic models were presented due to the effect of water on economic performance [80, 81]. Initially, hydroeconomic modeling incorporated simulation models in specific fields along with economic equations, yet it fell short of providing a comprehensive evaluation of adaptation options [82]. Consequently, these models were developed over time.

Recognizing the constraints of financial resources and time in implementing adaptation strategies, the integration of decision-making tools with developed hydroeconomic models became imperative, leading to the emergence of transdisciplinary tools. These analytical tools can be applied in combination with all three aforementioned methods.

Decision-making tools consist of two main groups: single decision criterion and more than one decision criterion. The development of transdisciplinary tools first has been done by incorporating common decision-making tools, including single-criteria decision tools (CBA, CEA) and multi-criteria analysis (MCA), in conjunction with hydroeconomic models [73, 83]. In the following, Robust decision-making tools were integrated with hydroeconomic models due to their relative superiority in addressing risks [84].

In summary, the evaluation of climate change adaptation strategies necessitates integrated dynamic models that account for climatic and non-climatic uncertainties affecting future conditions. These models should encompass various economic, social, and environmental factors, as well as mechanisms evolving over time. Additionally, decision-making tools alongside these models are imperative to the judicious selection of strategies, taking into account both risk factors and stakeholder participation. Below is an example of a suitable tool with the desired features.

6. Climate change adaptation through modern portfolio theory in risk management approach

The Modern Portfolio Theory (MPT) stands out as a robust decision-making tool suitable for reducing risk. Its distinctive capability lies in its capacity to calculate climate risk and the risk-taking of decision-makers. Rooted in economic theory, this model serves as a strategic planning methodology for investments in conditions of uncertainty. It simultaneously incorporates two criteria economic risk and economic return, within a unified mathematical model.

The portfolio optimization process involves selecting the optimal combination of financial assets, aiming to maximize investment returns while minimizing risk. At its core, MPT posits that by investing in assets with low correlation with each other, the inherent risks associated with each asset can be mitigated, resulting in a stable return with reduced overall risk [85].

By expanding the number and diversity of assets and allocating investments to assets with more return and less risk, MPT effectively manages both systematic risk (about changes affecting all activities) and unsystematic risk (related to changes specific to each activity). In actuality, MPT mitigates the system's susceptibility to fluctuations by leveraging the diversification feature. This, in turn, diminishes the system's vulnerability and risk.

According to the recommendations of the Intergovernmental Panel on Climate Change, the application of MPT in climate change adaptation planning has gained prominence in recent years across various sectors. Given the substantial impact of climate change on the water sector, the concept of portfolio risk distribution has been employed in investments to devise adaptation strategies and measures. For instance, within the domain of water supply, the portfolio approach integrates existing infrastructure with other options such as protective measures and water transfer. This method yields a diverse range of water supply portfolios, effectively reducing the required costs to meet demand and enhancing adaptability to evolving climatic conditions.

Moreover, recognizing the pivotal role of water in the agricultural sector, MPT has been instrumental in offering varied strategies to manage climate change-induced risks [86]. **Table 4** provides illustrative examples of these studies. In these investigations, diversification has been leveraged as a key feature to mitigate the risks associated with climate change.

In the realm of water allocation, wherein water functions both as a final and intermediate product within various economic sectors, water can be assumed as capital. The uncertainty associated with accessing this vital resource under climate change introduces a level of unpredictability to the economic performance of the system.

The MPT model, when applied to water allocation, strategically allocates this finite resource to activities characterized by higher economic efficiency and their economic performances are not affected by changing climate conditions simultaneously and to the same extent. Developing this set of activities with less sensitivity to climate fluctuations requires structural change. Within the revamped structure, enhanced adaptability is realized by minimizing uncertainty and fostering economic stability, resulting in a reduced vulnerability of the system. Consequently, the MPT, through the transformation adaptation process, serves as an effective strategy for risk management in climate change adaptation.

Since in the MPT model, a set of optimal points (returns and risks) is produced, the stakeholders with different utilities can choose one of the points based on their degree of risk tolerance, and in this way, the risk-taking of stakeholders is considered in the decision for water allocation. In this way, risk management by the MPT ensures a holistic evaluation that considers both the systemic vulnerabilities and the varying risk tolerance levels of involved stakeholders.

Surface water and groundwater resources management	[87, 88]
water quality	[89]
Coping with flood and drought extreme events	[90, 91]
Reducing environmental risks	[92]
Water distribution (reducing the risk of water distribution network)	[93]
Risk reduction in garden systems	[94]
Water supply (irrigation water, urban water, and areas with a common water source)	[95–101]
Combined planning of water supply and demand	[102]
Combination of various infrastructures of water resources	[103]
Land use change and cultivation pattern change	[88, 104, 105]
Diversity in products and sources of income for farmers	[106]
Diversification in products by examining the gross margins	[107]

Table 4.
The application of modern portfolio theory for adaptation to climate change in the water and agriculture sector.

Given the significant influence of water on both social and environmental systems, disregarding these dimensions may result in the ineffectiveness of climate change adaptation strategies [108]. Also, according to section (2–3), reversibility adaptation strategies can increase the adaptive capacity as a complement to the transformation adaptation strategy (allocation of water based on MPT). This dual approach, combining both reversible and transformation adaptation processes, presents a comprehensive strategy for risk management and ensuring successful adaptation to climate change, which was first proposed by the author of this chapter in the form of the water resources integrated dynamic model and MPT model were presented [109]. This research was conducted in a catchment of a lake. In this research, the MPT theory was developed with the social criterion, and water was reallocated as a decision variable between the economic activities of agriculture, services, and industry; in this way, this portfolio plane included optimal points produced. This reallocation by the transformation process reduced the basin's sensitivity to climate change and increased its adaptive capacity. In addition, reversible adaptation strategies were presented to moderate the effects of climate change, and these strategies were evaluated in the integrated model; then, the portfolio planes of these strategies were prioritized on the axes of return, risk, and Gini coefficient of the MPT model. Finally, an optimal point based on the risk tolerance of the beneficiaries was selected from the selected strategy screen for allocation in climate change conditions. The results of this research show the improvement of the economic performance of the basin (about 1.2 times) by reallocating more water to the service and industry sector in the transformation adaptation process, preserving the environment, and restoring the lake by saving water consumption in the agricultural sector in reversibility process.

7. Conclusions

The pressures out of system capacity brought to it due to environmental changes, including climate change, and the effects they have on different parts of the system take it out of balance. Finally, this situation leads to the failure of the system. In this situation, the only way to save and survive the system against pressures is to create adaptations so that the system adapts to new conditions. Achieving this adaptive system is possible through an efficient risk management approach. According to IPCC's definition, climate risk is a function of danger, exposure, and vulnerability, so the risk can be reduced through changes in vulnerability components, i.e., increasing adaptive capacity and reducing sensitivity.

Two processes are identified for dealing with risk of changes beyond the system's capacity: reducing vulnerability through adjustment measures (reversibility adaptation) or increasing resilience by establishing new equilibrium conditions (transformation adaptation). Research findings indicate that reversibility adaptation may not always restore the system to its initial state, while transformation adaptation leads to a new and sustainable equilibrium by altering the system's structure.

If structural changes in the transformation adaptation process are made based on the principle of decoupling, it not only enhances adaptive capacity but also establishes a stable (dynamic) balance over time, promoting resource preservation and promoting economic growth with reduced resource consumption.

In the decoupling method, resources are allocated to sectors with a higher economic benefit ratio to less resource consumption and minimized environmental impact. In this context, integrating considerations of wastewater and carbon emissions generated from economic activities to resource consumption leads to the realization of a green economy and sustainable development goals.

Contrary to the advantages offered by transformation adaptation, this approach encounters resistance from stakeholders due to the significant economic and social structural changes it introduces, particularly in regions where long-standing habits and dependencies on existing structures have developed. In contrast, reversible adaptation processes are more likely to gain social acceptance due to their minimal impact on existing structures. Consequently, for an effective climate change adaptation strategy within a risk management framework, it is advisable to implement both adaptation processes complementarily.

Given the interconnected nature of economic, social, and environmental subsystems and the inherent uncertainties in climate change adaptation outcomes, a comprehensive evaluation is imperative. Adopting a transdisciplinary approach through dynamic integrated models allows for a holistic assessment of these strategies. To aid decision-making and strategy selection, the use of robust decision-making tools, which consider current and future risks as well as stakeholders, is crucial.

The portfolio theory, renowned for its diversification feature, emerges as a fitting robust decision-making tool for risk management. By maximizing economic returns, minimizing risks, and adhering to the decoupling principle (maximizing return ratio to resource consumption), this theory allocates resources or capital strategically to a set of activities with high productivity, low risk, and less dependence on each other (reduction of system's sensitivity). This, in turn, enhances adaptive capacity and reduces vulnerability. His theory can also be used to choose appropriate strategies based on the risk-taking of stakeholders. In this way, it becomes possible to reduce risk through conversion and reversibility processes. Ultimately, integration of the portfolio model with dynamic integrated evaluation models further enhances the potential for comprehensive risk management under changing climate conditions.

Conflict of interest

The authors declare no conflict of interest.

Notes/thanks/other declarations

None.

Author details

Samin Ansari Mahabadi
Water Research Institute, Tehran, Iran

*Address all correspondence to: a_ansari25@yahoo.com

References

[1] IPCC. The Physical Science Basis, Technical Summary, the Working Group I contribution to the UN IPCC's Sixth Assessment Report (WG1 AR6). Cambridge, United Kingdom and New York, NY, USA: Cambridge University Press; 2021. pp. 35-144. DOI: 10.1017/9781009157896

[2] Winterhalder B. Environmental analysis in human evolution and adaptation research. Human Ecology. 1980;**8**:135-170

[3] McCarthy JJ. Climate Change 2001: Impacts, Adaptation, and Vulnerability: Contribution of Working Group II to the Third Assessment Report of the Inter-governmental Panel on Climate Change. Cambridge, United Kingdom and New York, NY, USA: Cambridge University Press; 2001. DOI: 10.1002/joc.775

[4] Parry ML. Climate Change 2007-Impacts, Adaptation and Vulnerability: Working Group II Contribution to the Fourth Assessment Report of the IPCC. Cambridge, United Kingdom and New York, NY, USA: Cambridge University Press; 2007. Available from: https://www.cabidigitallibrary.org/doi/pdf/10.5555/20083115494

[5] Lim B, Spanger-Siegfried E, Burton I, Malone E, Huq S. Adaptation Policy Frameworks for Climate Change: Developing Strategies, Policies and Measures. Cambridge: Cambridge University Press; 2005. Available from: https://www.adaptation-undp.org/sites/default/files/downloads/adaptation_policy_frameworks_for_climate_change_-_developing_strategies_policies_and_measures_0.pdf

[6] Limited M. Osting the impacts of climate change in the UK: Overview of guidelines. In: Technical Report. UK: UKCIP; 2004

[7] Orsato RJ, Ferraz de Campos JG, Barakat SR. Social learning for anticipatory adaptation to climate change: Evidence from a community of practice. Organization & Environment. 2019;**32**(4):416-440

[8] Rahman HT, Hickey GM. What does autonomous adaptation to climate change have to teach public policy and planning about avoiding the risks of maladaptation in Bangladesh? Frontiers in Environmental Science. 2019;**7**:2

[9] Holman IP, Brown C, Carter TR, Harrison PA, Rounsevell M. Improving the representation of adaptation in climate change impact models. Regional Environmental Change. 2019;**19**:711-721

[10] Kopke K, Lyons E, MacMahon E, O'Dwyer B, Gault J. Reflecting on Adaptation to Climate Change: International Best Practice Review and National MRE and Indicator Development Requirements. Ireland: Environmental Protection Agency; 2018. 59 p

[11] Lwasa S. A systematic review of research on climate change adaptation policy and practice in Africa and South Asia deltas. Regional Environmental Change. 2015;**15**(5):815-824

[12] Gain AK, Rouillard JJ, Benson D. Can Integrated Water Resources Management Increase Adaptive Capacity to Climate Change Adaptation? A Critical Review. Journal of Water Resource and Protection. 2013;**5**(4):11-20. DOI: 10.4236/jwarp.2013.54A003

[13] Huang-Lachmann J-T, Hannemann M, Guenther E. Identifying links between economic opportunities

and climate change adaptation: Empirical evidence of 63 cities. Ecological Economics. 2018;**145**:231-243

[14] Pahl-Wostl C. Adaptive and Sustainable Water Management: From Improved Conceptual Foundations to Transformative Change. Global Water Resources. United Kingdom: Routledge; 2021. pp. 175-193

[15] Ahmed K, Wei L. Adaptation as a response to climate change: A literature review. Available from: SSRN 2233070. 2012

[16] Berbel J, Gutiérrez-Martín C, Expósito A. Impacts of irrigation efficiency improvement on water use, water consumption and response to water price at field level. Agricultural Water Management. 2018;**203**:423-429

[17] Parry J-E, Terton A. Trends in Adaptation Planning: Observations from a Recent Stock-Taking Review. Canada: International Institute for Sustainable Development. 2016. 10 p. Available from: https://www.iisd.org/system/files/publications/trends-adaptation-planning-observations-review.pdf

[18] Expósito A, Beier F, Berbel J. Hydro-economic modelling for water-policy assessment under climate change at a river basin scale: A review. Water. 2020;**12**(6):1559

[19] Sa R. Climate change adaptation in SIDS: A systematic review of the literature pre and post the IPCC Fifth Assessment Report. Wiley Interdisciplinary Reviews: Climate Change. 2020;**11**(4):e653

[20] An A. An Australian Guide to the Science and Potential Impacts. Canberra, ACT: Australian Greenhouse Office; 2003. 239 p. ISBN: 1920840125

[21] Reduction ISfD. Living with Risk: A Global Review of Disaster Reduction Initiatives. Switzerland: United Nations; 2004;(1):429. Available from: http://www.unisdr.org/eng/about_isdr/bd-lwr-2004-eng.htm

[22] Willows R, Reynard N, Meadowcroft I, Connell R. Climate Adaptation: Risk, Uncertainty and Decision-Making. UKCIP Technical Report. Climate Impacts Programme. UK: Oxford; 2003. pp. 41-87

[23] Wu S, Bates B, Zbigniew Kundzewicz A, Palutikof J. Climate change and water. Technical Paper of the Intergovernmental Panel on Climate Change Geneva. Geneva: IPCC Secretariat; 2008. 210 p. Available from: https://archive.ipcc.ch/pdf/technical-papers/climate-change-water-en.pdf

[24] Bueno NP. Assessing the resilience of small socio-ecological systems based on the dominant polarity of their feedback structure. System Dynamics Review. 2012;**28**(4):351-360

[25] Nelson DR, Adger WN, Brown K. Adaptation to environmental change: Contributions of a resilience framework. Annual Review of Environment and Resources. 2007;**32**:395-419

[26] Pachauri RK, Allen MR, Barros VR, Broome J, Cramer W, Christ R, et al. Climate Change 2014: Synthesis Report. Contribution of Working Groups I, II and III to the Fifth Assessment Report of the Intergovernmental Panel on Climate Change. Geneva, Switzerland: IPCC; 2014. 151 p

[27] Pickett ST, Cadenasso ML, Grove JM. Resilient cities: Meaning, models, and metaphor for integrating the ecological, socio-economic, and planning realms. Landscape and Urban Planning. 2004;**69**(4):369-384

[28] Lister N-M. Resilience: Designing the new sustainability. Topos. 2015;**90**:14-21

[29] Farso M. Towards a New Landscape Architecture. Denmark: University of Copenhagen; 2010

[30] Ahern J. From fail-safe to safe-to-fail: Sustainability and resilience in the new urban world. Landscape and Urban Planning. 2011;**100**(4):341-343

[31] Gurr G, Johnson A, Liu J. Land use: Restoration and rehabilitation. Encyclopedia of Agriculture and Food Systems. UK: Academic Press; 2014;**8**(11):139-147

[32] Roni P, Beechie T. Stream and Watershed Restoration: A Guide to Restoring Riverine Processes and Habitats. West Sussex, UK: John Wiley & Sons; 2012. pp. 11-49

[33] Hong NB, Yabe M. Improvement in irrigation water use efficiency: A strategy for climate change adaptation and sustainable development of Vietnamese tea production. Environment, Development and Sustainability. 2017;**19**:1247-1263

[34] Tan Y, Liu X. Water shortage and inequality in arid Minqin oasis of Northwest China: Adaptive policies and farmers' perceptions. Local Environment. 2017;**22**(8):934-951

[35] Deligios PA, Chergia AP, Sanna G, Solinas S, Todde G, Narvarte L, et al. Climate change adaptation and water saving by innovative irrigation management applied on open field globe artichoke. Science of the Total Environment. 2019;**649**:461-472

[36] Rouabhi A, Hafsi M, Monneveux P. Climate change and farming systems in the region of Setif (Algeria). Journal of Agriculture and Environment for International Development (JAEID). 2019;**113**(1):79-954

[37] Lei Y, Liu C, Zhang L, Luo S. How smallholder farmers adapt to agricultural drought in a changing climate: A case study in southern China. Land Use Policy. 2016;**55**:300-308

[38] Wani MH, Baba SH, Bazaz NH, Sehar H. Climate change in Kashmir valley: Is it initiating transformation of mountain agriculture? Indian Journal of Economics and Development. 2015;**3**(2):142-154

[39] Barrett T, Feola G, Khusnitdinova M, Krylova V. Adapting agricultural water use to climate change in a post-soviet context: Challenges and opportunities in Southeast Kazakhstan. Human Ecology. 2017;**45**(6):747-762

[40] Faysse N. The rationale of the green Morocco plan: Missing links between goals and implementation. The Journal of North African Studies. 2015;**20**(4):622-634

[41] Morocco - Integrating Climate Change in the Implementation of the Plan Maroc Vert: P117081 - Implementation Status Results Report: Sequence 09 (English). Washington, D.C.: World Bank Group; 2016. Available from: http://documents.worldbank.org/curated/en/693371468186857140/Morocco-Integrating-Climate-Change-in-the-Implementation-of-the-Plan-Maroc-Vert-Project

[42] Lillo-Ortega G, Aldunce P, Adler C, Vidal M, Rojas M. On the evaluation of adaptation practices: A transdisciplinary exploration of drought measures in Chile. Sustainability Science. 2019;**14**(4):1057-1069

[43] Descheemaeker K, Nyssen J, Rossi J, Poesen J, Haile M, Raes D, et al. Sediment deposition and pedogenesis in exclosures in the Tigray highlands. Geoderma. 2006;**132**(3):291-314

[44] Mekuria W, Veldkamp E, Haile M, Nyssen J, Muys B, Gebrehiwot K.

Effectiveness of exclosures to restore degraded soils as a result of overgrazing in Tigray, Ethiopia. Journal of Arid Environments. 2007;**69**(2):270-284

[45] Mekuria W, Veldkamp E, Tilahun M, Olschewski R. Economic valuation of land restoration: The case of exclosures established on communal grazing lands in Tigray, Ethiopia. Land Degradation & Development. 2011;**22**(3):334-344

[46] Lowell EC, Maguire DA, Briggs DG, Turnblom EC, Jayawickrama KJS, Bryce J. Effects of silviculture and genetics on branch/knot attributes of coastal Pacific Northwest Douglas-fir and implications for wood quality—A synthesis. Forests. 2014;**5**(7):1717-1736

[47] Zhou H, Wang X, Ja W. A way to sustainability: Perspective of resilience and adaptation to disaster. Sustainability. 2016;**8**(8):737

[48] Faruque G, Sarwer RH, Karim M, Phillips M, Collis WJ, Belton B, et al. The evolution of aquatic agricultural systems in Southwest Bangladesh in response to salinity and other drivers of change. International Journal of Agricultural Sustainability. 2017;**15**(2):185-207

[49] Baird R. The Impact of Climate Change on Minorities and Indigenous Peoples. London: Briefing; 2008

[50] Kagunyu AW, Wanjohi J. Camel rearing replacing cattle production among the Borana community in Isiolo County of northern Kenya, as climate variability bites. Pastoralism. 2014;**4**(1):13

[51] Reenberg A, Rasmussen LV, Nielsen JØ. Causal relations and land use transformation in the Sahel: Conceptual lenses for processes, temporal totality and inertia. Geografisk Tidsskrift-Danish Journal of Geography. 2012;**112**(2):159-173

[52] Vermeulen SJ, Dinesh D, Howden SM, Cramer L, Thornton PK. Transformation in practice: A review of empirical cases of transformational adaptation in agriculture under climate change. Frontiers in Sustainable Food Systems. 2018:**2**:65

[53] Kates RW, Travis WR, Wilbanks TJ. Transformational adaptation when incremental adaptations to climate change are insufficient. Proceedings of the National Academy of Sciences. 2012;**109**(19):7156-7161

[54] Pringle P. Transforming how We Think About Adaptation [Blog]. UK: United Kingdom Climate Impacts Programme; 2013

[55] Belay GD, Mohammadi H, Ardalan A, Bavani AM, Hosseinzadeh-Attar MJ, Adera A. Climate change intervention and adaptation in Ethiopia: A critical appraisal of systematic review. 2017;**4**(3):12

[56] Fook CT, T. Transformational processes for community-focused adaptation and social change: A synthesis. Climate and Development. 2017;**9**(1):5-21

[57] Parmesan C, Morecroft MD, Trisurat Y. Climate Change 2022: Impacts, Adaptation and Vulnerability. GIEC; Cambridge, UK and New York, NY, USA: Cambridge University Press; 2022. pp. 197-377

[58] Cradock-Henry NA, Blackett P, Hall M, Johnstone P, Teixeira E, Wreford A. Climate adaptation pathways for agriculture: Insights from a participatory process. Environmental Science & Policy. 2020;**107**:66-79

[59] Hudson M, Mead ATP, Chagné D, Roskruge N, Morrison S, Wilcox PL, et al. Indigenous perspectives and gene editing in Aotearoa New Zealand. Frontiers in Bioengineering and Biotechnology. 2019;7:70

[60] Pearson LJ, Dare M. Farmer pathways to sustainability in the face of water scarcity. Environmental Science & Policy. 2021;**124**:186-194

[61] PM. G. A Guide to the Project Management Body of Knowledge. Newtown Square, PA: Project Management Institute (PMI); 2008

[62] Sanchez H, Robert B, Bourgault M, Pellerin R. Risk management applied to projects, programs, and portfolios. International Journal of Managing Projects in Business. 2009;**2**(1):14-35

[63] Jones RN, Preston BL. Adaptation and risk management. WIREs Climate Change. 2011;**2**(2):296-308

[64] Ludwig F, van Slobbe E, Cofino W. Climate change adaptation and integrated water resource management in the water sector. Journal of Hydrology. 2014;**518**:235-242

[65] Dessai S, Hulme M, Lempert R, Pielke R Jr. Do we need better predictions to adapt to a changing climate? Eos, Transactions American Geophysical Union. 2009;**90**(13):111-112

[66] Gonen AaZ N. Using risk management to increase the flexibility of transboundary water conflict resolutions. International Journal of Risk Assessment and Management. 2008;**10**(4):373-385

[67] Nottage RA. Climate Change Adaptation in New Zealand: Future Scenarios and some Sectoral Perspectives: New Zealand Climate. Change Centre; 2010

[68] Major DC. Climate change and water resources: The role of risk management methods. Journal of Contemporary Water Research and Education. 2011;**112**(1):8

[69] Kunreuther H, Heal G, Allen M, Edenhofer O, Field CB, Yohe G. Risk management and climate change. Nature Climate Change. 2013;**3**(5):447-450

[70] Gaaloul N, Eslamian S, Katlance R. Impacts of climate change and water resources management in the southern mediterranean countries. Water Productivity Journal. 2021;**1**(1):51-72

[71] Risk Management Framework for Water Planning and Management. In: Department of Environment WaNR, Government of South Australia, Australia. 2012

[72] Ray PA, Taner MÜ, Schlef KE, Wi S, Khan HF, Freeman SSG, et al. Growth of the decision tree: Advances in bottom-up climate change risk management. JAWRA Journal of the American Water Resources Association. 2019;**55**(4):920-937

[73] Bhave AG, Mishra A, Raghuwanshi NS. A brief review of assessment approaches that support evaluation of climate change adaptation options in the water sector. Water Policy. 2014;**16**(5):959-972

[74] Moss RH, Edmonds JA, Hibbard KA, Manning MR, Rose SK, Van Vuuren DP, et al. The next generation of scenarios for climate change research and assessment. Nature. 2010;**463**(7282):747-756

[75] Soboll A, Elbers M, Barthel R, Schmude J, Ernst A, Ziller R. Integrated regional modelling and scenario development to evaluate future water demand under global change conditions. Mitigation and Adaptation Strategies for Global Change. 2011;**16**:477-498

[76] Van Aalst MK, Cannon T, Burton I. Community level adaptation to climate change: The potential role of participatory community risk

assessment. Global Environmental Change. 2008;**18**(1):165-179

[77] Wende W, Bond A, Bobylev N, Stratmann L. Climate change mitigation and adaptation in strategic environmental assessment. Environmental Impact Assessment Review. 2012;**32**(1):88-93

[78] Larsen SV, Kørnøv L. SEA of river basin management plans: Incorporating climate change. Impact Assessment and Project Appraisal. 2009;**27**(4):291-299

[79] Yang Y, Xu H, Wang J, Liu T, Wang H. Integrating climate change factor into strategic environmental assessment in China. Environmental Impact Assessment Review. 2021;**89**:106585

[80] White JW, Hoogenboom G, Kimball BA, Wall GW. Methodologies for simulating impacts of climate change on crop production. Field Crops Research. 2011;**124**(3):357-368

[81] Niu X, Easterling W, Hays CJ, Jacobs A, Mearns L. Reliability and input-data induced uncertainty of the EPIC model to estimate climate change impact on sorghum yields in the US Great Plains. Agriculture, Ecosystems & Environment. 2009;**129**(1-3):268-276

[82] Divakar L, Babel MS, Perret S, Gupta AD. Optimal allocation of bulk water supplies to competing use sectors based on economic criterion–An application to the Chao Phraya River Basin, Thailand. Journal of Hydrology. 2011;**401**(1-2):22-35

[83] George B, Malano H, Davidson B, Hellegers P, Bharati L, Massuel S. An integrated hydro-economic modelling framework to evaluate water allocation strategies II: Scenario assessment. Agricultural Water Management. 2011;**98**(5):747-758

[84] Dittrich R, Wreford A, Moran D. A survey of decision-making approaches for climate change adaptation: Are robust methods the way forward? Ecological Economics. 2016;**122**:79-89

[85] Markowitz H. Portfolio selection. Journal of Finance. 1952;**7**(1):77-91

[86] Sewando PT. Efficacy of risk reducing diversification portfolio strategies among agro-pastoralists in semi-arid area: A modern portfolio theory approach. Journal of Agriculture and Food Research. 2022;**7**:100262

[87] Hua S, Liang J, Zeng G, Xu M, Zhang C, Yuan Y, et al. How to manage future groundwater resource of China under climate change and urbanization: An optimal stage investment design from modern portfolio theory. Water Research. 2015;**85**:31-37

[88] Marino G, Niso-Santano M, Baehrecke EH, Kroemer G. Self-consumption: The interplay of autophagy and apoptosis. Nature Reviews Molecular Cell Biology. 2014;**15**(2):81-94

[89] Marinoni O, Adkins P, Hajkowicz S. Water planning in a changing climate: Joint application of cost utility analysis and modern portfolio theory. Environmental Modelling & Software. 2011;**26**(1):18-29

[90] Aerts JC, Botzen W, van der Veen A, Krywkow J, Werners S. Dealing with uncertainty in flood management through diversification. Ecology and Society. 2008;**13**(1):41

[91] Buchecker M, Salvini G, Di Baldassarre G, Semenzin E, Maidl E, Marcomini A. The role of risk perception in making flood risk management more effective. Natural Hazards and Earth System Sciences. 2013;**13**(11):3013-3030

[92] Matthies BD, Jacobsen JB, Knoke T, Paul C, Valsta L. Utilising portfolio theory in environmental

research–new perspectives and considerations. Journal of Environmental Management. 2019;**231**:926-939

[93] Lee S, Shin S, Judi DR, McPherson T, Burian SJ. Criticality analysis of a water distribution system considering both economic consequences and hydraulic loss using modern portfolio theory. Water. 2019;**11**(6):1222

[94] Paut R, Sabatier R, Tchamitchian M. Reducing risk through crop diversification: An application of portfolio theory to diversified horticultural systems. Agricultural Systems. 2019;**168**:123-130

[95] Zeff HB, Kasprzyk JR, Herman JD, Reed PM, Characklis GW. Navigating financial and supply reliability tradeoffs in regional drought management portfolios. Water Resources Research. 2014;**50**(6):4906-4923

[96] Paydar Z, Qureshi M. Irrigation water management in uncertain conditions—Application of modern portfolio theory. Agricultural Water Management. 2012;**115**:47-54

[97] Kasprzyk JR, Reed PM, Characklis GW, Kirsch BR. Many-objective de novo water supply portfolio planning under deep uncertainty. Environmental Modelling & Software. 2012;**34**:87-104

[98] Kasprzyk JR, Reed PM, Kirsch BR, Characklis GW. Managing population and drought risks using many-objective water portfolio planning under uncertainty. Water Resources Research. 2009;**45**(12):18

[99] Characklis GW, Kirsch BR, Ramsey J, Dillard KE, Kelley CT. Developing portfolios of water supply transfers. Water Resources Research. 2006;**42**(5):14

[100] Kidson R, Haddad B, Zheng H, Kasower S, Raucher R. Optimising reliability: Portfolio modeling of contract types for retail water providers. Water Resources Management. 2013;**27**:3209-3225

[101] Kirsch BR, Characklis GW, Dillard KE, Kelley C. More efficient optimization of long-term water supply portfolios. Water Resources Research. 2009;**45**(3):12

[102] Zhang C. Embracing Uncertainty as the New Norm: A Risk-Based Portfolio Approach for Urban Water Investment Planning (Doctoral dissertation). United states: Harvard University, Graduate School of Arts & Sciences; 2016

[103] Shin S, Park H. Achieving cost-efficient diversification of water infrastructure system against uncertainty using modern portfolio theory. Journal of Hydroinformatics. 2018;**20**(3):739-750

[104] Gaydon D, Meinke H, Rodriguez D, McGrath D. Comparing water options for irrigation farmers using modern portfolio theory. Agricultural Water Management. 2012;**115**:1-9

[105] Bodin P, Olin S, Pugh TAM, Arneth A. Optimizing cropland cover for stable food production in sub-Saharan Africa using simulated yield and modern portfolio theory. Earth System Dynamics Discussions. 2014;**5**(2):1571-1606

[106] Sauer J, Finger R. Climate Risk Management Strategies in Agriculture–The Case of Flood Risk. Annual Meeting; July 27-29, 2014, Minneapolis, United states: Agricultural and Applied Economics Association; 2014

[107] Mitter H, Heumesser C, Schmid E. Spatial modeling of robust crop production portfolios to assess agricultural vulnerability and adaptation

to climate change. Land Use Policy. 2015;**46**:75-90

[108] Mahabadi SA, Bavani ARM, Bgheri A. Improving adaptive capacity of social-ecological system of Tashk-Bakhtegan Lake basin to climate change effects–A methodology based on Post-Modern Portfolio Theory. Ecohydrology & Hydrobiology. 2018;**18**(4):365-378

[109] Mahabadi SA, Bagheri A, Bavani ARM. Reducing vulnerability to the climate change-reversibility and transformation adopting in a hydro-economic model. Environmental Development. 2023;**47**:100893

Chapter 2

Perspective Chapter: Advancements in Disaster Risk Mitigation Strategies

Eleni Kalliontzi, Amalia Kouskoura, Evangelos Katsaros and Ioannis Bakouros

Abstract

Disasters, whether natural or human-induced, present persistent challenges, necessitating effective strategies for risk reduction and resilience enhancement. Recent advancements in disaster risk reduction (DRR) encompass dynamic risk assessment models, behavioral insights integration, and cutting-edge technologies like artificial intelligence and remote sensing. Cross-border collaborations further strengthen resilience efforts, exemplified by programs such as the European Union's Horizon 2020 initiative, Interreg and Directorate of Humanitarian Aid and Civil Protection (DG ECHO). Technological innovations, particularly in AI and machine learning, have revolutionized early warning and decision-making systems and rapid damage assessment, while policy frameworks increasingly advocate for integrated risk reduction and management approaches and climate resilience. Community engagement emerges as a cornerstone of that approach, incorporating local communities' values and priorities in DRR and DRM strategies. It is a necessity to empower communities through participatory and interactive tools, collaborative mapping, and planning processes and foster inclusive practices, crucial for vulnerable groups' protection. Interdisciplinary approaches and collaborative partnerships underscore the multifaceted nature of DRR, emphasizing the necessity of sustained investment and political commitment. Together, all these efforts can pave the way for building safer, more resilient societies equipped to confront the challenges of tomorrow's disasters.

Keywords: cross-border, project, risk management, disaster management, policies, natural disasters, macroregional development

1. Introduction

Disasters, whether natural or human-induced, have been an enduring challenge throughout human history, causing immense suffering, loss of life, and economic devastation [1]. From earthquakes and hurricanes to pandemics and industrial accidents, the impact of disasters reverberates across communities and nations, highlighting the urgent need for effective strategies to mitigate risk and enhance resilience.

Recent decades have witnessed significant advancements in our understanding of disaster risk and the development of innovative approaches to disaster risk reduction (DRR) [2]. From the emergence of dynamic risk assessment models to the integration of behavioral insights and the application of cutting-edge technologies, such as artificial intelligence and remote sensing, these innovations represent the latest results in DRR research and practice.

The climate crisis and its impacts demand proactive action. Prevention, Preparedness, and Immediate Response, focused on bolstering the country's resilience. These are the essential pillars for adapting our country to the new, unprecedented realities, which form the core of the ministry's approach to risk management arising from the climate crisis.

In this context, the development of a National Strategy for the Climate Crisis and Civil Protection is being pursued. This is a long-term strategy, not merely theoretical but based on scientific data and implemented through a specific action plan, founded on a national dialog on the climate crisis and its challenges.

Within the functions of the Ministry of Climate Crisis and Civil Protection, we rely on the established general plans for emergency needs and situations of emergency protection policy, as well as the management of their consequences. The institutional framework for handling urgent events, managing hazards, conducting environmental inspections, and related activities. Additionally, we rely on pollution mapping, drafting crisis management plans for civil protection phenomena, as well as plans formulated for emergency needs by municipalities and regions in Greece.

In addition to national efforts, cross-border projects have become increasingly significant in addressing transnational disaster risks. Initiatives such as the Horizon 2020 Programme, Interreg and DG ECHO prevention and preparedness, on Cross-Border Disaster Risk Reduction exemplify collaborative efforts that enhance regional resilience. These projects focus on sharing data, technology, and best practices across borders to effectively manage and mitigate risks posed by disasters that do not respect national boundaries.

In this chapter, we explore the latest developments in DRR and their implications for building safer, more resilient communities. We begin by examining the dynamics of risk, including the evolving nature of hazards and vulnerabilities, and the role of dynamic risk assessment models in providing more accurate and timely risk assessments. We then delve into the behavioral aspects of risk perception and decision-making, highlighting the importance of understanding human behavior in designing effective DRR strategies.

Advancements in technology have transformed the DRR landscape, offering new tools and capabilities for early warning, damage assessment, and spatial analysis [3]. The use of AI and remote sensing technologies, for instance, has revolutionized our ability to anticipate, prepare for, and respond to disasters with greater precision and efficiency.

Policy and governance frameworks play a crucial role in shaping DRR efforts, providing the legal and institutional framework necessary for effective risk management [4]. Integrated risk management approaches, which consider the interconnected nature of hazards and vulnerabilities, are increasingly being adopted to promote more coordinated and sustainable DRR strategies. Mainstreaming climate change adaptation measures into DRR policies and practices is also essential for building resilience to the growing impacts of climate change. Projects like the Integrated Risk Management Framework and the Climate Resilient Infrastructure Project highlight these integrated approaches.

Furthermore, international cooperation is vital for comprehensive disaster risk reduction. These collaborations enhance regional capacities to manage large-scale disasters through joint training exercises, shared resources, and harmonized policies. Such initiatives underscore the importance of a united approach to disaster resilience, transcending national limitations and fostering global solidarity.

At the center of DRR efforts are communities, whose resilience depends on their ability to anticipate, prepare for, and respond to disasters. Community engagement and empowerment initiatives, grounded in principles of inclusivity and participation, are essential for building resilience from the ground up [5]. The ROSES project aims to raise and enhance risk awareness, sharing of best practices, and risk communication by elaborating actions in the fields of Host Nation Support in cross-border areas, empowering local communities for joint disaster risk reduction and management, activating collaboration in bilateral agreements, public engagement, inclusion of vulnerable groups, risk awareness in educational structures, and innovative processes in the protection of cultural heritage, thus making disaster risk reduction inclusive and collaborative. Similarly, the SAILOR project focuses on cross-border risk assessment and action plans in the Georgia-Azerbaijan area, enriching the ongoing work on forest fires. Also, SOLVE project focuses on Cross Border Complex Floods And Forest Fires Prevention And Management between the cross-border area of North Macedonia and Greece, and the BALANCE project aims to address the pressing needs of effective multi-lateral response, reducing the disaster risk in highly prone regions and among social groups with inadequate institutional capacities for disaster management while enhancing response coordination to complex events and improving the skillsets and contingency plans and the bilateral and multilateral cooperation protocols in the field of civil protection.

The challenges posed by disasters are complex and multifaceted, requiring comprehensive and integrated approaches to risk reduction [6]. By harnessing the latest insights, innovations, and technologies, stakeholders can enhance their capacity to mitigate risk, build resilience, and create safer and more sustainable communities for future generations.

2. Behavioral insights in risk perception

Recent research in behavioral science has recognized different ways of thinking and mental shortcuts that affect how we see risks and make decisions [7]. By incorporating behavioral insights into risk communication and public engagement efforts, such as those undertaken by the Behavioral Risk Communication Project, policymakers can enhance community resilience and promote adaptive behaviors in disaster-prone areas.

Effective risk communication is essential for fostering public awareness and motivating behavioral change in disaster-prone areas [8]. Tailoring communication messages to specific audience segments and utilizing trusted channels, such as local community leaders and social media influencers, can enhance the effectiveness of risk communication efforts. Moreover, engaging communities in two-way communication processes, such as participatory risk mapping exercises and community forums, can facilitate dialog and build trust between stakeholders.

Behavioral nudges, such as default options, social norms, and incentives, can influence decision-making and encourage individuals to adopt risk-reducing behaviors [9]. By designing interventions that leverage behavioral insights, policymakers can promote actions that enhance community resilience and reduce vulnerability to disasters.

3. Technological innovations

Advancements in artificial intelligence (AI) and machine learning are revolutionizing disaster response and recovery efforts. AI-powered algorithms can analyze vast amounts of data to identify patterns and predict disaster impacts with unprecedented accuracy [10]. From optimizing evacuation routes to assessing infrastructure vulnerabilities, AI-driven solutions are transforming the way we prepare for and mitigate disasters.

Early warning systems play a crucial role in reducing the loss of life and property in disaster-prone areas. Machine learning algorithms can improve the accuracy and reliability of early warning systems by analyzing historical data and real-time observations to detect emerging threats. By continuously learning from new data and refining their predictive capabilities, these systems can provide timely alerts and actionable information to at-risk communities, enabling them to take proactive measures to mitigate disaster impacts [11].

Remote sensing technologies, such as satellite imagery and drones, offer invaluable insights into disaster-affected areas, allowing for rapid damage assessment and resource allocation. Coupled with geospatial analysis tools, these technologies enable responders to visualize spatial patterns of risk and prioritize interventions accordingly. The Remote Sensing Damage Assessment Initiative, for example, leverages high-resolution satellite imagery to identify and assess the impact of disasters on infrastructure and communities.

Following a disaster event, timely and accurate assessment of the extent and severity of damage is essential for coordinating response efforts and allocating resources effectively [12]. Satellite imagery provides a bird's-eye view of affected areas, allowing responders to identify damaged infrastructure, assess the impact on communities, and plan targeted interventions. By leveraging high-resolution imagery and automated image analysis techniques, responders can streamline the damage assessment process and expedite recovery efforts.

Unmanned aerial vehicles (UAVs), or drones, offer a flexible and cost-effective platform for conducting rapid assessments in disaster-affected areas [13]. Equipped with high-resolution cameras and sensors, UAVs can capture detailed imagery of inaccessible or hazardous locations, providing responders with real-time situational awareness and actionable intelligence. By deploying UAVs for damage assessment, search and rescue operations, and environmental monitoring, emergency responders can improve their operational efficiency and decision-making capabilities in dynamic disaster scenarios.

4. Policy and governance

Traditional approaches to DRR often focus on individual hazards in isolation, leading to fragmented and ineffective risk management strategies. Integrated risk management frameworks advocate for a holistic approach that considers the interconnected nature of hazards and vulnerabilities. By adopting a comprehensive risk governance approach, policymakers can enhance coordination among stakeholders and promote sustainable development practices that reduce overall risk exposure.

In regions prone to multiple hazards, such as earthquakes, tsunamis, and cyclones, hazard-specific approaches may overlook the compounding effects of concurrent or sequential disasters [14]. Multi-hazard risk assessment and planning frameworks integrate data from various sources to identify synergies and trade-offs between different risks, allowing policymakers to prioritize interventions that address multiple hazards simultaneously. The Integrated Risk Management Framework project exemplifies this multi-dimensional approach, aiming to build more resilient communities prepared to withstand complex disaster scenarios.

As climate change exacerbates the frequency and intensity of natural disasters, there is a growing recognition of the need to integrate climate change adaptation measures into DRR strategies [15]. By mainstreaming climate resilience considerations into policy formulation and infrastructure planning, governments can build more resilient communities better equipped to withstand future shocks and uncertainties. The Climate Resilient Infrastructure Project, for instance, focuses on incorporating climate-resilient design principles into new infrastructure developments.

Investing in climate-resilient infrastructure is essential for reducing vulnerability to climate-related hazards and enhancing long-term sustainability. Climate-resilient design principles, such as elevated structures, green infrastructure, and nature-based solutions, can help minimize the risk of damage and disruption caused by extreme weather events. By incorporating climate resilience into infrastructure planning and development processes, governments can future-proof critical assets and infrastructure systems, ensuring their continued functionality and serviceability in a changing climate.

Integrating climate adaptation considerations into development planning processes is essential for promoting sustainable and resilient growth [16]. By mainstreaming climate risk assessments into land use planning, infrastructure investment, and natural resource management decisions, governments can identify climate-related vulnerabilities and prioritize actions that enhance resilience. Moreover, incorporating climate adaptation measures into national policies and regulations helps build institutional capacity and fosters a culture of proactive risk management at all levels of governance.

5. Community engagement and empowerment

Empowering communities to actively participate in decision-making processes, co-create solutions, and take ownership of their resilience initiatives fosters a culture of preparedness, solidarity, and collective action. By nurturing partnerships between communities, governments, academia, and civil society organizations, we can leverage diverse perspectives, resources, and expertise to address the unique needs and priorities of each community and ensure inclusive and sustainable risk reduction outcomes. The ROSES project in the Western Balkans and the SAILOR project in the Georgia-Azerbaijan area are examples of cross-border initiatives that aim to enhance risk awareness, collaboration, and preparedness, thereby contributing to the resilience of communities across borders. The SOLVE project in the cross-border area of North Macedonia and Greece and BALANCE project in Western Balkans improve multilateral response, lower disaster risk in vulnerable areas and groups, boost coordination for complex events, enhance skills and contingency plans, and strengthen cooperation in civil protection.

Engaging communities in participatory risk mapping and planning exercises can help identify local hazards, vulnerabilities, and capacities, enabling residents to develop context-specific strategies for risk reduction and response. By combining traditional knowledge with scientific expertise, community-led risk mapping initiatives can generate valuable insights into local risk dynamics and inform the design of targeted interventions. Moreover, involving community members in decision-making processes fosters ownership and accountability, leading to more sustainable and effective risk reduction outcomes [17].

Building resilient communities requires strengthening social networks and fostering mutual support among residents. Community resilience networks, such as neighborhood associations and volunteer groups, play a crucial role in coordinating disaster preparedness and response activities at the local level [18]. By facilitating information sharing, skills training, and resource mobilization, these networks enhance community cohesion and adaptive capacity, enabling residents to cope with and recover from disasters more effectively. Moreover, leveraging existing social capital and cultural traditions can enhance the effectiveness and sustainability of community resilience initiatives.

Vulnerable and marginalized groups, such as women, children, and persons with disabilities, often bear the brunt of disasters due to pre-existing inequalities and discrimination. Inclusive approaches to risk reduction prioritize the needs and rights of these populations, ensuring their meaningful participation in decision-making processes and access to life-saving services [19]. By addressing underlying social and economic vulnerabilities, inclusive DRR strategies can enhance the resilience of the entire community and promote sustainable development outcomes.

Gender-responsive DRR approaches recognize the unique vulnerabilities and capacities of men, women, and gender-diverse individuals and seek to address gender inequalities in disaster preparedness, response, and recovery efforts [20]. By mainstreaming gender considerations into policy and programming, governments can ensure that women's voices are heard, their needs are met, and their contributions are valued throughout the disaster management cycle. Moreover, empowering women as agents of change enhances community resilience and fosters more inclusive and equitable societies.

Persons with disabilities often face significant barriers to accessing life-saving information, services, and support during disasters. Disability-inclusive DRR strategies prioritize the rights and needs of persons with disabilities, ensuring their full participation in disaster preparedness, response, and recovery efforts [21]. By mainstreaming disability considerations into policy development and program implementation, governments can remove physical, communication, and attitudinal barriers that prevent persons with disabilities from fully engaging in disaster risk reduction activities. Moreover, promoting disability-inclusive approaches fosters a culture of diversity, equity, and inclusion that benefits society as a whole.

6. Conclusions

The exploration of new insights and innovations in disaster risk reduction illuminates promising opportunities to fortify societies against the threats posed by disasters. The convergence of cutting-edge technologies, interdisciplinary approaches, and community-centered strategies has the potential to significantly bolster our collective resilience and sustainability in the face of diverse hazards.

The integration of advanced technologies, such as artificial intelligence, machine learning, and remote sensing, equips us with powerful tools for anticipating, assessing, and responding to disaster risks with unprecedented accuracy and efficiency. These technologies enable the development of early warning systems that provide timely alerts to at-risk populations, allowing for proactive measures to mitigate the impacts of impending disasters. Moreover, the utilization of remote sensing techniques facilitates rapid damage assessment and resource allocation, guiding targeted interventions and optimizing disaster response efforts.

Interdisciplinary approaches that draw upon diverse fields of expertise, including science, engineering, social sciences, and policy, foster a more comprehensive understanding of disaster risk dynamics and inform the design of effective risk reduction strategies. By synthesizing insights from multiple disciplines, we can develop holistic solutions that address the complex interplay of hazards, vulnerabilities, and societal factors contributing to disaster risk. This integrative approach enhances our ability to anticipate emerging threats, adapt to changing conditions, and build resilience across all levels of society.

Community engagement and empowerment lie at the heart of effective disaster risk reduction efforts, recognizing the invaluable role of local knowledge, capacities, and social networks in building resilience from the ground up. Empowering communities to actively participate in decision-making processes, co-create solutions, and take ownership of their resilience initiatives fosters a culture of preparedness, solidarity, and collective action. By nurturing partnerships between communities, governments, academia, and civil society organizations, we can leverage diverse perspectives, resources, and expertise to address the unique needs and priorities of each community and ensure inclusive and sustainable risk reduction outcomes.

Realizing the vision of a safer and more resilient future for all requires sustained investment, political commitment, and collaboration across sectors and stakeholders. It entails mobilizing financial resources to support research, innovation, and capacity-building initiatives aimed at strengthening our resilience to current and emerging threats. It also demands political leadership and governance structures that prioritize disaster risk reduction as a fundamental component of sustainable development, integrating risk considerations into planning, policymaking, and investment decisions at all levels.

Furthermore, achieving our shared goals necessitates fostering partnerships and fostering collaboration among governments, international organizations, academia, private sector entities, and civil society actors. By fostering an environment of cooperation, knowledge sharing, and mutual support, we can leverage the collective expertise and resources of diverse stakeholders to address complex challenges and maximize the effectiveness of our risk reduction efforts.

In the end, the journey toward building resilient and sustainable societies is complex and challenging. It is also filled with opportunities for innovation, collaboration, and transformative change. By embracing the latest insights and innovations in disaster risk reduction and working together across disciplines and sectors, we can build a safer, more resilient future for all, where communities are empowered to thrive in the face of adversity.

Acknowledgment

The authors acknowledge the use of trinka-grammar-checker tool for the language polishing of the manuscript. This tool provided invaluable assistance in refining the grammar, punctuation, and overall clarity of the text.

Funding

This chapter is funded by the DG ECHO, ROSES European project.

Author details

Eleni Kalliontzi, Amalia Kouskoura, Evangelos Katsaros and Ioannis Bakouros*
Management of Technology Research Lab (MaterLab), University of Western Macedonia, Kozani, Greece

*Address all correspondence to: ylb@uowm.gr

References

[1] Kalaycıoğlu M, Kalaycıoğlu S, Çelik K, Christie R, Filippi ME. An analysis of social vulnerability in a multi-hazard urban context for improving disaster risk reduction policies: The case of Sancaktepe, İstanbul. International Journal of Disaster Risk Reduction. 2023;**91**:101987-102005

[2] Dabral A, Bajwa S, Shioyama S, Chatterjee R, Shaw R. Social innovation hackathon for driving innovation in disaster risk reduction (DRR). Journal of Integrated Disaster Risk Management. 2021;**11**(1):12-28

[3] Girotto CD, Piadeh F, Bkhtiari V, Behzadian K, Chen AS, Campos LC, et al. A critical review of digital technology innovations for early warning of water-related disease outbreaks associated with climatic hazards. International Journal of Disaster Risk Reduction. 2024;**100**:107452-107472

[4] Iao-Jörgensen J. Networking in action: Taking collaborative capacity development seriously for disaster risk management. Progress in Disaster Science. 2024;**21**:100265-100282

[5] Pyke J, Law A, Jiang M, de Lacy T. Learning from the locals: The role of stakeholder engagement in building tourism and community resilience. Journal of Ecotourism. 2018;**17**(3):241-258

[6] Dimitrova M, Snair M. Classifying disaster risk reduction strategies: Conceptualizing and testing a novel integrated approach. Globalization and Health. 2024;**20**(1):101202-101215

[7] Gigerenzer G, Reb J, Luan S. Smart heuristics for individuals, teams, and organizations. Annual Review of Organizational Psychology and Organizational Behavior. 2022;**9**:135-155

[8] Abunyewah M, Gajendran T, Maund K. Conceptual framework for motivating actions towards disaster preparedness through risk communication. Procedia Engineering. 2018;**212**:433-440

[9] Beshears J, Kosowsky H. Nudging: Progress to date and future directions. Organizational Behavior and Human Decision Processes. 2020;**161**:3-19

[10] Liu S, Liu Y, Chu Z, Yang K, Wang G, Zhang L, et al. Evaluation of tropical cyclone disaster loss using machine learning algorithms with an explainable artificial intelligence approach. Sustainability. 2023;**15**(16):10876-10892

[11] Kamal, Paul S, Bhaumik P. Disaster management through integrative AI. In: ACM International Conference Proceeding Series. 2022;**17**:87-103

[12] Krichen M, Abdalzaher MS, Elwekeil M, Fouda MM. Managing natural disasters: An analysis of technological advancements, opportunities, and challenges. Internet of Things and Cyber-Physical Systems. 2024;**4**:100134-100147

[13] Bushnaq OM, Mishra D, Natalizio E, Akyildiz IF. Unmanned aerial vehicles (UAVs) for disaster management. In: Nanotechnology-Based Smart Remote Sensing Networks for Disaster Prevention. 2022;**13**:153-172

[14] De Angeli S, Malamud BD, Rossi L, Taylor FE, Trasforini E, Rudari R. A multi-hazard framework for spatial-temporal impact analysis. International Journal of Disaster Risk Reduction. 2022;**73**:102838-102854

[15] Farinós-Dasí J, Pinazo-Dallenbach P, Peiró Sánchez-Manjavacas E, Rodríguez-Bernal DC. Disaster risk management, climate change adaptation and the role of spatial and urban planning: Evidence from European case studies. Natural Hazards. 2024;**112**:349-372

[16] Kalogiannidis S, Kalfas D, Papaevangelou O, Chatzitheodoridis F, Katsetsiadou KN, Lekkas E. Integration of climate change strategies into policy and planning for regional development: A case study of Greece. Land (Basel). 2024;**13**(3):389-407

[17] Adabanya U, Awosika A, Moon JH, Reddy YU, Ugwuja F. Changing a community: A holistic view of the fundamental human needs and their public health impacts. Cureus. 2023;**15**:3823-3839

[18] Ganoe M, Roslida J, Sihotang T. The impact of volunteerism on community resilience in disaster management. Jurnal Ilmu Pendidikan dan Humaniora. 2023;**12**(3):143-162

[19] Nkombi Z, Wentink GJ. The role of public participation in disaster risk reduction initiatives: The case of Katlehong township. Jamba: Journal of Disaster Risk Studies. 2022;**14**(1):210-226

[20] Erman A, De Vries Robbe SA, Thies SF, Kabir K, Maruo M. Gender Dimensions of Disaster Risk and Resilience. 2021;**13**:159-176

[21] Sonpal D. Disability, disaster and the Law: Legislating disability inclusive disaster risk reduction. In: Routledge Readings on Law, Development and Legal Pluralism: Ecology, Families, Governance. 2022;**17**:290-305

Chapter 3

Risk Management in Mass Gatherings

Asghar Tavan, Abbasali Dehghani Tafti and Mahmood Nekoie-Moghadam

Abstract

In this section, risk management in mass gatherings will be addressed. Mass gatherings, due to their nature, fall into the category of unique events since they have the potential for incidents, injuries, and even fatalities. Identifying these potential risks and managing them can help event organizers to prevent accidents. This section of the book, by classifying and introducing threatening risks in mass gatherings, provides a better understanding of the organization of such events and ultimately facilitates the risk assessment process in mass gatherings.

Keywords: risk, risk management, mass gatherings, risk assessment, health

1. Introduction

According to the World Health Organization's definition, a mass gathering is an event, pre-planned or spontaneous, where the number of individuals is sufficient to strain the planning and response resources of the community at the local, state, or national level. Mass gatherings can occur for various reasons, including religious, social, cultural, political, and sporting events. Mass gatherings are broadly categorized into planned and unplanned (spontaneous) gatherings, each having its own subtypes. This categorization is crucial in the management of mass gatherings because the planning and execution of these events depend on the type of gathering and the conditions of the venue. Depending on the type of gathering, the conditions, and the facilities available, various risks may threaten the participants. Managing unplanned mass gatherings is notably more challenging, and these gatherings pose greater health risks to participants [1].

The safe organization of a mass gathering requires preparedness from various entities involved in its execution, such as the main organizing authority, different health departments, pre-hospital emergency services, regional hospitals, and collaborating sectors like the police and fire departments. Within these, the health sector must ensure readiness in various areas. As the first step in the preparedness plan, each sector should identify the risks associated with its role. After identifying the hazards and vulnerabilities of each event, we can analyze potential risks and design operational plans to eliminate or reduce them. Indeed, each of these events is susceptible to health-threatening risks and has its own specific incidents. Over the past decade, concerns about health-related hazards in mass gatherings have increased [2]. The primary health concerns in mass gatherings include infectious diseases such as respiratory and foodborne illnesses, injuries, traffic accidents, heat-related issues, insect bites, non-communicable diseases, and terrorism [3]. Many mass gatherings have reported high mortality rates due to communicable diseases, non-communicable diseases, injuries from stampedes, and terrorist attacks. In addition to infectious diseases, both endemic and seasonal (resulting from water, food, and respiratory sources), the health system faces challenges such as bruising, lacerations, limb amputations, muscle strains, headaches, dizziness, asthma and respiratory problems, heat-related issues, and abdominal pains that require specialized medical care [4]. In addition, participants in mass gatherings may undergo psychological changes that can have negative aspects, posing risks to their health [5]. Numerous statistics from around the world highlight the multiple injuries resulting from mass gatherings. These unfortunate incidents occur in various types of cultural, sports, and religious events. In addition to causing injuries, these events have also led to fatalities. The primary causes of mortality are often associated with stampedes and trauma [6]. The hazards associated with mass gatherings are related not only to the number of participants but also depend on various other factors. The risks have a direct relation to the existing hazards, the level of exposure, and the vulnerability of the population and the environment [7]. Organizers of mass gatherings must have access to reliable data from various aspects of the event to provide efficient resources and equipment and predict the required personnel in the health sector [8]. The most important step in this regard is assessing the risks of mass gatherings in the pre-event stage. Many injuries resulting from mass gatherings occur due to the lack of a risk management strategy. To prepare for managing mass gatherings and implement preventive measures to reduce risks, in the first step, it is necessary to identify risk factors and conditions that may contribute to health hazards in a mass gathering, creating the possibility for a comprehensive risk assessment. In essence, after identifying the hazards, we can analyze and evaluate them, assessing the likelihood of incidents and their impact on participants. To achieve maximum risks in mass gatherings, the best way is to categorize them based on a scientific methodology. With this approach, we are essentially aiming to develop effective management actions. One such classification is the 5-part categorization, which classifies various threatening risks into five main categories:

1. Environmental hazards

2. Hygiene hazards

3. Psychosocial hazards

4. Individual hazards

5. Management and facility hazards [9].

Understanding the hazards of each category in managing the risks of that category can be very beneficial. The collection of risks from all categories can provide event organizers with a comprehensive and complete overview of risks. In fact, after identifying risks in each category, a general risk assessment for the entire mass gathering

Figure 1.
Five domains of mass gathering hazards.

can be obtained. For a better understanding of each category and ultimately achieving an overall risk assessment, the threatening risks of each category will be fully explained in this section (**Figure 1**).

2. Environmental hazards

This area encompasses risks related to the event venue and weather conditions during the gathering. These risk factors include the number of participants and population density, moving crowds, hot weather air humidity, wide dispersion (large distances) in uneven environments, the likelihood of traffic accidents, rainfall, cold weather, outdoor gatherings, gatherings without clear boundaries, drowning incidents, insect bites, and sunburn. The larger the number of participants in a gathering, the greater the need for planning. As the population density increases, the likelihood of risks related to stampedes and associated traumas also rises. According to studies, between 1980 and 2007, 215 different human stampede incidents led to 7069 deaths and 14,000 documented injuries worldwide. With 965 people reportedly killed during a religious gathering in Baghdad in 2005, it was the deadliest stampede of its kind in the previous century. A total of 380 pilgrims lost their lives in the human stampede that occurred in Mina Valley during the 2006 Hajj. Only in the year 2015 occurred another human stampede during the Hajj, which tragically lost the lives of 769 pilgrims [10].

A study on the mass gathering of Hajj in 2018 showed that there were referrals to relief centers with different degrees of hyperthermia between 11:00 AM and 2:00 PM. In addition, people with old age and underlying diseases have a higher risk of hyperthermia [11]. A modeling study on 79 mass gatherings showed that the environmental heat index is one of the important factors in increasing the patient presentation rate (PPR) in mass gatherings [8].

Holding a mass gathering in extremely cold or hot weather increases the risks. In colder weather, the use of heating devices is necessary, which itself requires safety measures. In very hot weather, the risk of heat-related illnesses such as heatstroke increases. Additionally, participants' need for safe drinking water will be higher in these conditions.

2.1 Solutions and measures to reduce environmental hazards

One of the first steps organizers of mass gatherings should consider is estimating the number of participants based on the available space. Additionally, the entry of individuals into the gathering venue should be regulated through specific gates. Restriction on the entry of vehicles into the gathering area should be clearly defined at a specified boundary to minimize the risks associated with traffic accidents.

The facilities available in the venue or the environment of the gathering place should be proportional to the number of participants and accommodated at the event site. The population density needs to be estimated, and facilities should be strategically placed throughout the gathering location based on this density to ensure adequate coverage.

In gatherings held in open environments, especially those involving walking, it is essential to have designated rest areas with shade and access to clean water at specified intervals. If the gathering takes place near a river or lake, access to it should be restricted or prohibited to prevent the risks associated with drowning [9, 12].

3. Hygiene hazards

This area encompasses significant hazards that have the potential to pose risks to all participants and even the entire community. These risks include the outbreak of diseases, the hygiene quality of food and beverages, waste disposal, and wastewater discharge. In fact, this domain highlights risks that may have future consequences for the entire society, including epidemics of respiratory diseases that may occur after the mass gathering in the community.

In addition, due to the lack of adherence to hygiene requirements related to food and water, outbreaks of gastrointestinal epidemics may also occur. In gatherings, especially those lasting more than one day, negligence in proper waste disposal and sewage can pose environmental hazards to the region. Lack of health supervision in the preparation and distribution of food is another risk in this area. A 15-year retrospective study on the spread of meningitis in mass gatherings has shown that these gatherings have a high potential for the spread of infectious diseases, although the development of a vaccine to prevent this health problem has been helpful [13]. Also, in mass gatherings held in the United States, the spread and transmission of respiratory diseases such as influenza have been reported. This concern is also present in international gatherings about more diseases such as measles and mumps [14]. The outbreak of food poisoning is one of the concerns that has been observed in mass gatherings. In one case in India, 291 people with symptoms of diarrhea, abdominal pain, fever, chills, and vomiting went to the hospital due to food poisoning within a few hours after the gathering [15]. During the spread of the COVID-19 pandemic, it was shown that due to the possibility of the virus spreading in mass gatherings, the death rate caused by the virus increases in events [16].

3.1 Solutions and measures to reduce hygiene hazards

One of the most important solutions to reduce the risks related to the health sector is the design of a patient tracking and surveillance system. Based on the potential diseases that may cause an epidemic in mass gatherings, a surveillance system should be adjusted. The syndromic care system helps the organizers to identify people who have symptoms of respiratory, gastrointestinal diseases, etc. The distribution of food and water should be done under the supervision of health inspectors who are approved by the organizers, and the distribution of food and water from unreliable sources should be prevented. Health inspectors must have the necessary authority and responsibility to cancel the distribution of foods that are suspected from a health perspective. In planned gatherings, the venue for the gathering should be planned for standard collection of sewage and waste from a hygiene perspective. Another aspect to consider in this area is preventing the presence of street vendors in large gatherings because dealing with the distribution of food items by vendors can be very challenging. Additionally, the presence of street vendors in mass gatherings increases the possibility of terrorist and bioterrorist attacks [17, 18].

4. Psychosocial hazards

One of the factors that can turn a mass gathering into a severe incident with multiple casualties is the behavior of the crowd. Before organizing a large gathering, the crowd behavior should be determined based on the type of gathering. In gatherings accompanied by the excitement of participants, there is a higher possibility of risks. For example, confrontational gatherings where two opposing groups organize it together, election campaigns, or sporting events held in stadiums fall into this category.

In gatherings where alcohol consumption is possible, the risks in this area are highlighted because the crowd has the necessary behavioral potential for various risks. Another risk related to this area is the possibility of widespread rumors in gatherings. The spread of rumors among the crowd may lead to changes in crowd behavior, overcrowding, and pose health risks to the participants. Reported from sporting and music MGEs demonstrated that, drug and alcohol usage lead to intoxication, aggression, and serious injuries [19]. In addition to the behavior of the crowd, one of the other important factors in the psychological field of mass gatherings is the mood of the crowd. Identifying population mood is more difficult than identifying population behavior because it is often invisible. The mood of the population is divided into three levels: passive, active, and energetic. The more the mood of the population is toward the energetic level, the higher the probability of risks [20].

4.1 Solutions and measures to reduce psychosocial hazards

Some measures related to reducing these hazards need to be taken before holding the gathering, and some actions should also be carried out during the event.

Before the gathering, the type of event and the behavior of the crowd should be anticipated, and special facilities for it should be planned at the event venue accordingly. During the gathering, the behavior of the crowd, the level of excitement, and the potential for crowd confrontation should be monitored, and if any slight changes are observed, necessary interventions should take place. Before the gathering, laws related to alcohol consumption restrictions or prohibitions should be communicated to the participants. It is recommended that cameras responsible for monitoring the crowd during the gathering be equipped with operators trained in crowd behavior analysis. These operators should report any changes, congestion, or excitement observed during the event. To prevent the spread of rumors and the creation of anxiety among the participants, multi-layered communication should be carried out. This communication can be done through monitors, loudspeakers, or paper announcements. Timely and comprehensive communication will create confidence among participants and prevent emotional behaviors [9, 21].

5. Individual hazards

This area essentially reflects the general characteristics and demographics of participants in mass gatherings. The risks in this area increase as participants in mass gatherings come from vulnerable population groups. The presence of the elderly, children, and women in mass gatherings increases the required facilities for the event. Involving individuals with illnesses or disabilities and limited mobility in mass gatherings will face higher risks in this area. However, it is not possible to prevent the presence of these individuals in gatherings. Considering the nature of gatherings, which may involve recreational, social, or cultural activities, the presence of these individuals is unavoidable.

5.1 Solutions and measures to reduce psychosocial hazards

The first step in reducing such risks is to create a comprehensive profile of the population of individuals participating in the gathering. This action should be undertaken before the gathering by identifying and registering participants. Based on this profile, specific facilities may be added to the location of the mass gathering. Considerations such as wheelchair access, additional seating areas, and increased spacing for the crowd are among these measures. For gatherings that last more than a day or several days, providing medical and pharmaceutical needs for sick and elderly individuals should be mandatory.

6. Management and facility hazards

The risks arising from this area include improper planning, insufficient organizing personnel, inadequate facilities, and improper placement of facilities. For example, the lack or shortage of healthcare personnel at the gathering site (doctors, nurses, firefighters, police, and emergency medical workers), neglect of security and the potential for terrorist attacks, the absence of directional signs, failure in risk assessment, lack of communication, lack of pre-incident warning, absence of vaccination programs, lack of collaboration among agencies, difficult access to healthcare services, scarcity of medical service resources, challenging access to water and food, availability of alcohol, absence of free water, and the organization of gatherings for an extended period. Each of these risks has a high potential for causing incidents with a large number of casualties. Mass gatherings, given their nature, require planning similar to what happens in emergencies and disasters. In these gatherings, various organizations with different missions must work alongside each other. In this context, the unity of command and communication appropriate to the incident are crucial [9]. Statistics extracted from the Global Terrorism Database (GTD) over the last 50 years show that the number of terrorist attacks has increased in recent decades. These attacks have been carried out with various weapons in various music, sports, religious, and political festivals. These attacks in mass gatherings have killed hundreds of people and injured thousands of others [17].

6.1 Solutions and measures to reduce management and facility hazards

The most crucial step in this area is risk assessment before organizing a mass gathering. All potential risks, including those mentioned in previous areas, should be assessed, classified, and managed before conducting the mass gathering. Without conducting a risk assessment, any action taken will be blind and non-purposive.

In this area, it is essential to estimate the facilities and personnel required for organizing a mass gathering. This estimation depends on the results of the initial risk assessment. Another important category in this area is the placement of facilities and employees who provide services. This is particularly crucial, especially for emergency response personnel and healthcare service providers. In the event of an incident and the participants' need for medical services, both access to the target population and the evacuation of injured individuals are vital.

Access routes to the population and methods of transporting injured individuals to medical centers should be clearly defined and practiced on pre-designed maps. The safe conduct of mass gatherings requires the definition of multiple communication layers to establish effective communication among all organizing entities in the gathering. Due to population overload, the use of conventional communication systems such as mobile phones in mass gatherings may come with challenges. Therefore, organizers should equip themselves with alternative communication layers, such as using wireless communication, monitors, displays, and audio alerts at international gatherings should be compatible with the languages of the participants in the gathering.

In addition, in international gatherings, attention should be paid to the cultural differences among participants in mass gatherings. Warnings, educational programs, and executive plans should be designed in accordance with different cultures. In events that last more than a few days, a well-thought-out program for educating participants before the event is essential.

The Incident Command System (ICS) should be planned and practiced in advance. Individuals serving as responders in mass gatherings need to be familiar with their roles and responsibilities outlined in the ICS. This requires conducting multiple exercises and drills in the pre-event phase before the event takes place [9, 22]. In order to hold safe and create the necessary coordination between all service providers in mass gatherings, it is necessary to use an integrated management system, just like natural disasters. The implementation of the public health emergency operation center (PHEOC) is one of the necessities [23].

6.2 Opportunities for conducting mass gatherings

Despite all the risks that threaten the health of participants in mass gatherings and the potential for various incidents to occur, organizing these gatherings also presents numerous opportunities for the organizers. Many planned programs for handling incidents and disasters are executable and assessable during conducting the mass gatherings. Implementing an incident command system as the primary management structure during incidents at mass gatherings, making it feasible and evaluable.

Communication systems not commonly used in normal situations, such as wireless communication and satellite phones, can be utilized and evaluated during the execution of mass gatherings. Collaborative work and coordination between departments are inevitable during mass gatherings, enhancing organizations' capabilities in this regard. Some existing processes of organizations may require fundamental changes during incidents. These processes can be practiced, assessed, and revised during the execution of mass gatherings. Performing mass gatherings sometimes necessitates the establishment of additional infrastructures in the host community. These developed infrastructures can contribute to the overall development of the hosting community. This is particularly advantageous in cities and communities with lower levels of development, providing a golden opportunity for improvement.

7. Conclusion

Mass gatherings are organized for various social, cultural, and religious reasons, primarily aimed at providing entertainment in society. However, for various reasons, they have the potential to turn into an incident. Therefore, the organizers of these gatherings must conduct a comprehensive assessment of potential risks and, based on that, identify and reduce the identified hazards as much as possible. They should also anticipate the necessary facilities and human resources required for organizing mass gatherings. This section of the chapter provides a classification of threatening risks in mass gatherings and details each category, offering organizers the opportunity to identify these risks. In addition, each category includes proposed solutions to assist in managing and reducing these hazards.

Author details

Asghar Tavan[1]*, Abbasali Dehghani Tafti[2] and Mahmood Nekoie-Moghadam[3]

1 Health in Disasters and Emergencies Research Center, Institute for Futures Studies in Health, Kerman University of Medical Sciences, Kerman, Iran

2 Department of Health in Disasters and Emergencies, School of Public Health, Shahid Sadoughi University of Medical Sciences, Yazd, Iran

3 Department of Health in Disasters and Emergencies, School of Management and Medical Information Sciences, Kerman University of Medical Sciences, Kerman, Iran

*Address all correspondence to: at.tavan@yahoo.com

References

[1] World Health Organization. Public Health for Mass Gatherings: Key Considerations. World Health Organization; 2015;**25**:26. Available from: https://apps.who.int/iris/handle/10665/162109

[2] Aitsi-Selmi A, Murray V, Heymann D, McCloskey B, Azhar EI, Petersen E, et al. Reducing Tisks to health and wellbeing at mass gatherings: the role of the Sendai Framework for Disaster Risk Reduction. International Journal of Infectious Diseases. 2016;**47**:101-104

[3] Memish ZA, Steffen R, White P, Dar O, Azhar EI, Sharma A, et al. Mass gatherings medicine: Public health issues arising from mass gathering religious and sporting events. The Lancet. 2019;**393, 84**(10185):2073

[4] Hoang V-T, Gautret P. Infectious diseases and mass gatherings. Current Infectious Disease Reports. 2018;**20**:1-12

[5] Hopkins N, Reicher S. The psychology of health and well-being in mass gatherings: A review and a research agenda. Journal of Epidemiology and Global Health. 2016;**6**(2):49-57

[6] Schwartz B, Nafziger S, Milsten A, Luk J, Yancey A. Mass gathering medical care: resource document for the National Association of EMS Physicians Position Statement. Prehospital Emergency Care. 2015;**19, 68**(4):559

[7] Ranse J, Hutton A, Keene T, Lenson S, Luther M, Bost N, et al. Health service impact from mass gatherings: A systematic literature review. Prehospital and Disaster Medicine. 2017;**32, 7**(1):71

[8] Locoh-Donou S, Yan G, Berry T, O'Connor R, Sochor M, Charlton N, et al. Mass gathering medicine: Event factors predicting patient presentation rates. Internal and Emergency Medicine. 2016;**11, 52**:745

[9] TavanA,TaftiAD,Nekoie-MoghadamM, Ehrampoush M, Nasab MRV, Tavangar H, et al. Risks threatening the health of people participating in mass gatherings: A systematic review. Journal of Education and Health Promotion. 2019;**8**:209

[10] Ahmed QA, Memish ZA. From the "Madding Crowd" to mass gatherings-religion, sport, culture and public health. Travel Medicine and Infectious Disease. 2019;**28**:91-97

[11] Alkassas W, Rajab AM, Alrashood ST, Khan MA, Dibas M, Zaman M. Heat-related illnesses in a mass gathering event and the necessity for newer diagnostic criteria: a field study. Environmental Science and Pollution Research. 2021;**28**:16682-16689

[12] Arbon P. The development of conceptual models for mass-gathering health. Prehospital and Disaster Medicine. 2004;**19**(3):208-212

[13] Coudeville L, Amiche A, Rahman A, Arino J, Tang B, Jollivet O, et al. Disease transmission and mass gatherings: a case study on meningococcal infection during Hajj. BMC Infectious Diseases. 2022;**22**(1):1-10

[14] Rainey JJ, Phelps T, Shi J. Mass gatherings and respiratory disease outbreaks in the United States–should we be worried? results from a systematic literature review and analysis of the national outbreak reporting system. PLoS One. 2016;**11**(8):e0160378

[15] Bajaj S, Dudeja P. Food poisoning outbreak in a religious mass gathering.

Medical Journal Armed Forces India. 2019;75(3):339-343

[16] Ahammer A, Halla M, Lackner M. Mass gatherings contributed to early COVID-19 mortality: Evidence from US sports. Contemporary Economic Policy. 2023

[17] De Cauwer H, Barten DG, Tin D, Mortelmans LJ, Ciottone GR, Somville F. Terrorist attacks against concerts and festivals: a review of 146 incidents in the Global Terrorism Database. Prehospital and Disaster Medicine. 2023;**38**(1):33-40

[18] Tavan A, Tafti AD, Nekoie-Moghadam M, Ehrampoush M, Nasab MRV, Tavangar H. Public health risks threatening health of people participating in mass gatherings: A qualitative study. Indian Journal of Public Health. 2020;**64**(3):242-247

[19] Hutton A, Savage C, Ranse J, Finnell D, Kub J. The use of Haddon's matrix to plan for injury and illness prevention at outdoor music festivals. Prehospital and Disaster Medicine. 2015;**30**(2):175-183

[20] Hutton A, Ranse J, Gray KL, Turris S, Lund A, Munn MB. Psychosocial influences on patient presentations: considerations for research and evaluation at mass-gathering events. Prehospital and Disaster Medicine. 2020;**35**(2):197-205

[21] Hopkins N, Reicher S. Adding a psychological dimension to mass gatherings medicine. International Journal of Infectious Diseases. 2016;**47**:112-116

[22] Davarani ER, Tavan A, Amiri H, Sahebi A. Response capability of hospitals to an incident caused by mass gatherings in southeast Iran. Injury. 2022;**53**(5):1722-1726

[23] Elachola H, Al-Tawfiq JA, Turkestani A, Memish ZA. Public Health Emergency Operations Center-A critical component of mass gatherings management infrastructure. The Journal of Infection in Developing Countries. 2016;**10**(08):785-790

Section 2

Climate Data Analysis and Technological Tools

Chapter 4

Analysis of Climate Indices to Determine Global Climate Patterns: Techniques for Summarizing Complex Climate Data

Edgard Gonzales and Kenny Gonzales

Abstract

Large-scale atmospheric and oceanic circulation fluctuations have a strong impact on the global hydrological cycle and tropical cyclones (TC), which mainly generate intense precipitation and flooding. The primary objective of this research is to analyze the main climate indices (CI), which are techniques used to summarize complex climate information in simpler and more understandable forms. These indices are based on meteorological data, such as temperature, precipitation, humidity, and other atmospheric parameters to provide summary information about the climatic conditions in a particular region. Some common utilities and functions of climate indices are (i) climate monitoring; (ii) anomaly detection; (iii) agricultural planning; (iv) climate risk assessment; (v) scientific research; (vi) climate insurance; (vii) climate adaptation; and (viii) evaluation of water resources. CI play a crucial role in water management climate research and public policy planning, providing tools to understand and address challenges associated with climate conditions.

Keywords: global hydrological cycle, tropical cyclones, climate indices, precipitation, flooding

1. Introduction

It is now well-recognized that large-scale atmospheric and oceanic circulation fluctuations have a strong impact on the global hydrological cycle. Such relationships are helpful in the global understanding of the non-stationary relationships that exist between ocean and atmosphere mean conditions and freshwater discharge integrated at a continental scale [1]. TC dominate natural hazard losses, these losses have increased substantially in recent decades, driven largely by increased exposure, and studies that isolate the hazard component by normalizing the losses for exposure find no linear long-term trend [2].

CI are tools used to summarize complex climate information in simpler and more understandable forms. CI play a crucial role in various fields, from agriculture and water management to climate research and public policy planning, providing tools to understand and address challenges associated with weather conditions.

These indices are based on meteorological data, such as temperature, precipitation, humidity, and others, to provide summarized information about weather conditions in a particular region. Here are some common utilities and functions of climate indices: (i) climate monitoring; (ii) anomaly detection; (iii) agricultural planning; (iv) climate risk assessment; (v) scientific research; (vi) climate insurance; (vii) climate adaptation; and (viii) water resources assessment.

There is a diversity of climatic or oceanic indices developed by different researchers worldwide, however, for this research, only the most well-known and used indices are described.

2. North Atlantic Oscillation (NAO)

NAO, represents the dominant climate pattern, where a meridional shift of the atmospheric mass occurs over the North Atlantic area, Europe, and North America. The NAO measures the strength of westerly flow (positive with strong westerly winds and vice versa) and is an important source of seasonal to interdecadal variability, influenced by ocean and sea ice processes. The NAO and its recent counterpart, the Arctic Oscillation/Northern Hemisphere Annular Mode (AO/NAM), are the significant variability modes in the northern hemisphere winter climate [3].

NAO is based on the difference in the air pressure between the high area of the Azores (Azores Anticyclone) and the low area of Iceland (Iceland Anticyclone) and has two phases: positive and negative.

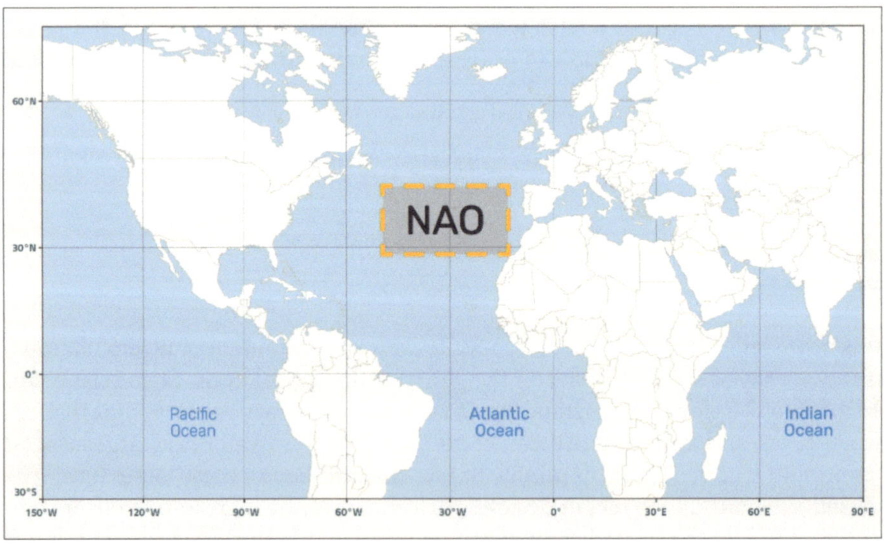

Figure 1.
NAO, based on the atmospheric pressure, differences between Iceland and the Azores represent the dominant climate pattern in the North Atlantic Region, and its teleconnection is characterized by a meridional displacement of atmospheric mass.

During positive phases of the NAO, weaker pressure occurs between Iceland and the Azores, generating above-average temperatures in the eastern US and throughout northern Europe and below-average temperatures in Greenland and throughout southern Europe, Europe, and the Middle East. This variability causes stronger wetter winds from the west toward Europe, causing milder, wetter winters in that region, as well as more frequent storms in the North Atlantic.

During the negative phase of the NAO, dominant processes occur for long periods, for example, in the mid-1950s until the winter of 1978/1979 (24 years), there were four periods of at least 3 years each in which the negative phase was dominant and the positive phase was absent.

Likewise, opposite patterns of temperature and precipitation anomalies occur (colder and drier winters in Europe). They are also associated with above-average rainfall in northern Europe and Scandinavia in winter, below-average rainfall in southern and central Europe, and a decrease in storms in the North Atlantic. Both phases of the NAO present prolonged periods of positive and negative phases generating changes in the location of the jet stream and the path of storms from the North Atlantic to Western and Central Europe and with large-scale modulations of the transport patterns of heat and humidity (**Figure 1**) [4].

NAO predicts winter weather patterns (temperature and precipitation) in Europe and North America, important for agriculture and water management.

3. Tropical North Atlantic Index (TNA)

The TNA index represents the damage potential of North Atlantic TCs due to the effect of winds and coastal surges using seasonal climate parameters of sea surface temperature and flow direction. These climatic parameters are indicators of the variables of intensity size and speed of advance of the TC. The potential damage from the cyclone is based on assessments of past periods expressed in actual damage and the characteristics of exposure and vulnerability [5].

It is the development of an index of cyclone damage potential formulated as [6]

$$\mathrm{CDP} = 4\frac{\left[\left(\frac{v_m}{65}\right)^3 + 5\left(\frac{R_h}{50}\right)\right]}{v_t} \tag{1}$$

where v_m is the maximum surface wind speed (knots), R_h is the radius of hurricane-force winds (nautical miles), and v_t is the forward translation speed (knots). The CDP applies to damage caused by wind, waves, and currents at sea as well as damage caused by intense winds on land and coastal flooding, excluding damage resulting from precipitation. The CDP coefficient scales from 0 to 10, and the damage is related to the duration of destructive winds (vt^{-1}), not just their maximum value (v_m).

TNA Positive: the SST in this region is above average, favoring the formation of hurricanes in the Atlantic.

TNA Negative: SSTs are below average; therefore, hurricane season is less active.

Applying the TNA index is important to evaluate winds, coastal waves, changes in SST, and forecast hurricane and TC activity in the tropical Atlantic region. Its use is essential to plan and reduce the adverse effects of climate (**Figure 2**).

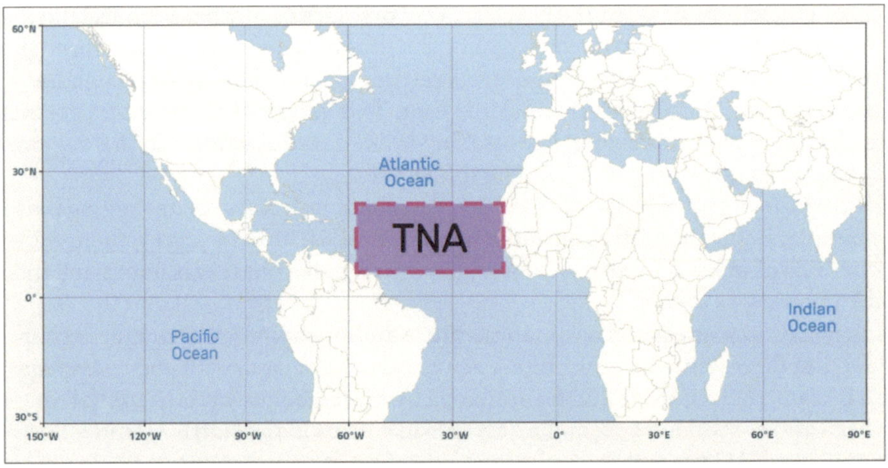

Figure 2.
TNA, based on SST variations of the tropical Atlantic and regions of the Gulf of Mexico and the west coast of Africa, monitors hurricane and tropical cyclone activity and is developed using seasonal climate variables of relative SST and steering flow.

4. South Atlantic Oscillation Index (SAM)

SAM describes the fluctuations in atmospheric pressure around the South Pole. It is the main mode of interannual atmospheric variability in the extratropics of the southern hemisphere (SH) (Atlantic Ocean, South America, Australia, New Zealand, southern Africa, and Antarctica), but is not consistent in the terrestrial areas of SH, indicating that the influence of the SAM index is modulated by regional effects. Teleconnections are stationary in regions with good proxy data coverage (South America, Tasmania, and New Zealand) [7].

Similarly, SAM is quantified by an empirical orthogonal function (EOF analysis) as EOF1 of the field or as the difference in mean sea level pressure (SLP) between mid and high latitudes [8]. Likewise, SAM also presents a barotropic vertical structure at the sea level. SAM and NAM (Northern Annular Mode) are used to determine the processes involving the troposphere and stratosphere in each hemisphere [9]. According to the EOF1 definition, SAM evolution would explain on average 30% of the climate variability of the hemisphere through the SLP or geopotential height [10].

Positive phase: it occurs during the summer, generating strengthening of the westerly winds associated with warming in the Peninsula. During the autumn, winter, and southern spring, warming occurs in the Antarctic Peninsula related to sea surface temperatures (SST) in the tropical Pacific and Atlantic.

Negative phase: it indicates a weakening of the processes generated in the positive phase.

The ENSO-Peninsula temperature relationship during autumn is weak on interannual time scales, and the tropical Pacific SST variability associated with the El Niño-Southern Oscillation (ENSO) is stronger mainly during winter and spring. These circulation changes dominate interannual temperature variability across the Antarctic Peninsula during summer and autumn.

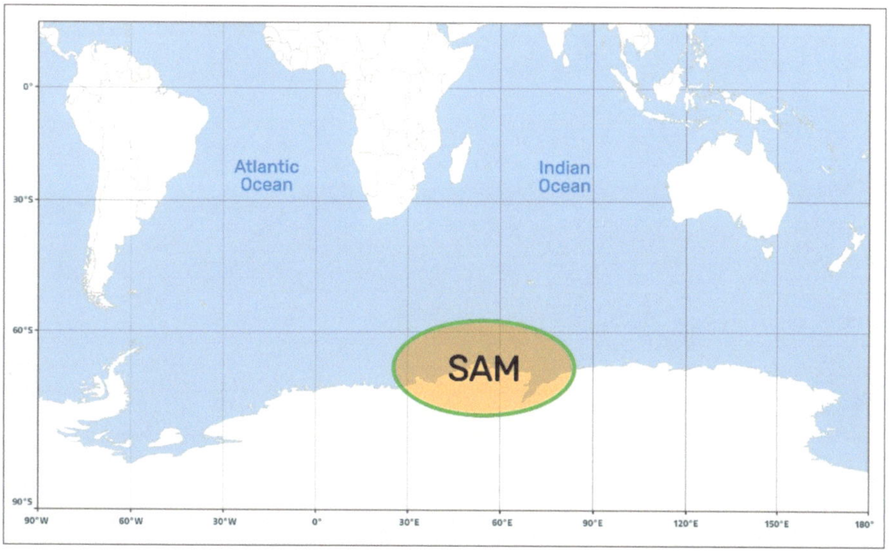

Figure 3.
SAM, based on fluctuations in atmospheric pressure over the South Atlantic; South America, Australia, New Zealand, Southern Africa, and Antarctica, presents positive and negative phases related to westerly winds associated with greater and lesser warming.

The SAM index can help foresee anomalous climatic conditions, such as heat waves, drought, or intense rains, floods, as well as the impact on ecosystems and biodiversity. Likewise, knowledge of the SAM index can help plan water resources management and natural risk management (**Figure 3**) [11].

5. Atlantic Multidecadal Oscillation Index (AMO)

The AMO operates on a longer time scale with North Atlantic SST fluctuations that can last decades, with impacts on atmospheric circulation and surrounding regions, influencing the distribution of precipitation, temperature, and extreme events such as hurricanes and storms. Similarly, the AMO monitors climate variability during alternating warm and cold phases in large parts of the Northern Hemisphere and is related to rainfall in northeastern Brazil, African Sahel, Atlantic hurricanes, and summer weather in North America and Europe.

During the positive phase of the AMO, the SST is warmer than average, in most of the North Atlantic, except for the east coast of the United States. The atmospheric patterns that appear are anomalous low pressures over the Atlantic between 20°S and 50°N, cyclonic surface winds around the low, and reduced wind speeds over the tropical Atlantic. Warm SST anomalies in the tropical Atlantic (via atmospheric teleconnections) generate anomalous cooling in the equatorial eastern-central Pacific.

During the negative phase of AMO, more or less opposite conditions occur and the SST is cold in most of the North Atlantic, producing an increase in precipitation in the eastern tropical Atlantic.

Anticyclonic anomalies in the North and South Pacific generate equatorward winds, distributing along the coasts of North and South America and contributing

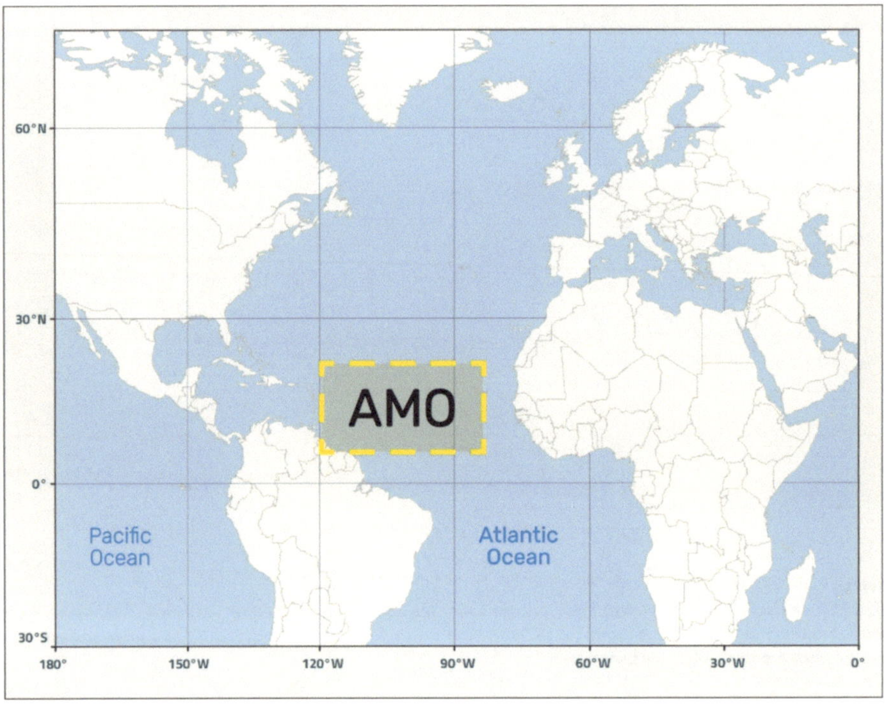

Figure 4.
AMO operates on a longer time scale (decades) with fluctuations in the North Atlantic SST, influencing atmospheric circulation, precipitation, temperature, and extreme events such as hurricanes and storms. It is a near global scale mode of observed multidecadal climate variability with alternating warm and cool phases of the northern hemisphere.

to further cooling [12]. However, there are differences between SST and atmospheric anomalies during periods of the same phase, especially in the extratropics. The correlation between the AMO and air temperature anomalies is positive over much of the world between 40°S and 50°N [13].

Global-scale Multidecadal Variability (GMV) could be generated by the AMO through atmospheric teleconnections and atmosphere-ocean coupling mechanisms. Its horizontal pattern resembles that of the Interdecadal Pacific Oscillation (IPO) in the Pacific and the AMO in the Atlantic Ocean, whose processes could affect precipitation and global temperature around the world. Observations show that there is a strong negative correlation when AMO outperforms GMV throughout approximately 4–8 years [14].

The usefulness of this index allows us to make forecasts of climate patterns in the North Atlantic on a multidecadal time scale, as well as planning in the management of the agricultural sectors, water resource management, and understanding of climate variability in the Atlantic North and its impacts on the global climate (**Figure 4**).

6. The Arctic Oscillation Index (AO)

The AO exerts a predominant influence on the variability of atmospheric circulation (atmospheric pressure) in the extratropical region of the northern hemisphere during the boreal winter, and the common feature is the oscillation of inverted sea level pressure (SLP) anomalies in high and mid-latitude regions [15].

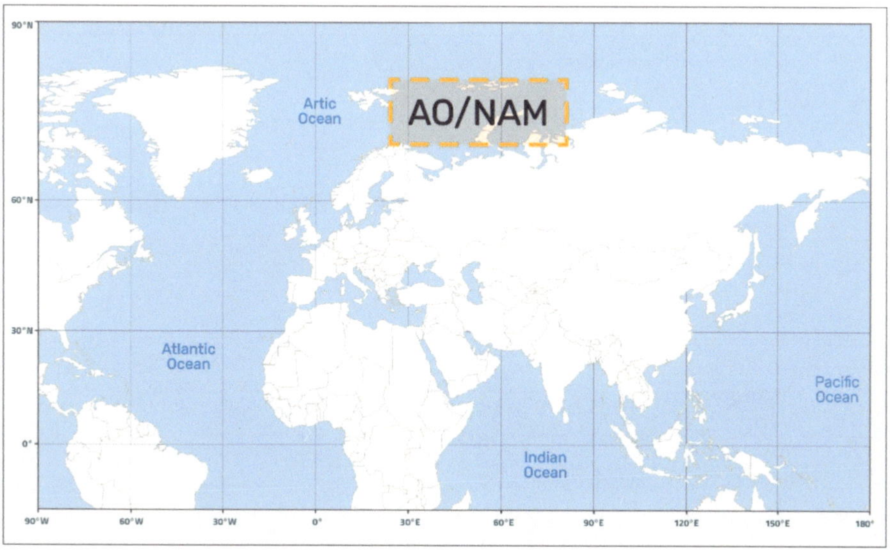

Figure 5.
AO describes the variability of atmospheric pressure at high latitudes in the northern hemisphere (Arctic) and presents positive and negative phases related to atmospheric pressure.

Positive phase: there is lower atmospheric pressure in the Arctic and higher pressure in the mid-latitudes of the northern hemisphere, leading to a strengthening of the polar jet stream and the generation of strong winds.

Negative phase: during this phase, atmospheric pressure is highest in the Arctic and lowest in mid-latitudes, and this process weakens the polar jet stream and causes greater variability in the climate.

The AO is also known as the near-surface mean sea level pressure (MSLP) pattern related to the northern annular modes (NAM). It has its application in climate models to predict climate patterns or changes in high latitudes of the northern hemisphere, as well as understanding climate variability in the Arctic and its effects on the global climate. It is also useful in short- and long-term planning in sectors such as agriculture, energy, and transportation (**Figure 5**) [16].

7. Oceanic Niño Index (ONI)

ONI is used to determine the occurrence and duration of El Niño episodes, based on monitoring SST in the central Pacific Ocean. It is used to detect a level of SST above the average that is maintained for several months, covering both the beginning and the end of an El Niño episode.

When a seasonal anomaly of 0.5°C in temperature is recorded for the first time, it indicates a high probability that an El Niño event will develop; although to confirm stronger alerts, it is necessary to wait several months. For this reason, it was suggested that an ONI value of 0.7°C marks a critical point at which the El Niño phenomenon stabilizes, providing an additional period for social decision-makers to implement preventive actions. This indicates that the initial detection of a value of 0.7°C could reliably indicate the stabilization phase of the El Niño phenomenon, providing greater reliability to the current El Niño onset indicator, set at 0.5°C, thus allowing

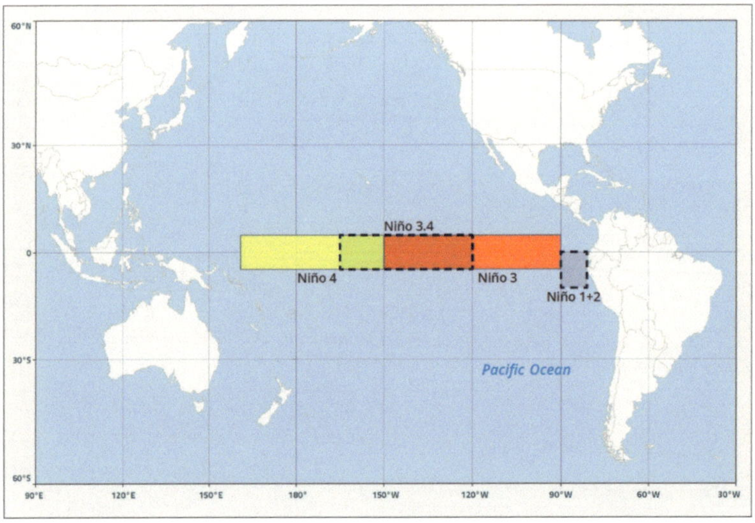

Figure 6.
ONI describes the occurrence and duration of El Niño and La Niña episodes, based on SST monitoring in the central Pacific Ocean.

vulnerable communities to prepare for the foreseeable social and ecological impacts of the phenomenon [17].

ENSO indicators can be evidenced through the Oceanic Niño Index (ONI), as well as through variations in SST that influence the intensity of precipitation. Consequently, the presence of El Niño and La Niña causes a reduction or increase in precipitation in Indonesia [18, 19].

Positive phase: known as El Niño, SSTs in the central and eastern equatorial Pacific are warmer than normal (weakening trade winds and a reduction in the resurgence of cold waters). There is an increase in precipitation over the equatorial Pacific and drought in some regions such as Australia and western South America. In addition, it could lead to an increase in the frequency of tropical cyclones in the Pacific, as well as hotter and more arid climatic conditions in certain areas of Asia and Africa.

Negative phase: known as La Niña, SSTs in the central and eastern equatorial Pacific are colder than normal (strengthening of the trade winds and an increase in the resurgence of cold waters). There is increased rainfall over the western Pacific and drier conditions in the eastern equatorial Pacific. Additionally, it may result in increased hurricane intensity in the Atlantic and cooler, wetter weather conditions in some regions of Asia and Australia.

These ONI phases, El Niño and La Niña, have important implications on global climate patterns and can affect sectors such as agriculture, fisheries, food security, natural disasters, and other aspects of society and the economy (**Figure 6**).

8. Southern Oscillation Index (SOI)

SOI, one of the world's most important climate indices, is a measure of the difference in air pressure in the Pacific Ocean, from Tahiti in the southeast to Darwin in the west. SOI is used to monitor and anticipate variations in both ENSO and the Walker Circulation (WC). During El Niño episodes, for example, the WC weakens, and the

SOI tends to show negative values. Climate fluctuations associated with alterations in the WC exert a significant influence on the climate, ecosystems, agriculture, and communities in various regions of the planet.

Related research has shown that (i) WC and SOI have weakened in recent decades, (ii) WC tends to decrease in climate models in response to higher concentrations of greenhouse gases in the atmosphere, as well as it has also been determined that the atmospheric pressure at the mean sea level (MSLP) between the eastern and western equatorial Pacific tends to increase during the twenty-first century, and not decrease.

Under global warming, the MSLP tends to increase in both the Darwin region and Tahiti which is located in a large region where the MSLP tends to increase in response to global warming.

The influence of global warming is already having a significant impact on the long-term variability of SOI, accounting for 45% of the standard deviation of the 30-year moving averages of the SOI. This proportion is estimated to increase to approximately 340% by the end of the twenty-first century. The implications of these findings for understanding recent climate change and seasonal prediction are discussed [20, 21].

Positive phase: during this phase, atmospheric pressure is higher than normal in the central and eastern Pacific region and lower than normal in the western Pacific region. This may indicate La Niña conditions and is associated with the cooling of sea surface waters in the central and eastern Pacific.

Negative phase: in this phase, atmospheric pressure is lower than normal in the central and eastern Pacific region and higher than normal in the western Pacific region. This would indicate El Niño conditions and is associated with a warming of sea surface waters in the central and eastern Pacific.

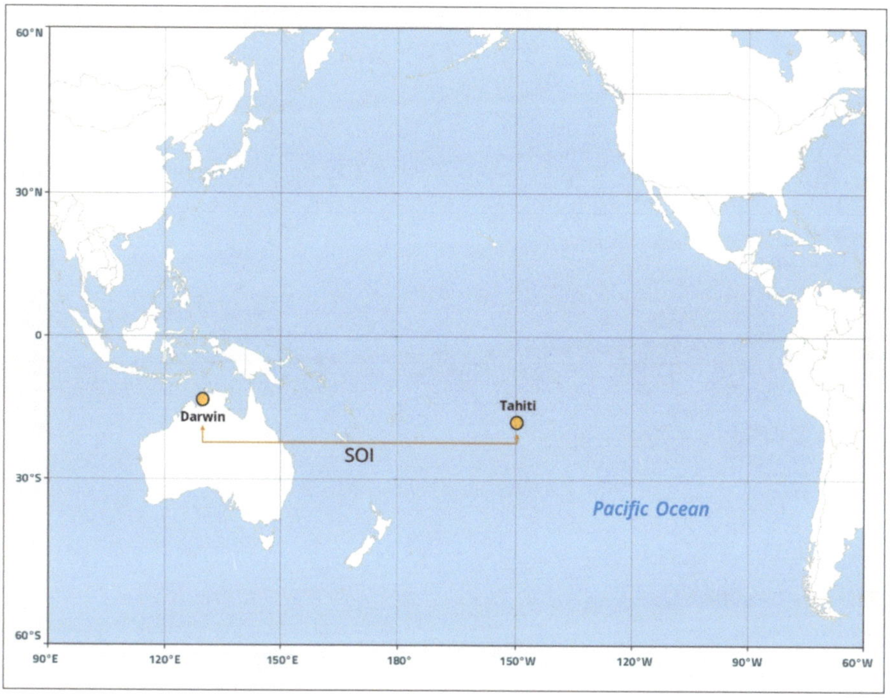

Figure 7.
SOI, monitors and anticipates variations in both ENSO and the Walker Circulation (WC) and presents positive and negative phases associated with cooling and warming of SST.

These phases of the SOI are fundamental to understanding the El Niño-Southern Oscillation (ENSO) phenomenon and its effects on global climate patterns. However, the SOI is not a good indicator of long-term equatorial changes related to global warming.

The SOI is used to monitor the evolution of ENSO and provide information on the phases of El Niño and La Niña, as well as used in climate models to predict the occurrence and intensity of El Niño and La Niña events. These predictions are important to assist in resource planning and management in sectors such as agriculture, water, and energy management, as well as preparing for potential climate impacts, such as drought and floods (**Figure 7**).

9. Multivariate ENSO Index (MEI)

MEI, uses several ocean-atmosphere variables (sea level atmospheric pressure (SLP), SST, zonal and meridional components of the surface wind, and outgoing long-wave radiation (OLR)) in the tropical Pacific basin, to monitor and evaluate the ENSO phenomenon more completely and flexibly. The MEI stands out as the only ENSO index that considers spatial variations of its main components concerning the seasonal cycle. Furthermore, it represents a weighted average of all these characteristics [22].

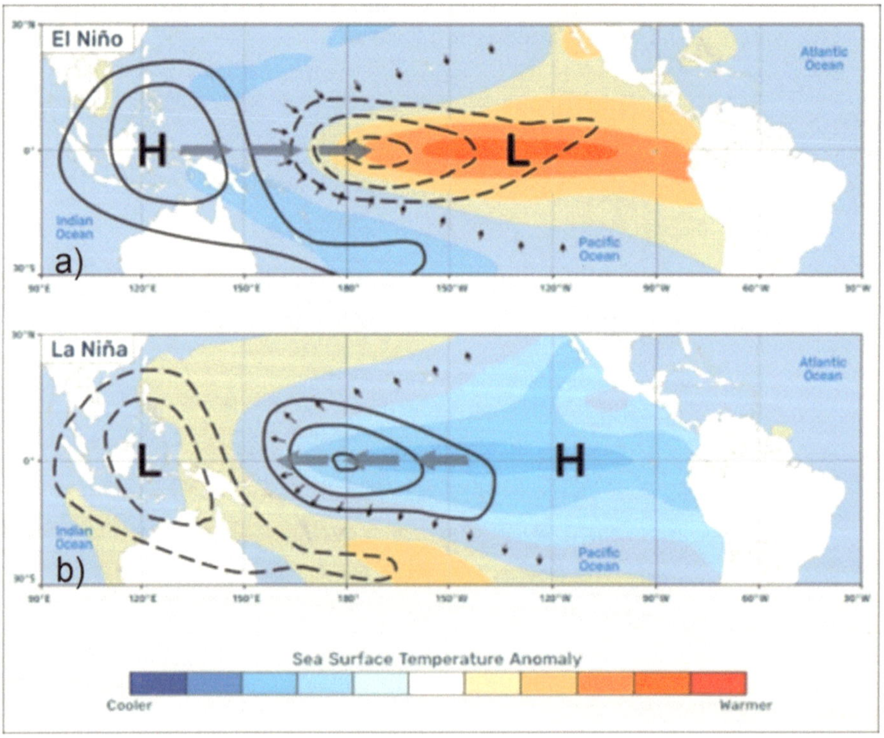

Figure 8.
MEI uses several ocean-atmosphere variables (SST, wind at different levels of the atmosphere, and cloudiness) in the tropical Pacific basin, to monitor and evaluate the ENSO phenomenon more completely and flexibly. (a) The highest values of MEI (typical of El Niño events). (b) The lowest values of MEI (typical of La Niña events). Source: Adapted https://psl.noaa.gov/enso/mei/. 2024.

The spectral analysis carried out using appropriate data reduction techniques on the monthly values of the MEI (1950–2008) has revealed the presence of a significant 60-month cycle, with a statistical reliability greater than 99%. The highest MEI values (typical of El Niño events, (**Figure 8a**)) and the lowest MEI values (typical of La Niña events, (**Figure 8b**)) coincide with the highest and lowest points of the 60-month cycle [23].

The extreme points of the MEI show a correlation with the extremes in the fluctuations of the atmospheric CO_2 concentration and are inversely related to the extremes of the variation in the length of day (LOD) rate. This indicates a strong connection between both local and global mechanisms in the Earth-atmosphere system, which are remarkably coupled and synchronized at various scales [24].

One of the applications of the MEI index is the prediction of the occurrence and intensity of El Niño and La Niña, which allows planning the management of risks associated with extreme climate phenomena, such as droughts, floods, and storms for the benefit of agriculture, as well as to implement water and soil management practices (**Figure 8**).

10. Rapid Intensity Index (RII)

RII has been designed to be applied in the Atlantic and Northeast Pacific basins, using large-scale predictor from the Statistical Hurricane Intensity Prediction Scheme (SHIPS), that is, to calculate the probability of rapid intensification (RI) occurring in the next 24 hours, using a linear discriminant analysis. The individual predictive factors used by RII show variations between the two basins. In the Atlantic basin, the most important factors include the change in intensity in the last 12 hours, the divergence at higher levels, and the vertical shear. In the case of the eastern basin of the North Pacific, the factors with the greatest weight are the change in intensity in the

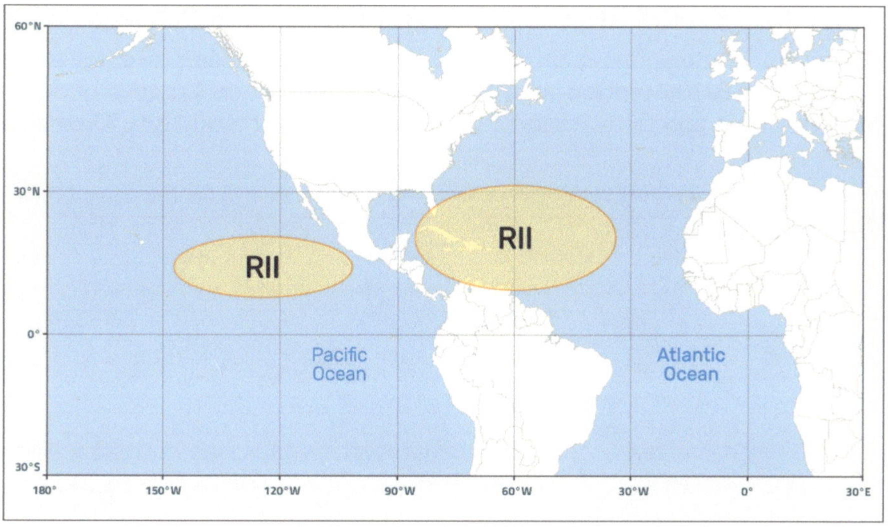

Figure 9.
RII, based on SST, atmospheric humidity, wind speed, and direction in different layers of the atmosphere, using mathematical models is designed for the prediction of hurricane intensity, that is, the probability of rapid intensification (RI) occurring in the next 24 hours (Atlantic and Northeast Pacific Basins).

last 12 hours, the symmetry of the internal convection areas, the difference between the current intensity, and the maximum potential of the system.

An analysis of forecast accuracy was performed by comparing climatological data for the 2006 and 2007 hurricane seasons, the results of which demonstrated that RII probabilistic forecasts are consistently accurate in both basins. Furthermore, using deterministic forecasts, RII was monitored to outperform all other available operational tools in terms of probability of detection (POD) and false alarm rate (FAR) [25]. Likewise, through the use of satellite images around the center of a TC, a characteristic pattern of precipitation rings has been found, which is associated with RI. By coupling the ring criterion with the RII index of the SHIPS, the probability of RI increased almost twofold. This suggests that both the ring and the SHIPS RII index contain independent information for predicting RI.

RII allows determining the probability that a tropical cycle will experience a rapid increase in its intensity. This information is valuable for forecasting agencies as well as for decision-making in terms of evacuations and emergency preparations. Additionally, RII can be combined with other criteria such as satellite-detected precipitation ring patterns to further improve the accuracy of predicting the rapid intensification of tropical cyclones (**Figure 9**) [26].

11. Genesis Potential Index (GPI)

GPI is an indicator to evaluate the probability of development of TC, its argumentation includes the processes that drive climate variability and the future change of the genesis of TC (TCG). The GPI of a TC estimates oceanic and atmospheric parameters, based on the understanding of oceanic impacts, where many factors are evaluated and discriminated, resulting in a GPI index. However, the precise determination of the contribution of thermodynamic environmental factors compared to dynamic factors in the formation of convective storms remains a challenge, especially in the context of global warming [27].

Through stepwise logarithmic regression analysis and considering 11 dynamic and thermodynamic factors, four main dynamic factors have been identified; the absolute vorticity at 850 hPa, the vertical velocity at 500 hPa, the vertical variation of the wind in the troposphere, and the vorticity of the zonal wind shear at 500 hPa. These four

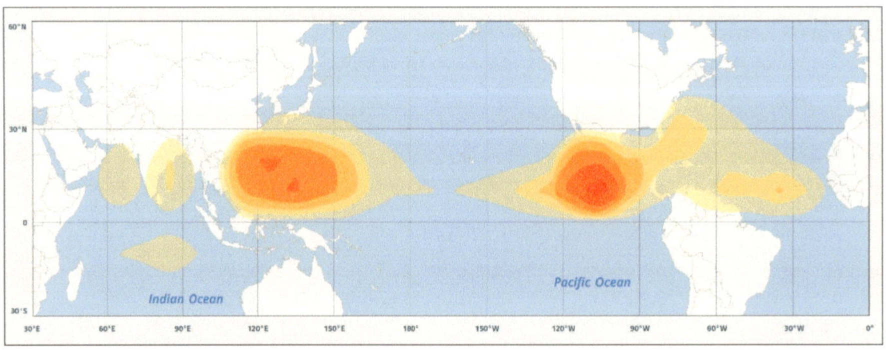

Figure 10.
GPI, based on atmospheric parameters such as SST, low-level cyclonic vorticity, atmospheric instability, vertical wind shear, and atmospheric humidity, evaluates the probability of TC development, based on the understanding of oceanic impacts, where several factors are evaluated and discriminated.

main dynamic factors have a significant influence on both the current simulation and the future projection of global warming. The dynamical GPI, which is composed of these four dynamical parameters, provides a valuable diagnostic tool for understanding future changes in tropical storm formation. At the same time, it improves the ability to represent interannual variations in the frequency of tropical storm formation in the oceans of the western Pacific and southern hemisphere [28].

Its application of the GPI helps to forecast and monitor the potential for the development of tropical cyclones and the intensity they could reach in a given region. It is important for preparation, mitigation, and response to possible extreme weather events (**Figure 10**).

12. The Pacific Decadal Oscillation Index (PDO)

The PDO is a crucial indicator of decadal climate variability in the Pacific, defined as the main empirical orthogonal function of North Pacific Sea Surface Temperature anomalies, which some experts have compared to a long-lasting pattern in the Pacific climate, similar to El Niño. Others have described it as a combination of two, sometimes independent, modes of climate variability in the Pacific, showing differences in terms of the location and temporal characteristics of variations in SST in the North Pacific. A growing body of evidence highlights the notable influence of the PDO in the southern hemisphere, with significant effects on climatic conditions in the mid-latitudes of the South Pacific Ocean, Australia, and South America.

PDO presents positive and negative phases. During positive phases of the PDO, SSTs in the North Pacific tend to be warmer than normal, while during negative phases, they tend to be colder than normal. These phases can persist for decades before changing, and their effects can be felt in different parts of the world, influencing climate patterns and weather phenomena. During the periods between 1890 and 1924, as well as between 1947 and 1976, a prevalence of "cold periods" of the PDO was observed. On the other hand, between 1925 and 1946, as well as from 1977 to the mid-1990s, "warm periods" of the PDO predominated.

Based on climate reconstructions from tree ring and coral records in the Pacific, it is suggested that variations in the PDO (along different time scales) can date back at least to the year 1600, although various indirect reconstructions show significant discrepancies, just as PDO fluctuations during the twentieth century showed two general patterns of periodicity (one from 15 to 25 years and another from 50 to 70 years), the mechanisms underlying this variability are still not fully understood [29].

It has been shown that the PDO can be reconstructed from a representation of sea surface temperature anomalies in the North Pacific, using a first-order autoregressive model and influenced by low-temperature variability in the Aleutians, the El Niño-Southern Oscillation (ENSO) phenomenon, and the zonal advection of oceanic anomalies in the Kuroshio-Oyashio extension. The latter originates from oceanic Rossby waves, driven by Ekman pumping in the North Pacific. The response patterns of sea surface temperature to these processes are not perpendicular and determine the spatial characteristics of the PDO. It is concluded that the PDO is not an independent dynamical mode but rather arises from the superposition of fluctuations in sea surface temperature with different dynamical origins [30].

PDO is used for the prediction of climate patterns such as droughts and floods, and the frequency and severity of events such as El Niño and La Niña, providing information on long-term SST trends in the North Pacific, is also an important topic

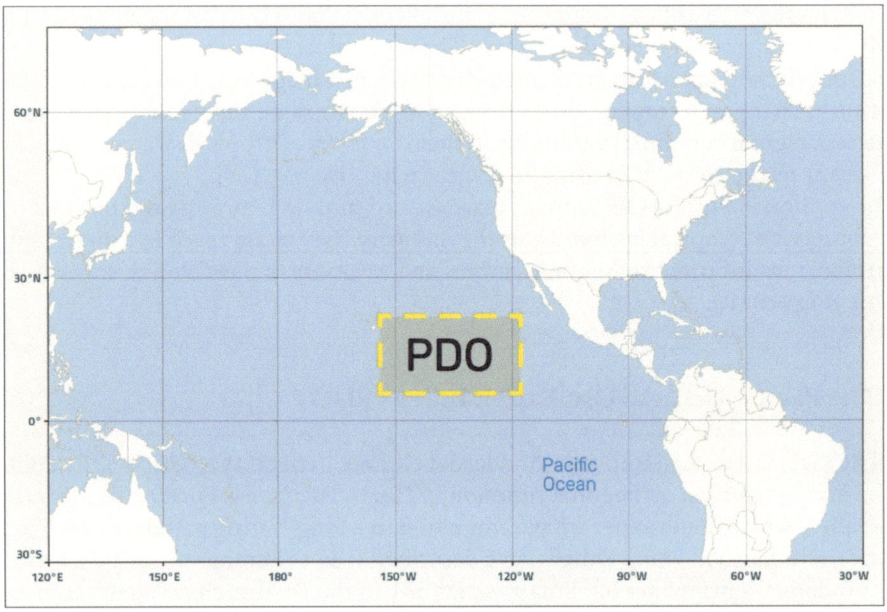

Figure 11.
PDO is an indicator of climate variability over decades with a long-lasting pattern in the Pacific climate similar to El Niño. PDO presents positive and negative phases. Positive phase: North Pacific SSTs are warmer than normal. Negative phase: the opposite of the positive phase. These phases can last decades before change, and their effects can be felt in different parts of the world.

of scientific research to better understand long-term climate variations and their impact on marine and terrestrial ecosystems. Likewise, understanding the positive and negative phases of PDO can help in the sustainable management of natural resources (**Figure 11**).

13. The Madden-Julian Oscillation Index (MJO)

The MJO index is a measure that describes the variability of rain and wind patterns in the tropical region of the Indian Ocean and the Pacific Ocean, allowing us to characterize the phase and intensity of the Madden-Julian Oscillation (natural cycle of changes in convection and winds in the equatorial region, characterized by two areas, one of suppressed rainfall and one of enhanced rainfall, each cycle of the oscillation lasts 30–60 days and is split into eight phases of equal length).

The MJO, like weather, has pulses of variability with a finite predictability time scale (predictability of at least 2–3 weeks), that is, future states of the MJO can be predicted up to a certain time in advance based on the knowledge of current and past states. Teleconnection links linked to MJO have a significant effect on circulations outside tropical regions, influencing meteorological and climate events such as monsoons in Southeast Asia, tropical storm formation in the Indian Ocean and the Pacific, the North Atlantic Oscillation, and the Pacific-North America pattern, and can also affect precipitation patterns in Australia and India. However, the accuracy of the MJO has been a considerable challenge in many general circulation models (GCMs), as these often introduce significant biases in the fundamental atmospheric fluxes. This makes it difficult to adequately reproduce the teleconnection patterns associated with

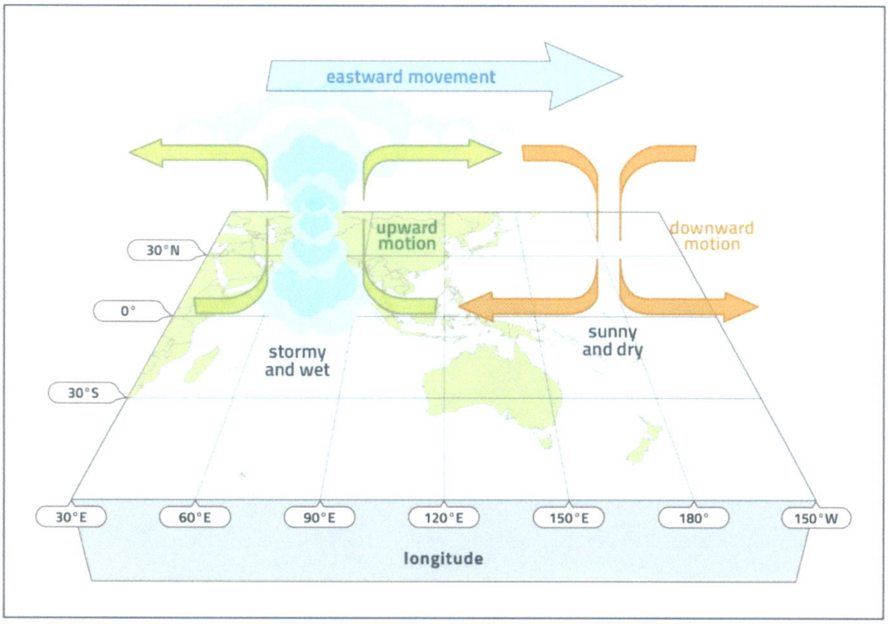

Figure 12.
MJO describes the variability of rainfall and wind patterns in the tropical Indian Ocean and Pacific Ocean, as well as to better understand climate variability on a global scale.

the MJO and the resulting impacts in the extratropics. Previous studies have revealed that the spatial arrangement of MJO-associated warming in the Indo-Pacific (MJO models) has a limited impact on the ability to predict teleconnection patterns [31].

These studies suggest that the sensitivity to the location of tropical heating near the subtropical jet is minimal. However, it has been observed that warming of the MJO east of the dateline could modify teleconnection trajectories in the North American region [32].

The MJO index is useful for short- and medium-term climate prediction in various regions of the world, improving predictions of extreme weather events, such as droughts, floods, and tropical cyclones, as well as to better understand climate variability on a global scale (**Figure 12**).

14. Indian Ocean Dipole Index (IOD)

IOD measures climate variability in the Indian Ocean region and its effects on regional and global climate. Determining the fluctuation of SST differences between two opposite regions in the western Indian Ocean and the eastern Indian Ocean plays an important role in predicting the El Niño Southern Oscillation in the tropical Pacific. IOD manifests itself as abnormal changes in SST, and atmospheric pressure in different parts of the Indian Ocean, and this process also generates positive and negative phases.

During the positive phase, the waters in the western Indian Ocean tend to be warmer than normal, while in the eastern Indian Ocean, they are colder than normal [33].

The Niño 3.4 index and IOD can be used to indicate drought phenomena, especially the 2-month main anomaly SST parameter in MJJ and NDJ [34].

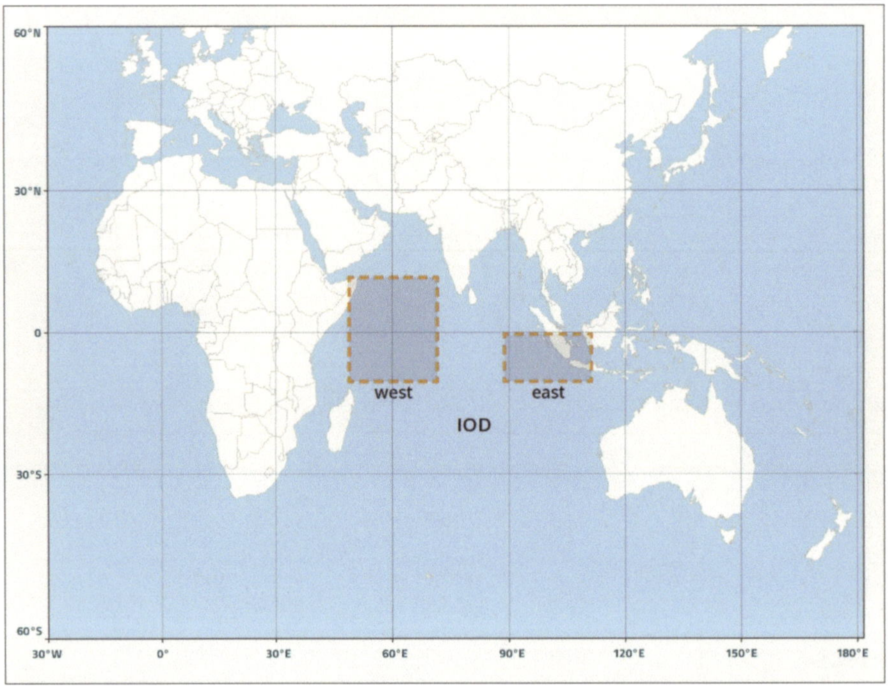

Figure 13.
IOD measures climate variability in the Indian Ocean region and its effects on regional and global climate (western Indian Ocean and eastern Indian Ocean) and plays an important role in predicting the El Niño Southern Oscillation in the tropical Pacific. IOD manifests itself as abnormal changes in SST and atmospheric pressure (it presents positive and negative phases), in different parts of the Indian Ocean.

The decade-long signal of the IOD index is closely related to the depth anomaly at which the 20°C isotherm is found in the ocean, suggesting that oceanic processes play an important role in the long-term variation of the IOD. Furthermore, this signal is also linked to anomalies in zonal winds, and in tropical regions, it is interpreted to have a decadal modulation of events occurring over single and multi-year periods [35].

The application of IOD allows the prediction of extreme weather events, droughts, floods, and storms, which can assist in disaster planning and mitigation, as well as natural resource management and agriculture in the Indian Ocean region (**Figure 13**).

15. Pacific Regional Equatorial Index (PREI)

This PREI Index was developed to determine the occurrence of a unique ENSO that occurred in 2017 specifically in the northern area of Peru, whose occurrence is not common. For example, a previous similar event occurred in early 1925. This type of ENSO was called Coastal ENSO by the Multisectoral Commission in Charge of the National Study of the El Niño Phenomenon (ENFEN).

The Coastal ENSO event that occurred in 2017 was one of the first recorded with Meteorological satellites, whose occurrence is due to a succession process of Modoki ENSO and Modoki La Niña events [36]. Likewise, it is also mentioned that the mechanism for the Coastal El Niño is the combined effect of local winds and equatorial Kelvin waves that caused the extreme coastal El Niño phenomenon of 2017, amplified by the positive Bjerknes feedback. Furthermore, pre-existing SST warming along the western coast of subtropical South America also favored anomalous northerly winds [37].

Its usefulness lies in differentiating the occurrence of a traditional ENSO and a Coastal ENSO, which have different formation processes, and allowing the prediction of extreme climate events such as precipitation and flooding in the northern area of Peru.

Table 1 presents a summary of the description and applications of the different climate indices developed in the present study.

Climate Index	Description	Applications
North Atlantic Oscillation (NAO)	Based on atmospheric pressure differences between Iceland and the Azores NAO positive: strong, moist winds, and more frequent storms in the North Atlantic. High rainfall in Scotland and southwestern Norway. Warm and wet winters in Western Europe. NAO negative: colder, drier winters in Europe and fewer storms in the North Atlantic	Prediction of weather patterns; temperature and precipitation, important for the management of agriculture and water resources
Tropical North Atlantic Index (TNA)	Based on SST variations of the tropical Atlantic, regions of the Gulf of Mexico and the west coast of Africa. Monitors hurricane and tropical cyclone activity. TNA positive: the SST is above average, favoring the formation of hurricanes in the Atlantic. TNA negative: the SST is below average. Therefore, the hurricane season is less active.	Predicts the intensity, size and activity of hurricanes and TCs in the tropical Atlantic region. Important for planning and mitigating the impacts of these climate events.
South Atlantic Oscillation Index (SAM)	Based on fluctuations in the atmospheric pressure over the South Atlantic, South America, Australia, New Zealand, Southern Africa, and Antarctica. Positive phase: it occurs during the summer, with west winds associated with warming in the Peninsula. Negative phase: indicates a weakening of the processes generated in the positive phase.	Prediction of heat waves, drought, heavy rain (floods), and storm. Important for the management of water resources and natural risks, as well as the mitigation of the impact on ecosystems and biodiversity.
Atlantic Multidecadal Oscillation Index (AMO)	It operates on a longer time scale (decades) with fluctuations in the North Atlantic SST, influencing atmospheric circulation, precipitation, temperature, and extreme events such as hurricanes and storms. Positive phase: SSTs in the North Atlantic are higher than normal. Negative phase: SSTs in the North Atlantic are lower than average.	Forecasting climate patterns in the North Atlantic on a multi-decadal time scale. It allows planning in the management of water resources, energy, and agriculture.
Arctic Oscillation Index (AO)	Describes the variability of atmospheric pressure at high latitudes in the northern hemisphere (Arctic) Positive phase: atmospheric pressure; lowest in the Arctic and highest in mid-latitudes of the northern hemisphere, strengthening of the polar jet stream and strong winds. Negative phase: atmospheric pressure; highest in the Arctic and lowest in mid-latitudes, weakening the polar jet stream and greater variability in climate.	Application in climate models, to predict climate patterns in high latitudes of the northern hemisphere (Arctic) and their effects on global climate. Short- and long-term planning in agriculture, energy, and transportation

Climate Index	Description	Applications
Oceanic Niño Index (ONI)	Describes the occurrence and duration of El Niño and La Niña episodes, based on SST monitoring in the central Pacific Ocean. Positive phase: known as El Niño, SSTs in the central and eastern equatorial Pacific are warmer than normal. Negative phase: known as La Niña, SSTs in the central and eastern equatorial Pacific are colder than normal.	Monitoring and forecasting the occurrence of El Niño and La Niña events, which can generate increased rainfall in the equatorial Pacific and drought in regions such as Australia and western South America and vice versa. Helps plan and mitigate its possible impacts.
Southern Oscillation Index (SOI)	Monitors and anticipates variations in both ENSO and the Walker Circulation (WC). Positive phase: indicates La Niña conditions, associated with cooling of sea surface waters in the central and eastern Pacific. Negative phase: indicates El Niño conditions, associated with a warming of sea surface waters in the central and eastern Pacific.	Monitors and predicts the evolution and occurrence of ENSO; El Niño y La Niña Resource planning and management; agriculture, water, and energy, as well as preparing for possible climate impacts, such as drought and floods.
Multivariate ENSO Index (MEI)	MEI uses several ocean-atmosphere variables (SST, wind at different levels of the atmosphere, and cloudiness) in the tropical Pacific basin, to monitor and evaluate the ENSO phenomenon more completely and flexibly.	Prediction of the occurrence and intensity of El Niño and La Niña, allowing planning for risk management associated with extreme climate events, such as droughts, floods, and storms.
Rapid Intensity Index (RII)	Based on SST, atmospheric humidity, wind speed, and direction in different layers of the atmosphere, using mathematical models. Designed for the prediction of hurricane intensity, that is, the probability of rapid intensification (RI) occurring in the next 24 hours (Atlantic and Northeast Pacific Basins)	Prediction that a tropical cycle experiences a rapid increase in intensity. This information is valuable for decision-making in terms of evacuations and emergency preparations.
Genesis Potential Index (GPI)	Based on atmospheric parameters, SST, low-level cyclonic vorticity, atmospheric instability, vertical wind shear, and atmospheric humidity. It evaluates the probability of TC development, based on the understanding of oceanic impacts, where several factors are evaluated and discriminated.	Forecast and monitor tropical cyclone development potential and intensity in a given region. Important for preparation, mitigation, and response to possible extreme weather events.
Pacific Decadal Oscillation Index (PDO)	PDO is an indicator of climate variability over decades with a long-lasting pattern in the Pacific climate similar to El Niño. PDO presents positive and negative phases. Positive phase: North Pacific SSTs are warmer than normal. Negative phase: the opposite of the positive phase. These phases can last decades before changing, and their effects can be felt in different parts of the world.	Predicting weather patterns such as droughts, floods, and the severity of events such as El Niño and La Niña. Understanding the positive and negative phases of PDO can help in sustainable natural resource management.
Madden-Julian Oscillation Index (MJO)	The MJO index describes the variability of rainfall and wind patterns in the tropical Indian Ocean and Pacific Ocean, as well as to better understand climate variability on a global scale.	It allows short- and medium-term climate prediction in various regions of the world of extreme weather events, such as droughts, floods, and tropical cyclones.

Climate Index	Description	Applications
Indian Ocean Dipole Index (IOD)	It measures climate variability in the Indian Ocean region and its effects on regional and global climate (western Indian Ocean and eastern Indian Ocean) and plays an important role in predicting the El Niño Southern Oscillation in the tropical Pacific. It presents positive and negative phases.	It allows the prediction of extreme weather events, droughts, floods, and storms in the Indian Ocean region.
Pacific Regional Equatorial Index (PREI)	PREI was developed to determine the occurrence of an ENSO, which occurred in 2017 specifically in the northern area of Peru (called coastal ENSO) whose occurrence is not common.	PREI allows the difference between the occurrence of a traditional ENSO and a Coastal ENSO, allowing the prediction of extreme precipitation and flooding in the northern area of Peru.

Table 1.
Summary of the description and applications of the different Climate Indices developed in the present study.

16. Conclusions

16.1 Tropical cyclones

The TNA is an indicator of the key parameters of TC intensity, size, and forward speed that constitute an index of existing cyclonic damage potential.

16.2 Southern hemisphere

SAM and SOI represent the atmospheric variability that involves the troposphere and stratosphere in the southern hemisphere and in the extratropical southern hemisphere.

16.3 North hemisphere

NAO, AMO, and RII represent dominant climate patterns in the North Atlantic and eastern North Pacific regions.

16.4 Extratropical

AO, MEI, GPI, and PDO. They generally represent the variability of atmospheric circulation and oceanic parameters in the extratropical region of the northern hemisphere.

16.5 Tropical

ONI and IOD represent climate variability in the central Pacific Ocean.

16.6 Tropical atmosphere

MJO is an index that represents the dominant mode of intraseasonal variability in the tropical atmosphere.

Conflict of interest

The authors declare no conflict of interest.

Author details

Edgard Gonzales[1*] and Kenny Gonzales[2]

1 National University of San Agustin of Arequipa, Arequipa, Peru

2 Advance Engineering and Research, Urb. Colegio de Ingenieros E3, Cerro Colorado, Arequipa, Peru

*Address all correspondence to: hgonzalesz@unsa.edu.pe

References

[1] Labat D. Cross wavelet analyses of annual continental freshwater discharge and selected climate indices. Journal of Hydrology. 2010;**385**(1-4):269-278. DOI: 10.1016/j.jhydrol.2010.02.029

[2] Pielke RA, Gratz J, Landsea CW, Collins D, Saunders MA, Musulin R. Normalized hurricane damage in the United States: 1900-2005. Natural Hazards Review. 2008;**9**(1):29-42. DOI: 10.1061/(asce)1527-6988(2008)9:1(29)

[3] Stenseth NC et al. Studying climate effects on ecology through the use of climate indices: The North Atlantic Oscillation, El Niño Southern Oscillation and beyond. The Royal Society. Aug 2003;**270**:2087-2096. DOI: 10.1098/rspb.2003.2415

[4] Wanner H, Brönnimann S, Casty C, Luterbacher J, Schmutz C, David B. North atlantic oscillation—Concepts and studies. Surveys in Geophysics. 2001;**22**(1984):321-382

[5] Done JM, Paimazumder D, Towler E, Kishtawal CM. Estimating impacts of North Atlantic tropical cyclones using an index of damage potential. Climatic Change. 2018;**146**:561-573. DOI: 10.1007/s10584-015-1513-0

[6] Holland GJ, Done JM, Douglas R, Saville GR, Ge M. Global Tropical Cyclone Damage Potential. Part of the book series: Hurricane Risk. 2019;**1**:23-42. DOI: 10.1007/978-3-030-02402-4_2

[7] Dätwyler C et al. Teleconnection stationarity, variability and trends of the Southern Annular Mode (SAM) during the last millennium. Climate Dynamics. 2017;**51**:2321-2339. DOI: 10.1007/s00382-017-4015-0

[8] Gong D, Wang S. Definition of Antarctic oscillation index. Geophysical Research Letters. 1999;**26**(4):459-462. DOI: 10.1029/1999GL900003

[9] Gillett NP, Kell TD, Jones PD. Regional climate impacts of the Southern Annular Mode. Geophysical Research Letters. 2006;**33**(23):1-4. DOI: 10.1029/2006GL027721

[10] Barrucand MG, Zitto ME, et al. Historical SAM index time series: Linear and nonlinear analysis. International Journal of Climatology. 2018;**38**:e1091-e1106

[11] Clem KR, Renwick JA, Mcgregor J, Fogt RL. The relative influence of ENSO and SAM on Antarctic Peninsula climate. Journal of Geophysical Research: Atmospheres. 2016;**121**(16):9324-9341. DOI: 10.1002/2016JD025305

[12] Knight JR, Folland CK, Scaife AA. Climate impacts of the Atlantic Multidecadal Oscillation. Geophysical Research Letters. 2006;**33**(May):2-5. DOI: 10.1029/2006GL026242

[13] Alexander MA, Kilbourne KH, Nye JA. Climate variability during warm and cold phases of the Atlantic Multidecadal Oscillation (AMO) 1871-2008. Journal of Marine Systems. 2014;**133**:14-26. DOI: 10.1016/j.jmarsys.2013.07.017

[14] Yang Y. A global-scale multidecadal variability driven by Atlantic multidecadal oscillation. National Science Review. 2020;**7**:1190-1197. DOI: 10.1093/nsr/nwz216

[15] Thompson DWJ, Wallace JM. The Arctic oscillation signature in

the wintertime geopotential height and temperature fields. Geophysical Research Letters. 1998;**25**(9):1297-1300. DOI: 10.1029/98GL00950

[16] Hill DJ, Csank AZ, Dolan AM, Lunt DJ. Pliocene climate variability: Northern Annular Mode in models and tree-ring data. Palaeogeography Palaeoclimatology Palaeoecology. 2011;**309**(1-2):118-127. DOI: 10.1016/j.palaeo.2011.04.003

[17] Glantz MH, Ramirez IJ. Reviewing the Oceanic Niño Index (ONI) to enhance societal readiness for El Niño's impacts. International Journal of Disaster Risk Science. 2020;**11**:394-403. DOI: 10.1007/s13753-020-00275-w

[18] Prasetyo Y, Nabilah F. Pattern analysis of El Niño and La Niña phenomenon based on sea surface temperature (SST) and rainfall intensity using Oceanic Niño Index (ONI) in West Java Area. In: IOP Conference Series: Earth and Environmental Science. Vol. 98, no. 1. 2017. p. 012041

[19] Tongkukut SHJ. El-Nino dan Pengaruhnya Terhadap Curah Hujan di Manado Sulawesi Utara. Jurnal Ilmiah Sains. 2011;**11**(1):102. DOI: 10.35799/jis.11.1.2011.51

[20] Power SB, Kociuba G. The impact of global warming on the Southern Oscillation Index. Climate Dynamics. 2011;**37**(9-10):1745-1754. DOI: 10.1007/s00382-010-0951-7

[21] Yan H, Sun L, Wang Y, Huang W, Qiu S, Yang C. A record of the Southern Oscillation Index for the past 2,000 years from precipitation proxies. Nature Geoscience. 2011;**4**(9):611-614. DOI: 10.1038/ngeo1231

[22] Wolter K, Timlin MS. El Niño/Southern Oscillation behaviour since 1871 as diagnosed in an extended multivariate ENSO index (MEI.ext). International Journal of Climatology. 2011;**31**(7):1074-1087. DOI: 10.1002/joc.2336

[23] Mazzarella A, Giuliacci A, Liritzis I. On the 60-month cycle of multivariate ENSO index. Theoretical and Applied Climatology. 2010;**100**(1):23-27. DOI: 10.1007/s00704-009-0159-0

[24] Mazzarella A, Giuliacci A, Scafetta N. Quantifying the Multivariate ENSO Index (MEI) coupling to CO2 concentration and to the length of day variations. Theoretical and Applied Climatology. 2013;**111**(3-4):601-607. DOI: 10.1007/s00704-012-0696-9

[25] Kaplan J, DeMaria M, Knaff JA. A revised tropical cyclone rapid intensification index for the Atlantic and eastern North Pacific basins. Weather and Forecasting. 2010;**25**(1):220-241. DOI: 10.1175/2009WAF2222280.1

[26] Kieper ME, Jiang H. Predicting tropical cyclone rapid intensification using the 37 GHz ring pattern identified from passive microwave measurements. Geophysical Research Letters. 2012;**39**(13):1-7. DOI: 10.1029/2012GL052115

[27] Zhang M, Zhou L, Chen D, Wang C. A genesis potential index for Western North Pacific tropical cyclones by using oceanic parameters. Journal of Geophysical Research: Oceans. 2016;**121**:7176-7191. DOI: 10.1002/2016JC011851

[28] Wang B, Murakami H. Dynamic genesis potential index for diagnosing present-day and future global tropical cyclone genesis. Environmental Research Letters. 2020;**15**(11):114008. DOI: 10.1088/1748-326/abbb01

[29] Mantua NJ, Harem SR. The Pacific decadal oscillation. Journal of Oceanography. 2002;**58**:35-44

[30] Schneider N, Cornuelle BD. The forcing of the Pacific Decadal Oscillation. Journal of Climate. 2005;**18**(21):4355-4373. DOI: 10.1175/JCLI3527.1

[31] Duane EW, William K.-ML. Intraseasonal Variability in the Atmosphere-Ocean Climate System. Vol. XVII. Berlin: Springer; 2005. 474 p 80 illus. 3-540-22276-6

[32] Henderson SA, Maloney ED, Son SW. Madden-Julian oscillation Pacific teleconnections: The impact of the basic state and MJO representation in general circulation models. Journal of Climate. 2017;**30**(12):4567-4587. DOI: 10.1175/JCLI-D-16-0789.1

[33] Li C, Feng Y, Sun T, Zhang X. Long term Indian Ocean Dipole (IOD) Index prediction used deep learning by convLSTM. Remote Sensing. 2022;**14**(3):523

[34] Chaowiwat W, Koontanakulvong S. Statistical Forecasting of Rainfall by ENSO/IOD Index in the Chao Phraya River Basin. 2nd World Irrigation Forum (WIF2). Chiang Mai, Thailand; 2016. pp. 6-8

[35] Ashok K, Le Chan W, Motoi T, Yamagata T. Decadal variability of the Indian Ocean dipole. Geophysical Research Letters. 2004;**31**(24):1-4. DOI: 10.1029/2004GL021345

[36] Gonzales E, Ingol E. Determination of a new coastal ENSO oceanic index for northern Peru. Climate. 2021;**9**(5):1-26. DOI: 10.3390/cli9050071

[37] Peng Q, Xie SP, Wang D, Zheng XT, Zhang H. Coupled ocean-atmosphere dynamics of the 2017 extreme coastal El Niño. Nature Communications. 2019;**10**(1):1-10. DOI: 10.1038/s41467-018-08258-8

Chapter 5

Synthesis of Tropical Cyclones: Understanding, Modeling, and Adapting to Climate Change Impacts

Jiayao Wang, Yu Chang and Kam Tim Tse

Abstract

Tropical cyclones, characterized by their destructive effects, pose significant threats to coastal regions worldwide. This review provides a comprehensive exploration of tropical cyclones, delving into their definitions, regional variations in nomenclature (hurricanes, typhoons, and tropical cyclones), and categorization based on intensity and core structural elements such as the eye, eyewall, and rainbands. Globally, the review meticulously analyzes the profound impacts of tropical cyclones, spanning environmental, social, and economic dimensions, and highlights the disproportionate vulnerability of coastal populations. A thorough literature review summarizes models, exploring their evolution and effectiveness in predicting cyclone behavior and impacts. Additionally, the review discusses emerging advancements in modeling techniques, including numerical simulations and machine learning algorithms, and their potential to enhance forecasting accuracy and risk assessment. Concluding with a critical discussion of current challenges, such as data limitations, model uncertainties, and the influence of climate change, the review underscores the pressing need for interdisciplinary collaborations and innovative solutions to mitigate the increasing risks posed by tropical cyclones in a changing climate.

Keywords: tropical cyclone, boundary layer, wind field modeling, climate change, risk assessment

1. Introduction

A tropical cyclone serves as the overarching descriptor for a non-frontal synoptic scale low-pressure system situated over tropical or sub-tropical waters. These meteorological phenomena exhibit organized convection, characterized by intense thunderstorm activity, and feature a well-defined cyclonic surface wind circulation [1]. The term "tropical cyclone" encompasses a broad category of weather systems, ranging from tropical depressions to hurricanes, and is rooted in the unique atmospheric conditions prevalent in tropical and sub-tropical regions. These dynamic and often powerful systems derive their energy from warm ocean waters, manifesting in a symphony of atmospheric processes that include convective updrafts, latent heat release, and the characteristic rotation around a central low-pressure center. During their formative stage, tropical cyclones characterized by maximum winds of 17 m/s or lower are referred to as tropical depressions. As these weather systems intensify and attain wind speeds ranging from 18 to 32 m/s, they are designated as tropical storms. Notably, tropical cyclones with maximum winds reaching 33 m/s or higher assume distinct regional names [2].

The terms "hurricane" and "typhoon" are regionally specific names for strong tropical cyclones. When tropical cyclones attain maximum sustained surface winds of 33 m/s (64 kt, 74 mph), they are designated by different names depending on their geographical location. In the North Atlantic Ocean, the Northeast Pacific Ocean east of the dateline, or the South Pacific Ocean east of 160E, these weather systems are termed "hurricanes." Across the Northwest Pacific Ocean west of the dateline, they adopt the name "typhoon." In the Southwest Indian Ocean, the term "tropical cyclone" is employed to describe storms with winds reaching this threshold. This regional nomenclature reflects the diverse terminology used globally to identify and characterize these meteorological phenomena based on their specific locations and the associated oceanic regions [3].

Tropical cyclones can exhibit considerable variability in their intensity, size, boundary layer composition, spiral band formation, eye characteristics, and overall symmetry, both between individual storms and over time [4]. The structure of a tropical cyclone is characterized by distinct features that contribute to its intensity and behavior. It can be visualized as a layered system of clouds and weather patterns.

Figure 1 illustrates a top view and a cross-section view of a TC. The size of the storm can be as wide as 300 km [5]. At the center of a mature TC lies the eye, where air descends instead of rising. This descending air suppresses cloud formation, resulting in

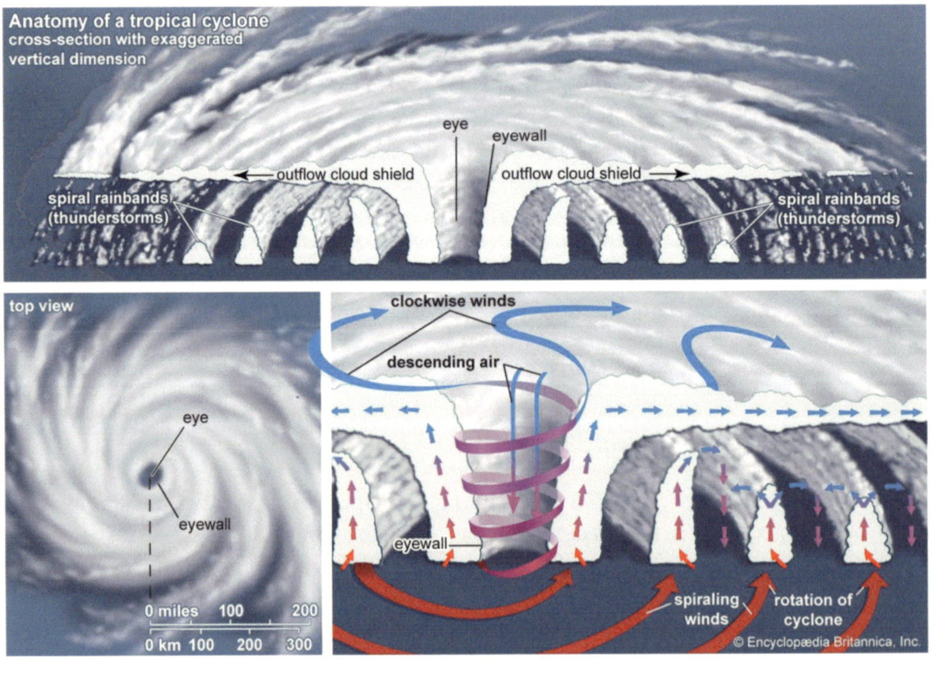

Figure 1.
The structure of a tropical cyclone (https://cdn.britannica.com/44/76744-050-E5B32AD5/view-cross-section-cyclone).

a clear and calm center. The eye is typically circular and can range in diameter from 30 to 65 kilometers (19–40 miles), although sizes as small as 3 kilometers (1.9 miles) and as large as 370 kilometers (230 miles) have been observed [6, 7].

Surrounding the eye is the "eyewall," which forms the cloudy outer edge and expands outward with height. Within the eyewall, the most intense weather conditions occur. This includes the highest wind speeds, most rapid ascent of air, tallest clouds, and heaviest precipitation. Consequently, the heaviest wind damage typically occurs when the eyewall of a tropical cyclone makes landfall. This is where most of the convection, or upward movement of air, takes place. This scattering is likely caused by sea spray generated by turbulent ocean surface waves due to strong winds. The eyewall can vary in steepness, ranging from almost vertical to about 45 degrees. Beyond the eyewall is a region with reduced precipitation surrounded by another ring of heavy rain known as the "outer eyewall." This layered structure of clouds and precipitation patterns defines the complex and dynamic nature of a tropical cyclone.

Associated winds in tropical cyclones exhibit distinct patterns and characteristics essential for understanding the dynamics of TCs. Typically, the strongest winds are located slightly outside the eye wall base, where they circulate counterclockwise in the northern hemisphere. This region, known as the radius of maximum winds, experiences the highest wind speeds, which can extend up to 100 kilometers from the center of the storm. Observations from drop-sondes reveal variations in wind profiles, with maximum wind speeds observed near the eyewall at altitudes ranging from 300 to 800 meters above the ground. However, in storms with larger radii, the strongest winds may occur at higher altitudes, typically between 1 and 2 kilometers above the ground [8].

Rainbands are important weather patterns in TCs, usually arranged in bands around the center of the storm. These rainbands often hover into the center of the storm, and the rainbands are sometimes stationary relative to the center, while in other cases they rotate around the center. This rotating band of clouds is usually associated with a pronounced swing in the storm's path. When a TC makes landfall, surface friction increases, leading to increased convergence of airflow toward the eyewall and the vertical movement of air that occurs there. The increased convergence and rise of moist air lead to the heavy rainfall associated with TCs, which can reach extremely high levels in a short period of time.

2. Categories of tropical cyclones

Tropical cyclones are classified into categories to communicate their potential impact and guide preparedness efforts. Tropical cyclones are categorized based on the wind speeds surrounding their circulation center and are assigned a rank. The choice of scale for a specific cyclone depends on its location, such as the utilization of the Saffir–Simpson hurricane wind scale and Australian tropical cyclone intensity scales in the western hemisphere. Regardless of the scale, all of them evaluate tropical cyclones by their maximum sustained winds. These winds are determined through observation, measurement, or estimation using various techniques, typically over a time span ranging from 1 to 10 minutes.

2.1 Saffir-Simpson hurricane wind scale (SSHWS)

The Saffir-Simpson Hurricane Wind Scale, originally developed by wind engineer Herb Saffir and meteorologist Bob Simpson, emerges as a pivotal tool for alerting the public about the possible impacts of various intensity hurricanes. The maximum sustained surface wind speed (peak 1-minute wind at the standard meteorological observation height of 10 m over unobstructed exposure) associated with the cyclone is the determining factor in the scale that determines a hurricane's category. With its five categories ranging from the minimal impact of Category 1 to the catastrophic consequences of Category 5, the scale provides a clear and concise framework for evaluating the severity of an impending storm. Categories 3–5 are specifically designated as major hurricanes. This widely recognized global system establishes a practical guide for meteorologists, emergency officials, and the public, offering a standardized means of communication crucial for effective preparedness, response, and mitigation efforts. The scale does not include hurricane-related damage from storm surges, rainfall-induced floods, etc. It should also be noted that the extent of damage caused by these winds can vary depending on factors such as local building codes in effect, the duration of high winds, changes in wind direction, and the age of structures. The criteria for each category are delineated in **Table 1** [9].

2.2 Hong Kong tropical cyclone classification

In Hong Kong, tropical cyclones are classified based on the World Meteorological Organization's guidelines, specifically focusing on their maximum sustained wind speeds near the center. The local classification system in Hong Kong, implemented since 2009, is distinctive as it defines classifications in terms of wind speeds averaged over a 10-minute period. This system introduces six categories, each delineating different levels of cyclonic intensity as in **Table 2**.[1]

2.3 Other criterions for tropical cyclone classification

The Beaufort scale, officially known as the Beaufort wind force scale, employed by meteorological agencies worldwide, evaluates wind speeds based on observational criteria related to the appearance of the sea surface and object behavior. Wind speed on the Beaufort scale is based on the empirical relationship between the equivalent wind

Category	Sustained wind speed (mph)	Damage	Example and year
1	74–95	Very dangerous winds will produce some damage	Humberto 2007
2	96–110	Extremely dangerous winds will cause extensive damage	Ike 2008
3	111–129	Devastating damage will occur	Alicia 1983
4	130–156	Catastrophic damage will occur	Harvey 2017
5	>156	Catastrophic damage will occur	Andrew 1992

Table 1.
Criteria for Saffir-Simpson hurricane wind scale.

[1] CHAN, C. Tropical Cyclone Classification. |Hong Kong Observatory(HKO)|Educational Resources. https://www.hko.gov.hk/en/education/tropical-cyclone/classification-naming-characteristics/00145-tropical-cyclone-classification.html.

Tropical cyclone classification	Maximum sustained winds near the center (km/h)
Tropical depression (TD)	<63
Tropical storm (TS)	63–87
Severe tropical storm (STS)	88–117
Typhoon (T)	118–149
Severe typhoon (ST)	150–184
Super typhoon (SuperT)	185 or above

Table 2.
Tropical cyclone classification in Hong Kong.

speed at 10 meters above the sea surface and Beaufort scale number B [10]. Extending from 0 (calm) to 12 (hurricane force), each level on the Beaufort scale is accompanied by specific descriptions of wind speed effects. The Beaufort scale was extended in 1946 when forces 13–17, with wind speeds more than 83mph, were added [11], which only applies to tropical cyclones. Internationally, the World Meteorological Organization (WMO) defined the Beaufort scale only up to force 12 and there was no recommendation on the use of the extended scale,[2] which is only used in China.

The China Meteorological Administration (CMA), the Hong Kong Observatory (HKO), the Philippine Atmospheric, Geophysical and Astronomical Services Administration (PAGASA), and the Japan Meteorological Agency (JMA) have established specific classifications for typhoons intended for domestic use.[3] The JMA employs a three-tier categorization system based on 10-minute maximum wind speeds, ranging from 33 m/s (64kt) to 44 m/s (85kt). A typhoon is categorized as a very strong typhoon when its maximum wind speeds fall within the range of 44 m/s (85kt) to 54 m/s (105kt). A violent typhoon, on the other hand, is characterized by maximum wind speeds of 54 m/s (105kt) or higher.[4]

The Indian Meteorological Department (IMD), for example, Regional Specialized Meteorological Center (RSMC), utilizing a *3-minute averaging for the sustained wind*, employs seven different classifications to grade systems.[5] In the southern hemisphere, a tropical cyclone is defined as having a clear organization of wind circulation with sustained wind speeds exceeding 34 knots (63 km/h or 39 mph) near the center for 10 continuous minutes. Once identified as a tropical cyclone, all centers proceed to name the system using the Australian tropical cyclone intensity scale. This scale, based on maximum mean wind speed and typical strongest gust, categorizes systems into five classes from 1 (weakest) to 5 (strongest).[6]

[2] https://library.wmo.int/viewer/41585?medianame=558-2012-2018_en_#page=29&viewer=picture&o=custom_bottom_Permalink&n=0&q=.

[3] Typhoon Committee (2015). Typhoon Committee Operational Manual 2015 (Report). World Meteorological Organization. Retrieved November 13, 2015.

[4] Tropical Cyclone Information: Scale and intensity of the tropical cyclone. https://www.data.jma.go.jp/multi/cyclone/cyclone_caplink.html?lang=en.

[5] https://rsmcnewdelhi.imd.gov.in/images/pdf/faq.pdf.

[6] RA V Tropical Cyclone Committee (2023). Tropical Cyclone Operational Plan for the South-East Indian Ocean and the Southern Pacific Ocean 2023 (Report). World Meteorological Organization. https://community.wmo.int/en/tropical-cyclone-operational-plans.

Meteorological agencies employ various scales to assess and categorize natural phenomena, providing crucial information for understanding and responding to potential threats. The Saffir-Simpson Hurricane Wind Scale is highly recognized internationally and is commonly referenced in media reports, public communications, and official statements during hurricane events. This scale is extensively used by the National Hurricane Center (NHC) in the Atlantic basin and the eastern North Pacific, as well as by other meteorological agencies globally.

3. Impacts of tropical cyclones on a global perspective

Tropical cyclones are powerful and devastating weather phenomena that have significant impacts on various parts of the world. In this section, we discuss the impact of typhoons on sustainability, looking at it from the economic, social, and environmental perspectives, through which, we can gain a clearer understanding of the challenges and impacts of TC events on various aspects of society. The assessment of the projected response of tropical cyclones to anthropogenic warming indicates that climate change will likely lead to change in the intensity and frequency of tropical cyclones, affecting different regions in unique ways [12–16]. The distribution of TCs on the planet has undergone clear changes since at least the 1970s. This is evident in a noticeable shift toward higher latitudes in the areas where TCs reach their peak intensity. Specifically, the latitudes where TCs are most intense have been moving toward the poles at a rate of approximately 0.5 degrees of latitude per decade [17]. With these changes, TCs have far-reaching impacts on a global scale, affecting different regions in diverse ways. In response to the environmental, social, and economic impacts caused by these storms, it is imperative to implement disaster preparedness, mitigation measures, and sustainable development strategies to strengthen resilience to disasters.

Figure 2 showcases the devastating impacts of tropical cyclones on various regions around the world, (a) depicts the catastrophic flooding that engulfed New Orleans, following the landfall of Hurricane Katrina. The storm surge breached levees, leading to extensive inundation and widespread destruction of property and infrastructure. (b) captures the consequence of Typhoon Haiyan on Tacloban City in the Philippines. The storm, accompanied by powerful earthquakes, caused widespread devastation, leaving behind a landscape of destruction and ruins. (c) lists the most economically costing tropical cyclones in the past half-century occurred in 2017. The combined economic damage from these storms amounted to roughly 225 billion U.S. dollars. Notably, Hurricane Katrina, recorded in 2005, remains the largest tropical cyclone in terms of economic losses since 1970, with damages exceeding 163 billion dollars.

Tropical cyclone activity impacts humans and the environment through strong winds, heavy rainfall, and storm surges, often resulting in landslides and floods [18–20]. Tropical cyclones have the potential to cause severe morbidity and mortality. It is estimated that these disasters have resulted in 1.9 million deaths over the past two centuries [21, 22]. The impact on health can be direct (due to adverse weather conditions) or indirect (due to infrastructure damage, environmental destruction, population displacement, and economic hardships). Historically, developing countries in the Asia-Pacific region have experienced the highest absolute and proportional mortality rates from tropical cyclones and other natural disasters [22]. The lack of advanced infrastructure and resources for effective early warning systems in developing

a[8] Flooding in New Orleans after Hurricane Katrina (August 2005)

b[9] Ruins of Tacloban in the Philippines after Typhoon Haiyan (November 2013),

coupled with earthquakes

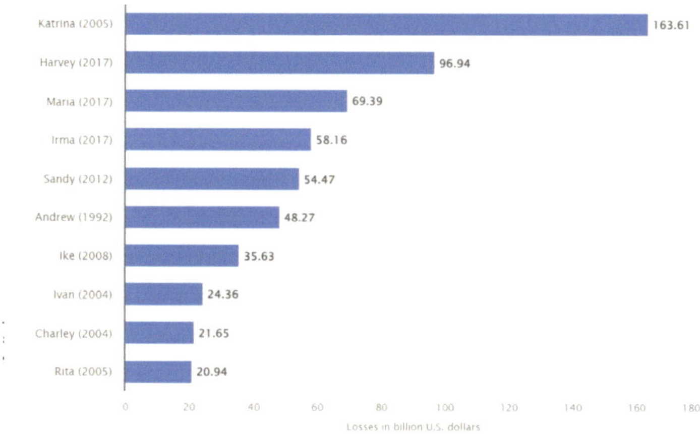

c[10] Economic Impacts of Tropical Cyclones in the Past Half-century

Figure 2.
Consequences of tropical cyclones. (a) flooding in New Orleans after hurricane Katrina (august 2005) (https://www.securitymagazine.com/articles/92876-us-high-tide-flooding-continues-to-increase). (b) ruins of Tacloban in the Philippines after typhoon Haiyan (November 2013), coupled with earthquakes (www.commondreams.org). (c) economic impacts of tropical cyclones in the past half-century (https://www.statista.com/statistics/1297538/global-leading-tropical-cyclones-economic-loss/).

countries, as compared to developed countries, exacerbates the severity of cyclone impacts, leading to increased social and economic burdens [20].

In addition to environmental impacts, the social and economic consequences of tropical cyclones are also substantial, with implications for loss of life, displacement, infrastructure damage, and economic burdens. Empirical data suggests that the effects of tropical cyclones can lead to a decrease in economic growth within the affected country for over a decade [23, 24].

Numerous research efforts have concentrated on the socio-economic impacts incurred by countries susceptible to tropical cyclones, including the United States, the Philippines, and China [25–30]. These studies have devised multiple damage functions specifically for tropical cyclones, aiming to analyze how the extent of damage resulting from these cyclones changes with fluctuations in global mean temperature and socio-economic advancement.

4. Literature review on tropical cyclone models

In recent years, the urgency of understanding and effectively mitigating the impacts of tropical cyclones has become increasingly evident. As the global community places greater emphasis on sustainable development and urban resilience [31, 32], the study of TCs and their associated boundary layer dynamics (TCBL) has garnered significant attention. The escalating trend of urbanization, characterized by the proliferation of high-rise structures and densely populated areas [33], underscores the critical need to comprehend the behavior of TCs within the context of a changing climate.

The rise in urbanization has led to the emergence of new challenges in disaster management and urban sustainability. The increasing frequency and intensity of TCs, exacerbated by climate change, pose significant threats to urban environments worldwide. As TCs evolve in response to changing climatic conditions, it becomes imperative to develop models that can accurately predict their behavior and associated impacts.

The important role of TC wind field modeling in disaster management and risk assessment provides critical support for building safer and sustainable urban environments. The application of wind field models can help to optimize economic losses and thus control or reduce the impacts caused by disasters. By collecting and analyzing disaster data, we can target redesign.

Tuan et al. [34] discussed the examination of the costs and benefits of implementing typhoon-resilient housing measures in Vietnam, which aims to assess whether using typhoon-resilient housing has a positive economic return, indicating a focus on redesigning buildings with typhoon-resilient features. In addition, the study identified positive returns while justifying the need to pursue policies related to this type of design. Pantua et al. [35] evaluated the structural integrity of current roofing designs in response to extreme environmental events, such as typhoons. The results highlight structural weaknesses in the current design, considering factors such as wind angle, structural frame, and materials. Then they specifically focused on the optimization of photovoltaic (PV) rooftop installations. It highlights the vulnerability of solar installations to typhoon-force winds and proposes a framework that integrates fluid–structure interaction modeling and building energy simulation to evaluate the structural and energy performance of roof-mounted solar panels [36, 37]. Mata et al. [38] highlighted the need to find suitable building orientation and roof angles for single-family residential houses using simulations, especially in countries highly prone to typhoons like the Philippines. The objective is to optimize these factors to minimize drag forces during typhoon conditions, thereby enhancing the resilience of buildings against such extreme weather events.

Such designs can reduce the damage to buildings and infrastructure caused by disasters and lower the cost of repair and reconstruction. The application of wind field modeling can enhance personal safety and security, especially through the connection with storm warning systems [39]. Early warning systems can provide timely alerts so that people have enough time to take action to avoid disaster areas or take other countermeasures, thus minimizing casualties and property damage. City governments should invest in effective and timely early warning systems and integrate climate change adaptation and resilience into planning and policy development [40, 41]. Wind field models can also be coupled with other hazards such as typhoon-induced rainfall and secondary hazards such as floods and mudslides [42, 43]. By establishing an integrated coastal hazard coupling field, we can provide comprehensive disaster data support and provide a scientific basis for disaster prevention and risk assessment in coastal areas.

A model focusing on the boundary layer within a tropical cyclone serves as a valuable tool for examining the mechanisms behind primary and secondary air circulations within the TC, which are directly linked to much of the associated damages [44]. Therefore, it holds significant importance to develop models and subsequently predict the wind field within the tropical cyclone boundary layer (TCBL). This is particularly crucial for forecasting wind speeds at specific locations of interest, and these models find practical applications in various real-world scenarios, including disaster preparedness and urban planning [45, 46].

4.1 Foundation of the models

The TCBL models constitute a crucial tool utilized for understanding and predicting the behavior of these meteorological phenomena. They leverage a combination of direct observations garnered from within the TCBL itself and a deep understanding of the intricate mechanisms underpinning energy and momentum exchanges occurring within this layer. The TCBL models rely on both direct observations collected within the TCBL and knowledge of the mechanisms governing energy and momentum exchanges within this layer [47]. With observational insights alongside theoretical constructs, the TCBL models strive to furnish a comprehensive depiction of the evolving structure and intensity of tropical cyclones, thereby enhancing our ability to anticipate their impacts and facilitate more effective mitigation and response strategies.

4.1.1 Observation data

Regarding observation, data is primarily gathered through ground soundings, meteorological observation towers, and radar systems. Ground sounding stands out as the predominant technique for uncovering wind flow characteristics within the TCBL. Additionally, surveillance aircraft deploy drop-down sounding instruments, known as Global Positioning System (GPS) dropsondes, equipped with GPS technology to track their positions as they descend through the TCBL. These GPS dropsondes offer invaluable direct measurements of wind speeds and other meteorological variables, allowing for a comprehensive understanding of the thermodynamic and dynamical structures within the TCBL [48, 49]. Since their inception, advancements in measurement techniques and data processing of GPS dropsondes have significantly enhanced their capabilities, leading to substantial contributions to wind profile modeling within the TCBL [50–52].

4.1.2 Theoretical foundation

The original Navier-Stokes equation is modified through the introduction of the balance between the gravity and static pressure of the air. When the mean wind velocities replace the instantaneous wind velocities, the turbulent exchange coefficient of K is used to show the turbulent mixing effect due to the shear of mean wind velocities. Considering the horizontal mixing of the turbulence is generally less strong than the vertical mixing by at least an order of magnitude, only the vertical mixing coefficient is commonly presented in the simplified Navier-Stokes equation. In addition, the variations in mean vertical wind velocities are significantly less than those in horizontal wind velocities. Consequently, the transportation equation of vertical winds is generally neglected. The azimuthal gradients for the variations in horizontal wind velocities are generally neglected considering the azimuthal size of the TC.

It is noted that both the original and the simplified Navier-Stokes equation are nonlinear in nature due to the connective term. Consequently, the analytical solutions are not available and numerical treatments are required to solve the governing equation for the sake of modeling the wind field inside the TCBL. According to whether employing the numeric treatment of linearization, the modeling of the TCBL wind field is categorized into the linear models and nonlinear simulation.

4.2 Research progress of TCBL models

In the realm of meteorological research, understanding and predicting the behavior of TCs is a paramount challenge with far-reaching implications for disaster preparedness, response, and risk mitigation. Various TCBL wind field models are all based on the same set of governing equations, which is mentioned in Section 4.2.1, depicting the dynamics or thermodynamics of the TCBL, and hence the differences among the models are mainly attributed to the modifications/simplifications made to the original set of equations. **Figure 3** gives an overview of TCBL models available.

Since the analytical solutions could be derived for the governing equation of the TCBL wind field once the nonlinear terms are linearized. The linear models of the TCBL wind field are the mainstream tools for the investigation of the meteorology corresponding to a TC in the era with weak computational capacity.

4.2.1 Semi-empirical linear models

Initially, the slab model is introduced, offering a simplified yet insightful representation of the main vortex of a tropical cyclone in a two-dimensional plane. The general idea of the slab model is to average the vertical variations in the TCBL wind field, and hence to focus on the main circulation of the TC wind flow in the horizontal plane. Specifically, the TCBL is assumed as a slab at the constant depth, which leads to the analytical solution to the governing equation under an axis-symmetric pressure distribution.

Smith [53] proposed a pioneering description of the TCBL wind field, modeling it as a steady, axis-symmetric potential vortex, simplifying the analytical solutions but

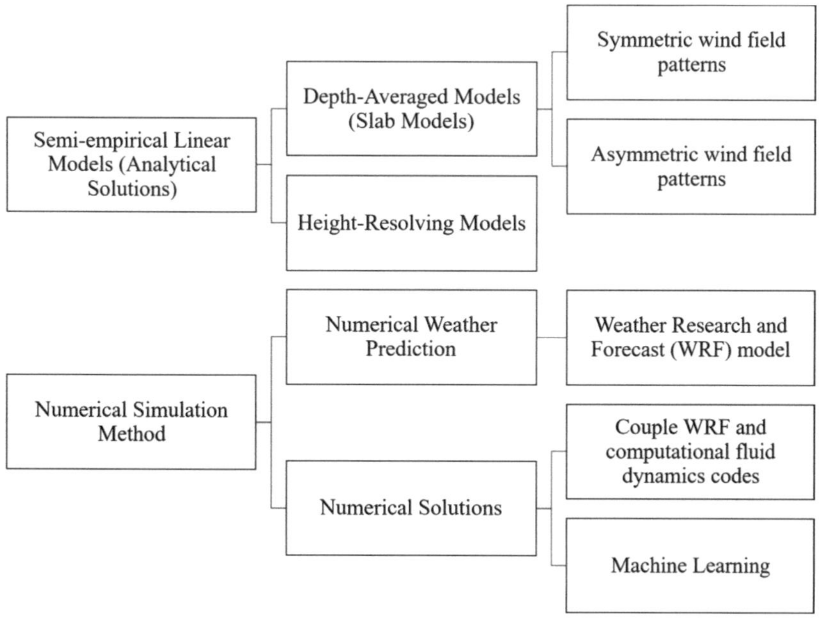

Figure 3.
Categories of TCBL wind field models.

lacking realism due to an inadequate understanding of TCBL turbulence. Subsequent refinements by Bode and Smith [54] and Smith [55] improved the model's accuracy, predicting spatial oscillations in vertical wind velocity at the TCBL top.

While the symmetric slab model provides satisfactory surface wind estimates, due to insufficient knowledge of turbulent structures at the time, a more realistic understanding of the turbulent layer was not possible. Recent studies [56, 57] identified inconsistencies in predictions due to neglect of vertical momentum exchanges and terrain characteristics.

Recognizing the limitations of two-dimensional modeling, recent advancements focus on height-resolving models to incorporate three-dimensional TCBL wind field dynamics. Meng et al. [58] proposed an analytical model considering underlying terrains, showing promise in predicting TCBL wind fields under complex conditions. Subsequent enhancements by Huang and Xu [59] and Snaiki and Wu [60] incorporated temperature variations and increasingly emerged effects of climate change in the prediction of extreme wind speeds, respectively.

Batts et al. [61] proposed a linear model within the slab category to estimate maximum surface wind speed based on gradient-level winds, applied in Monte-Carlo simulations for U.S. coastal hazard assessment. Shapiro [62] employed a slab boundary layer model of constant depth, analyzing the steady flow under a specified translating symmetric vortex in gradient balance. To validate the model's findings, comparisons are made with observations from Hurricanes Frederic (1979) and Allen (1980), as well as other relevant observational and theoretical data, which reveals that the simple boundary layer formulation accurately captures the qualitative features of the wind field observed in Hurricane Frederic. The asymmetric slab model, widely used in wind engineering, assesses TC hazards with acceptable computational burden. For instance, Vickery et al. [63] applied Shapiro's model to estimate extreme TC wind speeds along the U.S. coast with specific recurrence periods, favoring it over Batts' model. However, recent studies [56, 57] found inconsistency between slab model predictions and observational data on radial TCBL wind variations, due to deliberate neglect of vertical momentum exchanges. While Shapiro's model provides computational efficiency for risk analysis of large insurance portfolios, its simplifications and assumptions may introduce limitations in accurately estimating hurricane-related losses [64, 65].

The average-depth slab model for the TC boundary layer is a simplified yet insightful approach used to understand and predict the dynamics of the boundary layer within TCs, however, all of the above are modeled in two dimensions; the actual boundary layer wind field is three-dimensional [60, 66–68], and not taking height into account may lead to inaccurate results.

The analytical model by Meng et al. [58] for calculating the wind field in a moving typhoon boundary layer incorporates two distinct layers: An upper inviscid layer and a lower friction layer. By integrating these layers, the model accurately predicts wind speeds and directions, including variations observed from elevated vantage points during typhoons. Moreover, it accounts for rapid wind speed changes caused by surface features. Although the model obtained an analytical solution and was verified [59, 68], the structural characteristics of the TCBL wind field are not reliably estimated because the gradient balance is not always satisfied for real TCs. Huang and Xu [59] improved upon Meng's model by considering temperature effects and removing the assumption of a constant central pressure difference with height. Through a decomposition method, it provides more accurate wind speed and direction predictions, validated against field measurements and compared to other models like Meng's and Shapiro's. Results show the refined model offers better accuracy in wind speed, direction, and profile estimations, as confirmed by spatial wind speed distribution and mean wind profile analyses.

As a milestone in the development of the height-resolving model, Kepert [69] proposed a linear model with an analytical solution available by neglecting the vertical advection, and the horizontal advection is linearized. This model uses a linear analytical model and a numerical model to describe the boundary layer flow, highlighting the role of vertical diffusion, vertical advection, and horizontal advection in maintaining inflow against outward acceleration. Therefore, the vertical turbulent diffusivity and surface drag coefficient are important as they parameterize the momentum exchanges from the bottom boundary upwards. Subsequent scholars have studied the parameterization of coefficients [52, 66, 70–72].

A key shortcoming with current height-resolving models is that the super-gradient winds predicted by the model are generally weaker than the observations because it disregards the vertical advection in the TCBL [73]. Yang et al. [68] proposed a height-resolving model that simultaneously considers horizontal and vertical advections, whose better performance confirmed the importance of a height-resolving model to include the vertical advection.

Since climate change cannot be ignored and the effects on TCs have been observed, the new generation of mods is now moving toward considering the effects. Fang et al. [70] incorporate various elements to develop a typhoon velocity field model suitable for the gradient layer and boundary layer, considering multi-field parameters correlation and terrain effects. Overall, the model demonstrates the ability to illustrate specific wind environment characteristics with satisfactory precision, including non-exponential wind profiles in typhoon boundary layers. Kim and Lee [74], on a Monte Carlo Simulation approach, incorporate various components such as genesis, intensity, tracks, and wind field to estimate extreme wind speeds of future typhoons. The method considers climatological factors, and it incorporates sea surface temperature (SST) and oceanic occupation ratio (OOR) for intensity modeling, and a modified Batts' model for wind field estimation.

4.2.2 Numerical simulation method

As computational capacity increased, the linear model's limitations led to the rise of numerically solving the nonlinear governing equation to estimate the TCBL wind field, becoming mainstream. Numerical weather prediction packages are commonly employed to depict TC structures, including the TCBL wind field. Additionally, the direct solution of the nonlinear partial differential equation governing TCBL dynamics is reported in the literature.

Examples of TCBL wind models include the Weather Research and Forecasting (WRF) model, which is widely used for weather and climate research. Researchers utilize WRF simulations of historical TCs, known as hindcasting, to analyze TCBL wind fields [75–79]. For instance, Nolan, McNoldy, and Yunge [80] explored the impact of different planetary boundary layer schemes on TCBL numerical simulations by comparing WRF-simulated tracks, intensities, and sizes of a historical TC with observational data. They generated two WRF simulations with different boundary layer parameterizations and verified their accuracy against observed best track records. Another example is the Large Eddy Simulation (LES) model, which resolves the large-scale turbulent structures in the atmosphere and is used to study boundary layer dynamics. Zhu [79] introduced a LES framework within the WRF model, revealing TCBL boundary layer processes and coherent large eddy circulations during TC landfall.

While numerical weather prediction software can offer detailed representations of the TCBL wind field based on large-scale distributions of meteorological variables like wind velocities, air pressures, and temperatures, its use in TCBL investigations is constrained by boundary and initial condition preparation. Moreover, detailed TCBL wind field structures provided by numerical weather simulations come at the cost of significant computational resources, making it challenging to use them in Monte Carlo simulations to quantitatively assess TC hazards. Three-dimensional numerical models like WRF need continual adjustments based on an evolving understanding of meteorological processes within the TCBL. However, progress in TC meteorology research is slow and limited, hindering accurate simulation of TCBL wind fields with models like WRF ([81]; Wang and Wu [82]). Consequently, simulations of TCBL wind fields with the WRF model may suffer from inaccuracies in TC intensity and size.

In addition to the numerical weather prediction, directly solving the nonlinear governing equation of the TCBL wind field is suggested using general-purpose computational fluid dynamic tools. For example, Ma and Sun [83] showed the structure of the TCBL wind field via the help from the general computational fluid dynamics software of OpenFOAM. However, the numerical models combining the numerical weather prediction tool and computational fluid dynamics code still have difficulties in predicting winds in the strong convective inner core region. Recent modeling has focused on how to integrate various models, to study the impact of climate change on the dynamics of TCBL wind fields. Through Monte Carlo simulations incorporating climate change scenarios, the study reveals potential changes in wind profiles, suggesting that existing wind profile models may be overly conservative when evaluating structural safety under the influence of climate change, especially in tall buildings, when compared to current standards [72].

As various models play critical roles in TC forecasting, it is essential to assess their performance and accuracy in capturing different aspects of TC dynamics is very important, thus facilitating improvements in forecasting accuracy and resilience-building efforts. The wind fields simulated by the WRF model show an overall good agreement with observations, while noting discrepancies such as smaller simulated inflow angles compared to observations and mixed results in comparisons with Doppler radar wind profiles [80, 84]. Models include the OpenFOAM code lacks the possibility to describe the mesoscale process of TCs and their boundary layer processes [85–87]. Therefore, to replicate the actual meteorological system of a TC, many studies coupled WRF tool and computational fluid dynamics codes due to the limitations of mesoscale WRF tool and small-scale computational fluid dynamics code [88–91]. Computational fluid dynamics code coupled with WRF can have the highest accuracy in estimating the wind field among other mesoscale meteorology models [92]. The accuracy of such numerical models has improved significantly in recent years [93], making their wind fields more suitable for operational use.

In recent years, machine learning (ML) holds promise for parameterizing boundary layer simulations due to the highly nonlinear nature of TCBL processes, which traditional parametrization struggles to address. ML algorithms, characterized by high dimensions and nonlinearities, offer a potential solution to the challenges in TCBL parameterization [94]. In the realm of numerical modeling, the most active research on ML application pertains to convective parameterization in atmospheric circulation models within climate research [95, 96] and unified physics parameterization [97, 98]. O'Gorman and Dwyer [96] discuss the use of ML to develop new parameterizations for moist convection directly from high-resolution model output, which is a significant aspect of convective parameterization within climate research. The study focuses on understanding the behavior of these ML-based parameterizations when integrated into a general circulation model (GCM) and their effectiveness in simulating climate change or extreme events. Han et al. [95] The study discussed focuses on developing a new moist physics parameterization scheme using deep learning techniques, specifically a residual convolutional neural network. This scheme aims to address biases in simulated precipitation and atmospheric circulation caused by current moist physics parameterization schemes in GCMs. It demonstrates superior performance in accurately reproducing simulations and predicting convective events in various atmospheric conditions.

The application of ML in boundary layer parameterization remains relatively limited. For example, McGibbon and Bretherton [99] describe the training of an artificial neural network to reproduce thermodynamic tendencies and boundary layer properties from high-resolution reanalysis data over a specific region. It demonstrates the use of ML techniques to emulate and parameterize boundary layer processes, which is a crucial aspect of atmospheric modeling. Wang et al. [100] describe the development of an emulator using deep neural networks for a planetary boundary layer parameterization in the WRF climate model. Parameterizations are crucial for representing diurnal variations in the atmospheric boundary layer. The research indicates that neural networks can successfully simulate vertical profiles within the boundary layer and produce accurate predictions of wind speed, temperature, and water vapor profiles. Chen et al. [101] discuss using ML for both pure data-driven models and for enhancing numerical models by incorporating ML techniques. The passage also highlights successful cases of ML methods in various aspects of tropical cyclone forecasting, such as genesis forecasts, track forecasts, intensity forecasts, extreme weather forecasts, and storm surge forecasts. This implies that ML is being applied to improve the accuracy of numerical forecast models, which includes parameterization of boundary layer processes within the context of tropical cyclones.

The limited utilization of ML in TCBL parameterization is primarily attributed to factors such as the intricate nature of TCBL processes, insufficient training data, challenges in interpretability of ML models, as well as difficulties in integrating ML techniques with current parameterization frameworks, underscoring the critical importance of ensuring compatibility and maintaining model accuracy during integration efforts.

Artificial intelligence (AI) has great potential in integrating with climate change models. In the parameterization of climate models, traditional physical parameterization methods may have limitations in accurately describing complex nonlinear processes in the climate system. ML techniques, on the other hand, are able to discover patterns in the climate system by learning from massive amounts of data, and also taking climate change into account, thus providing more accurate parameterization schemes.

5. Exploration of the challenges and frontiers

5.1 Summary of research progress

This review has delved into the multiple realms of tropical cyclones, providing an in-depth exploration of their characteristics, classification systems, and the extensive impacts of tropical cyclones on a global scale, emphasizing their implications for sustainability and the evolving landscape of climate change. This discussion underscored the urgent need for comprehensive research and action to mitigate the adverse effects of TCs.

Tropical cyclone modeling represents a frontier in meteorological research, where the quest for accuracy and reliability is constantly evolving. Current approaches to hurricane and typhoon modeling face many challenges and also show vast frontiers, including the impacts of climate change and the integration of AI technologies.

Traditional linear models offer quick calculations and reveal certain physical mechanisms, yet their comprehensiveness and reliability fall short compared to nonlinear simulations. Nonlinear methods, while slower, provide indispensable accuracy in predicting real tropical cyclone trajectories. However, their computational demands remain a barrier, limiting applicability to TCs with specific boundaries and initial conditions. The integration of AI and ML presents a transformative avenue in tropical cyclone modeling. While wind field models have limitations, AI-enhanced simulations have the potential to revolutionize TC prediction by significantly reducing computational efforts. Nonetheless, challenges persist, including the need for extensive training data and overcoming the limitations of existing AI models in producing accurate predictions.

5.2 Challenges and future directions

The complexities of physical processes within TCs pose formidable challenges for both AI and numerical models. Boundary layer wind field modeling encounters obstacles such as insufficient data quality and quantity, the intricate influence of boundary layer turbulence, and the generalization capability of AI models across diverse scenarios.

To address these challenges, a balanced approach that integrates AI and numerical modeling is paramount. This entails leveraging the strengths of each approach, incorporating uncertainty analysis, enhancing observational data quality, and regularly updating models to adapt to changing meteorological conditions and geographic environments. By embracing this holistic strategy, the future of hurricane and typhoon modeling holds promise for more accurate predictions and improved resilience in the face of these powerful natural phenomena.

As we navigate the complex terrain of hurricane and typhoon modeling, it becomes evident that collaboration, innovation, and adaptability are key to advancing our understanding and preparedness in a changing climate landscape.

For the practical application of TC models, they can be integrated into a coupled multi-hazard spatial field model to support disaster warning and risk assessment of coastal infrastructure. Such an integrated model can provide multifaceted hazard data, including TC wind field information, to provide a more comprehensive information base for relevant agencies and decision makers.

In addition, the application of TC modeling can be extended to the development of advanced warning systems. By forecasting TC risks in advance, the system can provide reasonable countermeasures for coastal residents and infrastructure, such as advising residents in high-rise buildings to take measures such as evacuating, moving to a safer floor, or closing the windows, in order to minimize potential casualties and property damage. TC models can also be used to study vulnerability and analyze the resilience of building structures and coastal infrastructure. By simulating the performance of buildings and infrastructures under the influence of typhoons, potential weak points can be identified and recommendations for strengthening and improvement can be made to enhance their resilience.

Through concerted efforts and interdisciplinary collaboration, we can forge a path toward more effective modeling techniques and enhanced resilience to tropical cyclones worldwide.

Author details

Jiayao Wang[1], Yu Chang[2*] and Kam Tim Tse[3*]

1 Department of Civil and Environmental Engineering, The Hong Kong Polytechnic University, Kowloon, Hong Kong

2 Institute for Ocean Engineering, Tsinghua Shenzhen International Graduate School, Shenzhen, Guangdong, China

3 Department of Civil and Environmental Engineering, The Hong Kong University of Science and Technology, Hong Kong, China

*Address all correspondence to: changy23@mails.tsinghua.edu.cn and timkttse@ust.hk

References

[1] Holland GJ. Ready reckoner. Global Guide to Tropical Cyclone Forecasting. 1993;**9**(9):26

[2] Emanuel K. Tropical cyclones. Annual Review of Earth and Planetary Sciences. 2003;**31**(1):75-104

[3] Neumann CJ. Global Overview: Global Guide to Tropical Cyclone Forecasting. Geneva, Switzerland: World Meteorological Organization (WMO); 1993

[4] Chan JC, Kepert JD. Global Perspectives on Tropical Cyclones: From Science to Mitigation. Singapore: World Scientific Publishing Co. Pte Ltd.; 2010

[5] Emanuel K. Divine Wind: The History and Science of Hurricanes. New York, United States: Oxford University Press; 2005

[6] Annamalai H, Slingo J, Sperber K, Hodges K. The mean evolution and variability of the Asian summer monsoon: Comparison of ECMWF and NCEP–NCAR reanalyses. Monthly Weather Review. 1999;**127**(6):1157-1186

[7] Pasch RJ, Pasch RJ, Blake ES, Cobb HD III, David P. Tropical Cyclone Report Hurricane Wilma 15-25 October 2005. Miami, USA: Roberts National Hurricane Center; 2006

[8] Kepert JD. Observed boundary layer wind structure and balance in the hurricane core. Part I: Hurricane Georges. Journal of the Atmospheric Sciences. 2006;**63**(9):2169-2193

[9] Taylor HT, Ward B, Willis M, Zaleski W. The Saffir-Simpson Hurricane Wind Scale. Washington, DC, USA: Atmospheric Administration; 2010

[10] Beer T. Environmental Oceanography. Boca Raton, USA: CRC Press; 2017

[11] Saucier WJ. Principles of Meteorological Analysis. Vol. 438. Chicago, IL: University of Chicago Press; 1955

[12] Knutson T, Camargo SJ, Chan JC, Emanuel K, Ho C-H, Kossin J, et al. Tropical cyclones and climate change assessment: Part II: Projected response to anthropogenic warming. Bulletin of the American Meteorological Society. 2020;**101**(3):E303-E322

[13] Masson-Delmotte V, Zhai P, Pörtner H-O, Roberts D, Skea J, Shukla PR. Global Warming of 1.5 C: IPCC Special Report on Impacts of Global Warming of 1.5 C above Pre-Industrial Levels in Context of Strengthening Response to Climate Change, Sustainable Development, and Efforts to Eradicate Poverty. New York, United States: Cambridge University Press; 2022. DOI: 10.1017/9781009157940

[14] Mendelsohn R, Emanuel K, Chonabayashi S, Bakkensen L. The impact of climate change on global tropical cyclone damage. Nature Climate Change. 2012;**2**(3):205-209. DOI: 10.1038/nclimate1357

[15] Middelanis R, Willner SN, Otto C, Levermann A. Economic losses from hurricanes cannot be nationally offset under unabated warming. Environmental Research Letters. 2022;**17**(10):104013. DOI: 10.1088/1748-9326/ac90d8

[16] Neukom R, Steiger N, Gómez-Navarro JJ, Wang J, Werner JP. No evidence for globally coherent warm and cold periods over the

preindustrial common era. Nature. 2019;**571**(7766):550-554. DOI: 10.1038/s41586-019-1401-2

[17] Studholme J, Fedorov AV, Gulev SK, Emanuel K, Hodges K. Poleward expansion of tropical cyclone latitudes in warming climates. Nature Geoscience. 2022;**15**(1):14-28

[18] Keim ME. Cyclones, tsunamis and human health. Oceanography. 2006;**19**(2):40-49

[19] Noji EK. Analysis of medical needs during disasters caused by tropical cyclones: Anticipated injury patterns. The Journal of Tropical Medicine and Hygiene. 1993;**96**(6):370-376

[20] Shultz JM, Russell J, Espinel Z. Epidemiology of tropical cyclones: The dynamics of disaster, disease, and development. Epidemiologic Reviews. 2005;**27**(1):21-35. DOI: 10.1093/epirev/mxi011

[21] Nicholls RJ, Mimura N, Topping J. Climate change in south and south-East Asia: Some implications for coastal areas. Journal of Global Environmental Engineering. 1995;**1**:137-154

[22] Prevention, U. N. D. P. B. F. C. Reducing Disaster Risk: A Challenge for Development-a Global Report. New York, United States: United Nations; 2004

[23] Berlemann M, Wenzel D. Hurricanes, economic growth and transmission channels: Empirical evidence for countries on differing levels of development. World Development. 2018;**105**:231-247

[24] Krichene H, Geiger T, Frieler K, Willner S, Sauer I, Otto C. Long-term impacts of tropical cyclones and fluvial floods on economic growth–empirical evidence on transmission channels at different levels of development. World Development. 2021;**144**:105475

[25] Bakkensen LA, Mendelsohn RO. Risk and adaptation: Evidence from global hurricane damages and fatalities. Journal of the Association of Environmental and Resource Economists. 2016;**3**(3):555-587

[26] Baldwin JW, Lee C-Y, Walsh BJ, Camargo SJ, Sobel AH. Vulnerability in a tropical cyclone risk model: Philippines case study. In: Weather, Climate, and Society. Boston, USA: AMS Publications; 2023

[27] Cinco TA, de Guzman RG, Ortiz AMD, Delfino RJP, Lasco RD, Hilario FD, et al. Observed trends and impacts of tropical cyclones in the Philippines. International Journal of Climatology. 2016;**36**(14):4638-4650

[28] Elliott RJ, Strobl E, Sun P. The local impact of typhoons on economic activity in China: A view from outer space. Journal of Urban Economics. 2015;**88**:50-66

[29] Emanuel K. Global warming effects on US hurricane damage. Weather, Climate, and Society. 2011;**3**(4):261-268

[30] Schmidt S, Kemfert C, Höppe P. The impact of socio-economics and climate change on tropical cyclone losses in the USA. Regional Environmental Change. 2010;**10**:13-26

[31] Agbedahin AV. Sustainable development, education for sustainable development, and the 2030 agenda for sustainable development: Emergence, efficacy, eminence, and future. Sustainable Development. 2019;**27**(4):669-680

[32] Holden E, Linnerud K, Banister D. The imperatives of

sustainable development. Sustainable Development. 2017;**25**(3):213-226

[33] Ali MM, Al-Kodmany K. Tall buildings and urban habitat of the 21st century: A global perspective. Buildings. 2012;**2**(4):384-423

[34] Tuan TH, Tran P, Hawley K, Khan F, Moench M. Quantitative cost-benefit analysis for typhoon resilient housing in Danang city, Vietnam. Urban Climate. 2015;**12**:85-103

[35] Pantua CAJ, Calautit JK, Wu Y. A novel fluid-structure interaction modelling and optimisation of roofing designs of buildings for typhoon resilience. In: IOP Conference Series: Materials Science and Engineering. Vol. 556. Chengdu, China: IOP Publishing Ltd.; 2019

[36] Pantua CAJ, Calautit JK, Wu Y. A fluid-structure interaction (FSI) and energy generation modelling for roof mounted renewable energy installations in buildings for extreme weather and typhoon resilience. Renewable Energy. 2020;**160**:770-787

[37] Pantua CAJ, Calautit JK, Wu Y. Sustainability and structural resilience of building integrated photovoltaics subjected to typhoon strength winds. Applied Energy. 2021;**301**:117437

[38] Mata JL, Orejudos JN, Opon JG, Guirnaldo SA. Optimizing building orientation and roof angle of a typhoon-resilient single-family house using genetic algorithm and computational fluid dynamics. Buildings. 2022;**13**(1):107

[39] Tan C, Fang W. Mapping the wind hazard of global tropical cyclones with parametric wind field models by considering the effects of local factors. International Journal of Disaster Risk Science. 2018;**9**:86-99

[40] Nhuan MT, Tue NT, Quy TD. Enhancing resilience to climate change and disasters for sustainable development: Case study of Vietnam coastal urban areas. In: Resilient Asia: Fusion of Traditional and Modern Systems for a Sustainable Future. Tokyo, Japan: Springer; 2018. pp. 63-79

[41] Sajjad M, Li Y, Li Y, Chan JC, Khalid S. Integrating typhoon destructive potential and social-ecological systems toward resilient coastal communities. Earth's Future. 2019;**7**(7):805-818

[42] Kim J-S, Chen A, Lee J, Moon I-J, Moon Y-I. Statistical prediction of typhoon-induced rainfall over China using historical rainfall, tracks, and intensity of typhoon in the Western North Pacific. Remote Sensing. 2020;**12**(24):4133

[43] Yin J, Lin N, Yang Y, Pringle WJ, Tan J, Westerink JJ, et al. Hazard assessment for typhoon-induced coastal flooding and inundation in Shanghai, China. Journal of Geophysical Research: Oceans. 2021;**126**(7):e2021JC017319

[44] Li T-H, Wang Y. The role of boundary layer dynamics in tropical cyclone intensification. Part I: Sensitivity to surface drag coefficient. Journal of the Meteorological Society of Japan. Ser. II. 2021;**99**(2):537-554. DOI: 10.2151/jmsj.2021-027

[45] Kijewski-Correa T, Taflanidis A, Vardeman C, Sweet J, Zhang J, Snaiki R, et al. Geospatial environments for hurricane risk assessment: Applications to situational awareness and resilience planning in New Jersey. Frontiers in Built Environment. 2020;**6**:549106

[46] Powell MD, Houston SH. Hurricane Andrew's landfall in South Florida.

Part II: Surface wind fields and potential real-time applications. Weather and Forecasting. 1996;**11**(3):329-349

[47] Emanuel K. 100 years of progress in tropical cyclone research. Meteorological Monographs. 2018;**59**:15.11-15.68. DOI: 10.1175/AMSMONOGRAPHS-D-18-0016.1

[48] Franklin JL, Black ML, Valde K. GPS dropwindsonde wind profiles in hurricanes and their operational implications. Weather and Forecasting. 2003;**18**(1):32-44. DOI: 10.1175/1520-0434(2003)018< 0032:GDWPIH>2.0.CO;2

[49] Hock TF, Franklin JL. The ncar gps dropwindsonde. Bulletin of the American Meteorological Society. 1999;**80**(3):407-420. DOI: 10.1175/1520-0477(1999)080< 0407:TNGD>2.0.CO;2

[50] Powell MD, Vickery PJ, Reinhold TA. Reduced drag coefficient for high wind speeds in tropical cyclones. Nature. 2003;**422**(6929):279-283. DOI: 10.1038/nature01481

[51] Uhlhorn EW, Black PG, Franklin JL, Goodberlet M, Carswell J, Goldstein AS. Hurricane surface wind measurements from an operational stepped frequency microwave radiometer. Monthly Weather Review. 2007;**135**(9):3070-3085. DOI: 10.1175/MWR3454.1

[52] Vickery PJ, Wadhera D, Powell MD, Chen Y. A hurricane boundary layer and wind field model for use in engineering applications. Journal of Applied Meteorology and Climatology. 2009;**48**(2):381-405. DOI: 10.1175/2008JAMC1841.1

[53] Smith RK. The surface boundary layer of a hurricane. Tellus. 1968;**20**(3):473-484. DOI: 10.1111/j.2153-3490.1968.tb00388.x

[54] Bode L, Smith RK. A parameterization of the boundary layer of a tropical cyclone. Boundary-Layer Meteorology. 1975;**8**:3-19. DOI: 10.1007/BF02579390

[55] Smith RK. A simple model of the hurricane boundary layer. Quarterly Journal of the Royal Meteorological Society: A Journal of the Atmospheric Sciences, Applied Meteorology and Physical Oceanography. 2003;**129**(589):1007-1027. DOI: 10.1256/qj.01.197

[56] Kepert JD. Slab-and height-resolving models of the tropical cyclone boundary layer. Part I: Comparing the simulations. Quarterly Journal of the Royal Meteorological Society. 2010;**136**(652):1686-1699. DOI: 10.1002/qj.667

[57] Khare S, Bonazzi A, West N, Bellone E, Jewson S. On the modelling of over-ocean hurricane surface winds and their uncertainty. Quarterly Journal of the Royal Meteorological Society: A Journal of the Atmospheric Sciences, Applied Meteorology and Physical Oceanography. 2009;**135**(642):1350-1365. DOI: 10.1002/qj.442

[58] Meng Y, Matsui M, Hibi K. An analytical model for simulation of the wind field in a typhoon boundary layer. Journal of Wind Engineering and Industrial Aerodynamics. 1995;**56**(2):291-310. DOI: 10.1016/0167-6105(94)00014-5

[59] Huang W, Xu YL. A refined model for typhoon wind field simulation in boundary layer. Advances in Structural Engineering. 2012;**15**(1):77-89. DOI: 10.1260/1369-4332.15.1.77

[60] Snaiki R, Wu T. Modeling tropical cyclone boundary layer: Height-resolving pressure and wind fields. Journal of

Wind Engineering and Industrial Aerodynamics. 2017b;**170**:18-27. DOI: 10.1016/j.jweia.2017.08.005

[61] Batts ME, Simiu E, Russell LR. Hurricane wind speeds in the United States. Journal of the Structural Division. 1980;**106**(10):2001-2016. DOI: 10.1061/JSDEAG.000554

[62] Shapiro LJ. The asymmetric boundary layer flow under a translating hurricane. Journal of Atmospheric Sciences. 1983;**40**(8):1984-1998. DOI: 10.1175/1520-0469(1983)040<1984:TABLFU>2.0.CO;2

[63] Vickery PJ, Skerlj P, Steckley A, Twisdale L. Hurricane wind field model for use in hurricane simulations. Journal of Structural Engineering. 2000;**126**(10):1203-1221. DOI: 10.1061/(ASCE)0733-9445(2000)126:10(1203)

[64] Daneshvaran S, Yao T, Morden R, Wen Y, Zadeh M. A parametric model for efficient hurricane risk analysis of large portfolios. In: International Conference of Structural Safety and Reliability (ICOSSAR). Rotterdam, The Netherlands: Balkema; 1997

[65] Vickery PJ, Twisdale LA. Wind-field and filling models for hurricane wind-speed predictions. Journal of Structural Engineering. 1995;**121**(11):1700-1709. DOI: 10.1061/(ASCE)0733-9445(1995)121:11(1700)

[66] Li W, Hu Z, Pei Z, Li S, Chan PW. A discussion on influences of turbulent diffusivity and surface drag parameterizations using a linear model of the tropical cyclone boundary layer wind field. Atmospheric Research. 2020;**237**:104847 %@ 100169-104847 %@ 108095. DOI: 10.1016/j.atmosres.2020.104847

[67] Snaiki R, Wu T. A linear height-resolving wind field model for tropical cyclone boundary layer. Journal of Wind Engineering and Industrial Aerodynamics. 2017a;**171**:248-260. DOI: 10.1016/j.jweia.2017.10.008

[68] Yang J, Chen Y, Zhou H, Duan Z. A height-resolving tropical cyclone boundary layer model with vertical advection process. Natural Hazards. 2021;**107**:723-749. DOI: 10.1007/s11069-021-04603-1

[69] Kepert J. The dynamics of boundary layer jets within the tropical cyclone core. Part I: Linear theory. Journal of the Atmospheric Sciences;**58**(17):2469-2484. DOI: 10.1175/1520-0469(2001)058<2469:TDOBLJ>2.0.CO;2

[70] Fang G, Zhao L, Cao S, Ge Y, Pang W. A novel analytical model for wind field simulation under typhoon boundary layer considering multi-field correlation and height-dependency. Journal of Wind Engineering and Industrial Aerodynamics. 2018;**175**:77-89 %@ 0167-6105. DOI: 10.1016/j.jweia.2018.01.019

[71] Hong X, Hong HP, Li J. Solution and validation of a three dimensional tropical cyclone boundary layer wind field model. Journal of Wind Engineering and Industrial Aerodynamics. 2019;**193**:103973 %@ 100167-106105. DOI: 10.1016/j.jweia.2019.103973

[72] Wang J, Tse TK, Li S, Fung JC. A model of the sea–land transition of the mean wind profile in the tropical cyclone boundary layer considering climate changes. International Journal of Disaster Risk Science. 2023;**14**:1-15. DOI: 10.1007/s13753-023-00488-9

[73] Kepert J, Wang Y. The dynamics of boundary layer jets within the tropical cyclone core. Part II: Nonlinear enhancement. Journal of the Atmospheric Sciences. 2001;**58**(17):2485-2501 %@ 0022-4928.

DOI: 10.1175/1520-0469(2001)058<2485:TDOBLJ>2.0.CO;2

[74] Kim GY, Lee S. Prediction of extreme wind by stochastic typhoon model considering climate change. Journal of Wind Engineering and Industrial Aerodynamics. 2019;**192**:17-30

[75] Davis C, Wang W, Chen SS, Chen Y, Corbosiero K, DeMaria M, et al. Prediction of landfalling hurricanes with the advanced hurricane WRF model. Monthly Weather Review. 2008;**136**(6):1990-2005. DOI: 10.1175/2007MWR2085.1

[76] Done J, Davis CA, Weisman M. The next generation of NWP: Explicit forecasts of convection using the weather research and forecasting (WRF) model. Atmospheric Science Letters. 2004;**5**(6):110-117. DOI: org/10.1002/asl.72

[77] Jiménez PA, González-Rouco JF, García-Bustamante E, Navarro J, Montávez JP, De Arellano JV-G, et al. Surface wind regionalization over complex terrain: Evaluation and analysis of a high-resolution WRF simulation. Journal of Applied Meteorology and Climatology. 2010;**49**(2):268-287. DOI: 10.1175/2009JAMC2175.1

[78] Lo JCF, Yang ZL, Pielke RA Sr. Assessment of three dynamical climate downscaling methods using the weather research and forecasting (WRF) model. Journal of Geophysical Research: Atmospheres. 2008;**113**(D9). DOI: 10.1029/2007JD009216

[79] Zhu P. A multiple scale modeling system for coastal hurricane wind damage mitigation. Natural Hazards. 2008;**47**:577-591. DOI: 10.1007/s11069-008-9240-8

[80] Nolan DS, McNoldy BD, Yunge J. Evaluation of the surface wind field over land in WRF simulations of hurricane Wilma (2005). Part I: Model initialization and simulation validation. Monthly Weather Review. 2021a;**149**(3):679-695. DOI: 10.1175/MWR-D-20-0199.1

[81] Marks FD, Shay LK, PDT-5. Landfalling tropical cyclones: Forecast problems and associated research opportunities. Bulletin of the American Meteorological Society. 1998;**79**(2):305-323

[82] Wang Y-Q, Wu C-C. Current understanding of tropical cyclone structure and intensity changes–a review. Meteorology and Atmospheric Physics. 2004;**87**(4):257-278

[83] Ma T, Sun C. Large eddy simulation of hurricane boundary layer turbulence and its application for power transmission system. Journal of Wind Engineering and Industrial Aerodynamics. 2021;**210**:104520

[84] Nolan DS, McNoldy BD, Yunge J, Masters FJ, Giammanco IM. Evaluation of the surface wind field over land in WRF simulations of hurricane Wilma (2005). Part II: Surface winds, inflow angles, and boundary layer profiles. Monthly Weather Review. 2021b;**149**(3):697-713

[85] Durán P, Meißner C, Rutledge K, Fonseca R, Martin-Torres J, Adaramola MS. Meso-microscale coupling for wind resource assessment using averaged atmospheric stability conditions. Meteorologische Zeitschrift. 2019;**28**(4):273-291. DOI: 10.1127/metz/2019/0937

[86] Hristov Y, Oxley G, Žagar M. Improvement of AEP predictions using diurnal CFD modelling with site-specific stability weightings provided from mesoscale simulation. Journal of Physics: Conference Series. 2014;**524**(1):012116

[87] Luo X, Cao Y. Simulation of the wind fields over complex terrain with coupling of CFD and WRF. Journal of Computational Methods in Sciences and Engineering. 2021;**21**(5):1155-1166. DOI: 10.3233/JCM-204759

[88] Huang M, Wang Y, Lou W, Cao S. Multi-scale simulation of time-varying wind fields for Hangzhou Jiubao bridge during typhoon Chan-hom. Journal of Wind Engineering and Industrial Aerodynamics. 2018;**179**:419-437. DOI: 10.1016/j.jweia.2018.06.020

[89] Mughal MO, Lynch M, Yu F, Sutton J. Forecasting and verification of winds in an east African complex terrain using coupled mesoscale-and micro-scale models. Journal of Wind Engineering and Industrial Aerodynamics. 2018;**176**:13-20. DOI: 10.1016/j.jweia.2018.03.006

[90] Nakayama H, Takemi T, Nagai H. Coupling of WRF and building-resolving urban CFD models for analysis of strong winds over an urban area. In: Proceedings of the 14th Conference on Mesoscale Processes. Los Angeles, CA, USA: American Meteorological Society; 2011

[91] Talbot C, Bou-Zeid E, Smith J. Nested mesoscale large-eddy simulations with WRF: Performance in real test cases. Journal of Hydrometeorology. 2012;**13**(5):1421-1441. DOI: 10.1175/JHM-D-11-048.1

[92] Mohan M, Sati AP. WRF model performance analysis for a suite of simulation design. Atmospheric Research. 2016;**169**:280-291. DOI: 10.1016/j.atmosres.2015.10.013

[93] Duan YH, Fang J, Cheng ZQ, Xu J, Li QQ, Zhan RF. Advances and trends in tropical cyclone research and forecasting: An overview of the ninth World Meteorological Organization International Workshop on tropical cyclones (IWTC-9). Acta Meteorologica Sinica. 2020:537-550

[94] Goodfellow I, Bengio Y, Courville A. Deep Learning. Cambridge, USA: MIT Press; 2016

[95] Han Y, Zhang GJ, Huang X, Wang Y. A moist physics parameterization based on deep learning. Journal of Advances in Modeling Earth Systems. 2020;**12**(9):e2020MS002076

[96] O'Gorman PA, Dwyer JG. Using machine learning to parameterize moist convection: Potential for modeling of climate, climate change, and extreme events. Journal of Advances in Modeling Earth Systems. 2018;**10**(10):2548-2563

[97] Brenowitz ND, Bretherton CS. Spatially extended tests of a neural network parametrization trained by coarse-graining. Journal of Advances in Modeling Earth Systems. 2019;**11**(8):2728-2744

[98] Gentine P, Pritchard M, Rasp S, Reinaudi G, Yacalis G. Could machine learning break the convection parameterization deadlock? Geophysical Research Letters. 2018;**45**(11):5742-5751

[99] McGibbon J, Bretherton CS. Single-column emulation of reanalysis of the Northeast Pacific marine boundary layer. Geophysical Research Letters. 2019;**46**(16):10053-10060

[100] Wang J, Balaprakash P, Kotamarthi R. Fast domain-aware neural network emulation of a planetary boundary layer parameterization in a numerical weather forecast model. Geoscientific Model Development. 2019;**12**(10):4261-4274

[101] Chen R, Zhang W, Wang X. Machine learning in tropical cyclone forecast modeling: A review. Atmosphere. 2020;**11**(7):676

Chapter 6

Current and Future Multirisk Analysis in Climate Change Scenarios with Riskcoast WebGIS

Nelson Mileu and José Luís Zêzere

Abstract

Several regions in Europe are exposed to multiple climate hazards, although their integrated understanding is still limited. The Riskcoast WebGIS platform, developed in the context of the project with the same name, aims to identify the exposed elements and carry out a current and future multirisk mapping assessment in climate change scenarios, for a set of climate hazards: landslides, flash flooding, estuarine flooding, coastal flooding, and coastal erosion. The main objective of this chapter is to present the main functionalities of the Riskcoast WebGIS platform and the multi-risk assessment capabilities for different future risk scenarios arising from climate change applied to the case study of the municipality of Setúbal, Portugal.

Keywords: WebGIS, multirisk, climate change, coastal areas, scenarios

1. Introduction

Human-caused climate change is already generating weather and climate extremes in every region across the globe, leading to widespread adverse impacts on food and water security, human health, and on economies and society and related losses and damages to nature and people [1]. The increase in the frequency, intensity and geographical extent of hazardous events associated with climate extremes has been analyzed in various studies [2, 3], denoting a growing concern in understanding the processes of climate evolution and adaptation. In recent years, research on the adverse impacts associated with extreme events has linked climate change to the amplification of the risk of various phenomena such as floods, coastal floods, heat waves, droughts, storms, or forest fires, studying this relationship for single specific climate or weather hazard [2, 4, 5]. Moreover, the development of comprehensive approaches to assess natural disaster risks, and in particular those related to climate change with a multiple risk perspective, has been addressed by few authors [4, 6, 7], considering all aspects that contribute to increased hazard, exposure, and vulnerability. Recognizing and integrating the manifold hazards linked to climate change is crucial, given that threats become more pronounced in regions exposed to various climate risks [2]. Effective risk management can only be achieved through the integration of this comprehensive knowledge.

IntechOpen

According to the "Report of the open-ended intergovernmental expert working group on indicators and terminology relating to disaster risk reduction" [8], the term "multi-hazard" means the selection of multiple major hazards that a region faces and the specific contexts where hazardous events may occur simultaneously, cascadingly, or cumulatively over time, taking into account the potential interrelated effects. Despite the importance of analyzing multiple hazards for describing future disaster events in terms of their magnitude and probability [5], such an approach presents several challenges. According to Forzieri et al. [2], the study of multiple hazards poses two major challenges: hazards are not directly comparable, as their processes and description metrics differ, and hazards can interact, triggering cascading effects and coupled dynamics. The significance of examining multiple hazards is especially pronounced in coastal regions, as they are exacerbated by climate change [9, 10]. These areas are particularly exposed and susceptible to climate-related threats.

As with research centered on a single specific climate or weather hazard, the development of stand-alone applications for modeling risks was also common in recent years, particularly for earthquakes, tsunamis, landslides, and floods. More recently, there has been the development of a new generation of platforms in the field of natural risks based on open source code and WebGIS technologies [11]. The evolution of platforms toward open-source solutions is justified in research and development projects due to the principles of open research, where it is possible to access freely modifiable source code [12], while the adoption of WebGIS platforms has become commonplace because this technology deals with geographic information, including geospatial analysis, within the online environment [13].

To overcome all these challenges, a WebGIS application called Riskcoast was developed as part of the project with the same name. The project aimed to develop risk prevention cartographic tools to be applied in spatial planning and emergency planning. The main tools developed included the creation and updating of hazard, vulnerability and risk mapping in the SUDOE coastal regions, adapted to different future risk scenarios arising from climate change. The WebGIS platform aims to identify the exposed elements and carry out a current and future multirisk mapping assessment in climate change scenarios, for a set of processes that respond to climate drivers: landslides, flash flooding, estuarine flooding, coastal flooding, and coastal erosion. The main objective of this chapter is to present the main functionalities of the WebGIS Riskcoast platform and the multirisk assessment capabilities for different future risk scenarios arising from climate change applied to the case study of the municipality of Setúbal, Portugal.

This chapter is divided into five sections, besides this introduction. Firstly, the case study is presented, followed by the data used in the WebGIS platform. The methodology section includes the conceptual approach and multirisk workflow analysis, the conceptualization of climate change scenarios, and the platform architecture description. The results section illustrates its application to the municipality of Setúbal, demonstrating its value to address current and future multirisk analysis in climate change scenarios. The chapter ends with conclusions and notes for future developments.

2. Case study: Setúbal, Portugal

The study area used to test the WebGIS tool is the municipality of Setúbal, Portugal (**Figure 1**). The municipality of Setúbal is located in the Lisbon Metropolitan

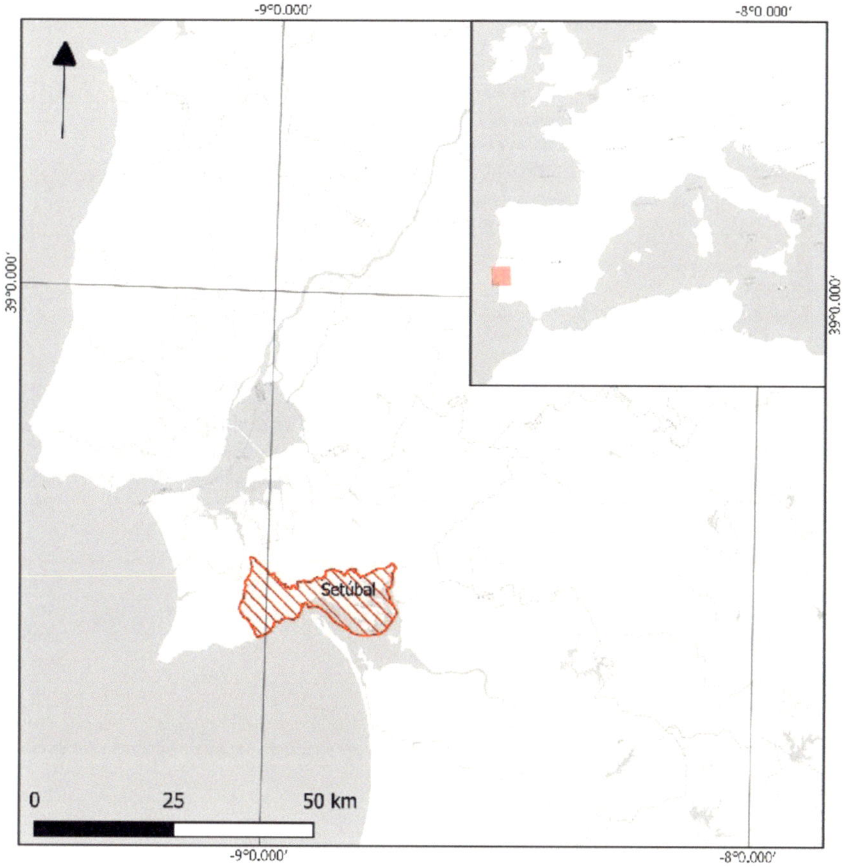

Figure 1.
Case study location.

Area, Portugal. With a total area of 230 km^2, the municipality comprises five parishes: Azeitão, Gâmbia-Pontes-Alto da Guerra, Sado, Setúbal, and São Sebastião.

The municipality of Setúbal is located on the north bank of the mouth of River Sado, bordering the Atlantic Ocean and flanked to the west by the Arrábida Mountain. In 2021, the municipality of Setúbal had around 123,000 inhabitants, about 16% of the total population of the Setúbal Peninsula. Due to its riverside location, its high population density (536 inhabitants/km^2), the existence of a historic urban area, and strong industrial development, it is particularly vulnerable to the consequences of climate change.

3. Datasets

The variables used in the calculations of the Multihazard Risk Index for implementing current and future multirisk mapping assessment in climate change scenarios are listed in **Table 1** and are available in the GitHub repository at https://github.com/nmileu/riskcoast/tree/main/data. The geographical unit of analysis was the census block, obtained from the National Statistics Institute [15]. All layers used in the application are in geoJSON format.

Dataset	Description	Source
Susceptibility | Flash floods	The layers corresponding to the climate hazards analyzed (flash floods, landslides, estuarine flooding, coastal flooding, and coastal erosion) were obtained in Zêzere et al.'s study [14] and constitute single layers in the system according to the current climate scenario (present-day; 2100 – RCP 4.5 and 2100 – RCP 8.5 and have been converted to GeoJSON format.	Zêzere et al. [14]
Susceptibility | Landslides		
Susceptibility | Estuarine flooding		
Susceptibility | Coastal flooding		
Susceptibility | Coastal erosion		
Exposure | Census blocks	The exposure analysis in the Multihazard Risk Index analysis component is based on calculating the population and number of buildings for each census block. These layers were also converted in the system to GeoJSON format.	National Statistics Institute - INE [15]
Exposure | Building centroids		
Social vulnerability	The vulnerability component is a layer defined from the census blocks. This layer was also converted in the system to GeoJSON format.	Zêzere et al. [14] Santos and Ferreira [16]

Table 1.
Source datasets used in Riskcoast WebGIS application. All data are available at https://github.com/nmileu/riskcoast/tree/main/data.

4. Methods

4.1 Risk analysis approach

The risk analysis carried out is based on the Multihazard Risk Index that combines the three main components of risk: susceptibility, exposure, and vulnerability. Zêzere et al. [14] presented a comprehensive description of susceptibility mapping in the technical report "Actual and Future Risk in Climate Change Scenarios Setúbal Test Site," including detailed methodological expositions for the considered set of climate hazards: landslides, flash flooding, estuarine flooding, coastal flooding, and coastal erosion. As the above report includes all technical details, we will present here only the definitions and methodological outline of each component of the current and future multirisk mapping assessment model in climate change scenarios, describing, when appropriate, how the calculations were adapted for the WebGIS framework.

The analysis of climate hazards, now and in the future, was carried out for the set of physical processes with a relevant impact on the municipality of Setúbal and was supported by technical-scientific methods, adjusted to the municipal scale and to the available data. The susceptibility assessment was carried out independently for each type of climate hazard [14]. The mapping of actual hazards was carried out based on the direct delimitation of the areas affected by the hazardous processes considered or using indirect zoning methods, that is, quantitative or semi-quantitative methods based on analyzing the causes of the hazardous processes. Future climate hazards were estimated, whenever possible, quantitatively, based on the territorial incidence of the current hazard and its foreseeable evolution obtained from the projections adjusted to two state-of-the-art climate scenarios (Representative Concentration Pathways in case of limited (4.5) or no action (8.5) until 2100).

The exposure to hazards was considered to be the situation of people, infrastructure, housing, production capacities, and other tangible human assets located in hazard prone areas [8]. The exposure was considered on the platform through roadways, railways, residential buildings, resident population, and strategic, vital, and/or critical equipment. However, for calculating multihazard exposure, it was an option to consider only the number of buildings and the resident population. The buildings are represented geometrically by their centroid, corresponding to all buildings with total or partial residential function. The resident population per building was estimated by dasymetric analysis between the building layers and the census blocks of the 2011 census [17]. The assessment of current exposure in the calculation of the Multihazard Risk Index was carried out by intersecting the buildings and resident population with the hazardous areas corresponding to landslides, flash flooding, estuarine flooding, coastal flooding, and coastal erosion. In the context of assessing future exposure and calculating the Multihazard Risk Index, we decided to intersect the existing exposed elements (buildings and resident population) with the areas identified as hazardous at the end of the twenty-first century, in both the RCP 4.5 and RCP 8.5 scenarios. The results obtained through this approach emphasize the exposure of the elements currently present in the study area to the climate projected for 2100. This underscores the potential consequences for hazardous processes driven by climate factors if such conditions were to occur in the present day.

With regard to the vulnerability component, we chose to consider only the Criticality of Social Vulnerability. Criticality of Social Vulnerability is defined as the set of characteristics and behaviors of individuals that condition their propensity to suffer damage after the occurrence of a disastrous event [14]. These characteristics can contribute to the breakdown of the system and community resources that enable them to respond to or deal with catastrophic scenarios. In this work, we use the results obtained for the complete Metropolitan Area of Lisbon [16], using data from the 2011 census for the statistical census block disaggregation. Following the original work, it was possible to define an initial set of 45 variables, from which population density and building density were removed as they reflect exposure rather than vulnerability, leaving a total of 43 variables [14], representing the following domains: demographics, social support, the condition of built heritage, the economy, education, housing, family structure, employment, and health. In order to ensure comparison between territorial units of analysis, most of the variable data was expressed as a proportion, and social vulnerability was calculated using a principal component analysis (PCA).

4.2 Multihazard Risk Index

The Multihazard Risk Index (MRI) is dimensionless and results from the product of susceptibility (S), exposure (E), and vulnerability (V), using Eq. (1).

$$MRI = S^{\frac{1}{a}} \times E^{\frac{1}{a}} \times V^{\frac{1}{a}} \tag{1}$$

where ^{a} corresponds to the exponent of the Multihazard Risk Index.

The MRI, with due regard for differences in scale, risk components, and input data, is based on the INFORM risk index, which is an international reference risk index that combines data from 16 components describing hazards, exposure, vulnerability, and lack of coping capacity [18]. The formulation adopted in this work has recently been successfully applied at the municipal scale in Portugal, for flood risk [19], landslides [20], and wildfires [21].

The risk analysis, with the calculation of the MRI, was carried out for the present day and for the climate at the end of the twenty-first century, considering the RCP 4.5 and RCP 8.5 scenarios. The Territorial Unit (TU) used for analysis was the census block, as defined in the Geographical Base for Referencing Information of the National Statistics Institute [21]. Multihazard susceptibility was calculated using the intersection of the TUs with the areas prone to be affected by each of the considered climate hazards: landslides, flash flooding, estuarine flooding, coastal flooding, and coastal erosion. The exercise was carried out for three scenarios, corresponding to the actual situation (scenario 1) and the climate at the end of the twenty-first century, for RCP 4.5 (scenario 2) and RCP 8.5 (scenario 3). For each UT, and for each scenario, multihazard susceptibility was calculated by adding up the percentage of area affected by each of the processes considered, using Eq. (2).

$$STUi = \sum_{i=1}^{n} ZiCHi \tag{2}$$

where CHi is climate hazard i (landslides, flash flooding, estuarine flooding, coastal flooding, and coastal erosion) and Zi is the percentage of Territorial Unit i intersected by each hazard.

As the processes involved in landslides and coastal erosion along coastal cliffs are the same, the corresponding values in Eq. (2) have only been counted once in cases where there is a spatial overlap between these two types of hazard.

Multihazard exposure was calculated from the intersection of the TUs with the buildings exposed to each of the climate hazards considered landslides, flash flooding, estuarine flooding, coastal flooding, and coastal erosion. The population living in these buildings was estimated by dasymetric mapping based on data from the 2011 censuses. As in the case of susceptibility, the exercise was carried out for three scenarios: current situation (scenario 1) and the climate at the end of the twenty-first century, for RCP 4.5 (scenario 2) and RCP 8.5 (scenario 3).

For each TU, and for each scenario, multihazard exposure was calculated by adding up the resident population exposed to each of the physical processes considered, according to the Eq. (3).

$$ETUi = \sum_{i=1}^{n} PiCHi \tag{3}$$

where Ri is the resident population in the buildings in Territorial Unit i intersected by each climate hazard.

The vulnerability considered in Eq. (1) corresponds to criticality and was calculated for each TU independent of susceptibility and exposure, as described in Section 2.1.

In the end, before integration into the Multihazard Risk Index, the three components of the MRI (susceptibility, exposure and vulnerability) were scaled to the interval [0, 1] using the min-max method (4).

$$X_{i,norm}^{m} = \frac{X_i^m - X_{i,\min}}{X_{i,\max} - X_{i,\min}} \tag{4}$$

The values of $X_{i,\max}$ and $X_{i,min}$ were determined for each component of the MRI, taking into account the full range of values obtained for the three scenarios considered.

Finally, it should be noted that, as the integration process is multiplicative, the MRI is equal to zero whenever any of the three components that define it (susceptibility, exposure, and vulnerability) is equal to zero.

4.3 Combined hazard selection

Multihazard susceptibility and exposure are calculated by intersecting the TUs with the areas prone to be affected by each of the climate hazards: landslides, flash flooding, estuarine flooding; coastal flooding; and coastal erosion. As this is a multihazard approach, the selection of the processes to be considered in the MRI calculation is a user option in the WebGIS framework, with several possible combinations. The user can calculate the MRI using a single climate hazard or any combination of several climate hazards. The number of possible combinations without repetition for the five processes was determined from expression (5).

$$C_r^n = \frac{n!}{r!(n-r)!} \tag{5}$$

where C_r^n = number of combinations, n = total number of objects in the set, and r = number of choosing objects from the set.

In this way, the user can choose between 32 combinations of processes for one actual and two future climate change scenarios, making it possible to obtain a total of 96 multirisk mapping assessments.

4.4 Riskcoast WebGIS development and architecture

The Riskcoast WebGIS application was written in Javascript, using HTML, CSS, and Javascript files (**Figure 2**) directly from a repository on GitHub (https://github.com/nmileu/riskcoast), with instructions in this version only in Portuguese. The mapping component was implemented using Leaflet open-source JavaScript library, with base maps provided by OpenStreetMap. The spatial analysis tasks, namely, the spatial intersections between hazard prone-areas and exposed elements, were carried out using the Turf Javascript library. This is a modular library for geospatial analysis in browsers, using the GeoJSON data format, which has the advantage of speed and does not require to send data to a WebGIS or database server.

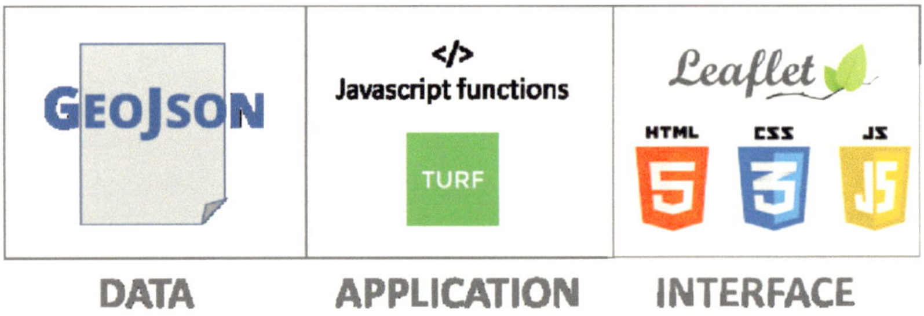

Figure 2.
Riskcoast WebGIS architecture.

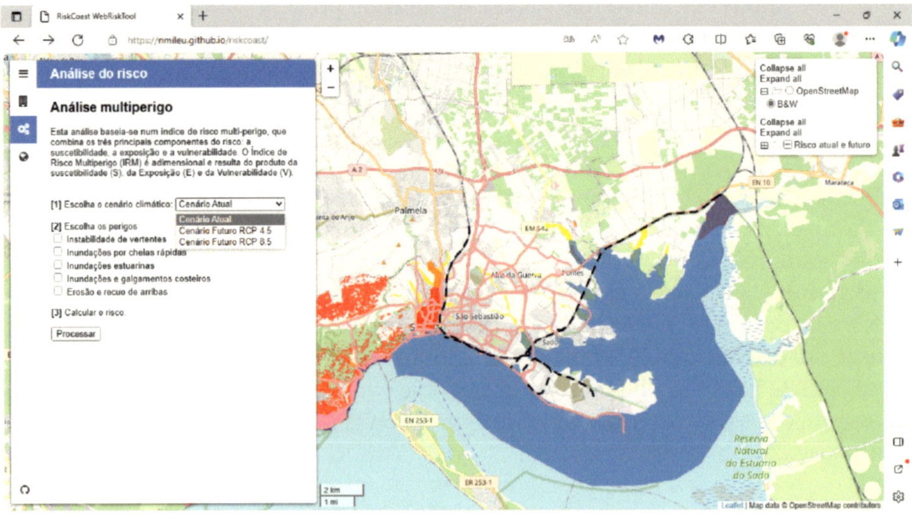

Figure 3.
Riskcoast WebGIS interface.

For the Multihazard Risk Index calculation, the processing is done through Javascript functions. The choice of Javascript in the processing was guided by the ease of implementation and the processing speed as most of the code runs in the user's browser. To represent geographic layers with their non-spatial attributes, GeoJSON was the chosen format. Despite the limitations, the format allows the representation and parsing of simple geographical features like the layers used in Riskcoast project in a simple and human-readable format.

The WebGIS tool can be accessed via a web browser on a desktop or mobile device at https://nmileu.github.io/riskcoast/. The site has the main map area where the user can view the layers available in the Tree Layers Control, namely, the layers of social vulnerability, exposure, hazard (susceptibility), and risk. In the main map, the user can zoom or pan around the map and visualize the layer classification for each census block (**Figure 3**).

The main map has four main tabs ('Project description', 'Exposure analysis', 'Multi-Hazard Risk analysis', and 'Legends'). The first tab gives the description of the "Riskcoast" research project. The next tab, the Exposure analysis tab, controls the exposition scenarios. The first step is to select the climate scenario (current or future), followed by one of the climate hazards for which the user want to intersect with the exposed elements (roads, railways, buildings, population, and strategic, vital, and/or critical equipment). In the next tab, the user can select the climate scenario (current or future, RCP 4.5, RCP 8.5) and the combination of climate hazards for which the MRI is computed at the census block scale. The last tab is dedicated to showing the legends for the layers.

5. Results

At this point, we present two applications (current and future multirisk analysis in climate change scenarios) in which we have calculated the risk using different process combinations: climate hazards associated with precipitation (landslides and floods), climate hazards associated with the rise of sea level (estuarine flooding, coastal flooding, and coastal erosion), and all climate hazards.

5.1 Current multirisk analysis in climate change scenarios

Figure 4 shows the mapping of the Multihazard Risk Index in the municipality of Setúbal, obtained using the Riskcoast WebGIS application, for current landslides and flash floods. In this figure, it is possible to identify a large number of territorial units with a high risk (MRI > 0.4), all located in the Setúbal city center and essentially exposed to the flash flooding hazard.

Figure 5 shows the mapping of the Multihazard Risk Index in the municipality of Setúbal, obtained using the Riskcoast WebGIS application, for current estuarine flooding, coastal flooding, and coastal erosion. In this figure, it is possible to identify a group of territorial units with a medium risk (MRI > 0.2), located in the parish of Sado (Mitrena Industrial Zone) and essentially exposed to estuarine flooding.

Figure 6 shows the mapping of the Multihazard Risk Index in the municipality of Setúbal, obtained using the Riskcoast WebGIS application, for current landslides, flash floods, estuarine flooding, coastal flooding, and coastal erosion, where the city center and the Mitrena industrial zone stand out as risk areas.

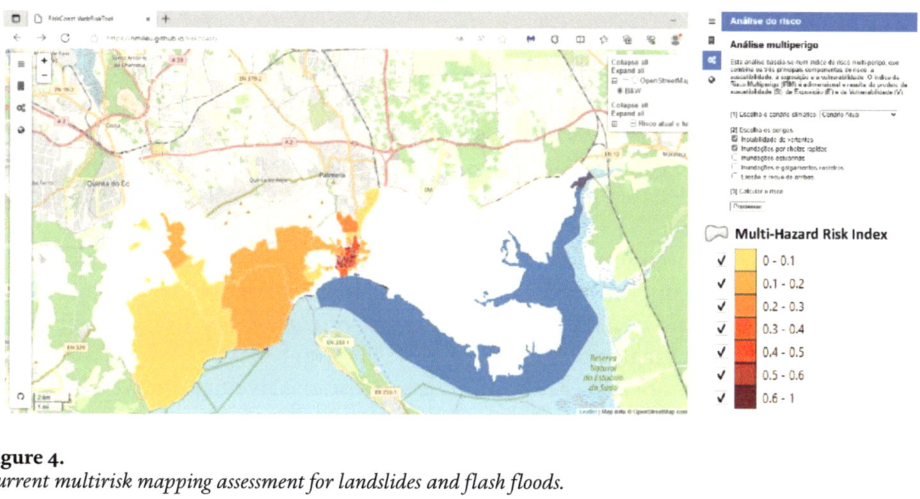

Figure 4.
Current multirisk mapping assessment for landslides and flash floods.

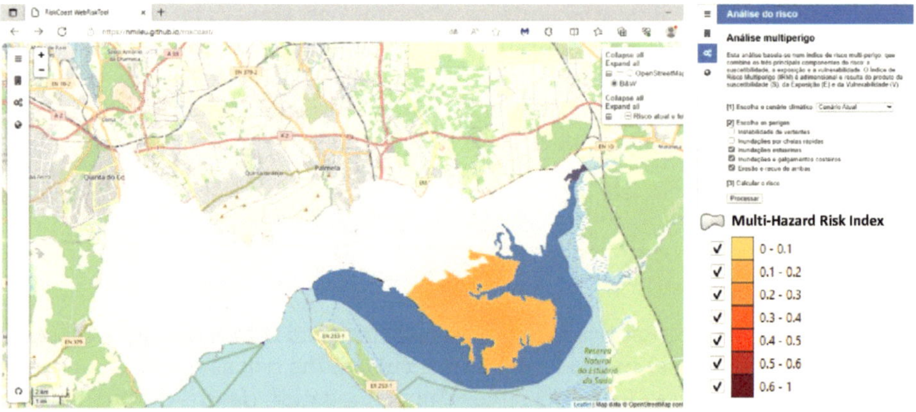

Figure 5.
Current multirisk mapping assessment for estuarine flooding, coastal flooding, and coastal erosion.

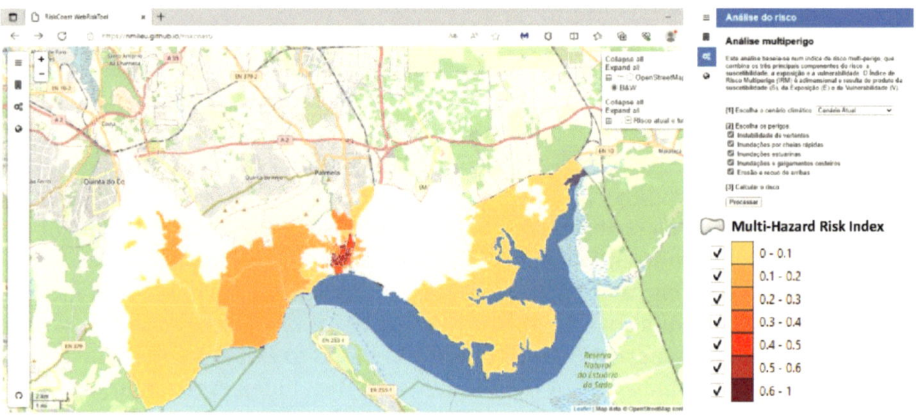

Figure 6.
Current multirisk mapping assessment for all climate hazards.

5.2 Future multirisk analysis in climate change scenarios

Figure 7 shows the mapping of the Multihazard Risk Index in the municipality of Setúbal, obtained using the Riskcoast WebGIS application, for landslides and flash floods for the end of the twenty-first century and considering the worst climate scenario (RCP 8.5). The figure highlights a large number of territorial units with a high risk (MRI > 0.4), located in Setúbal city center and essentially exposed to the flash flooding hazard.

Figure 8 shows the mapping of the Multihazard Risk Index in the municipality of Setúbal, obtained using the Riskcoast WebGIS application for estuarine flooding, coastal flooding, and coastal erosion for the end of the twenty-first century and considering the worst climate scenario (RCP 8.5). The figure highlights a group of territorial units with a high risk (MRI > 0.4), located in the riverside area and in the inner area of the Sado River estuary.

Figure 9 shows the mapping of the Multihazard Risk Index in the municipality of Setúbal, obtained using the Riskcoast WebGIS application, for the end of the twenty-first century (RCP 8.5) and accounting the complete set of considered climate

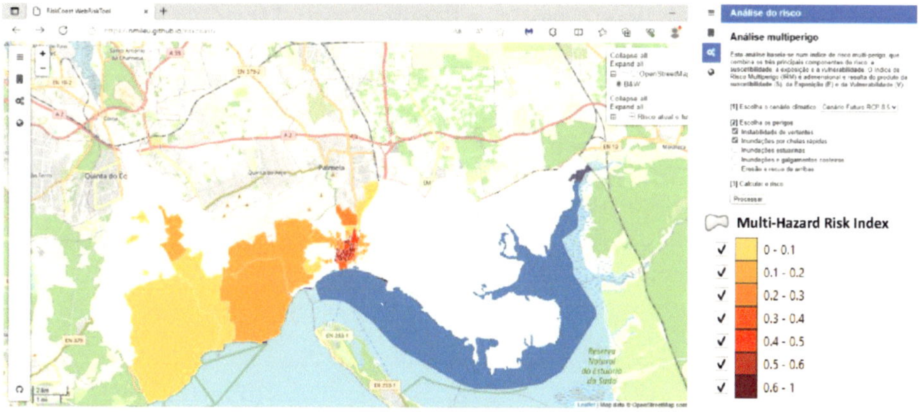

Figure 7.
Future (RCP 8.5) multirisk mapping assessment for landslides and flash floods.

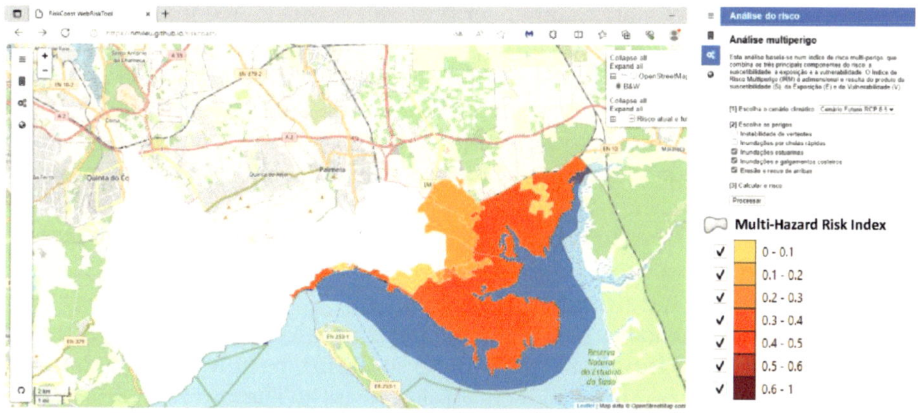

Figure 8.
Future (RCP 8.5) multirisk mapping assessment for estuarine flooding, coastal flooding, and coastal erosion.

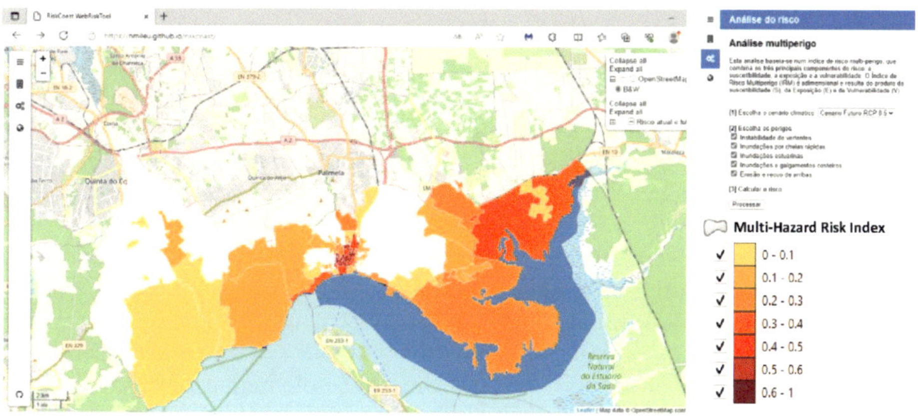

Figure 9.
Future (RCP 8.5) multirisk mapping assessment for all climate hazards.

hazards: landslides, flash floods, estuarine flooding, coastal flooding, and coastal erosion. The multihazard risk analysis for the end of the century, with the climate conditions defined by the RCP 8.5 scenario and considering the elements currently exposed in the municipality, shows an increase in risk. There was no reduction in the Multihazard Risk Index in any of the municipality's territorial units. In addition to those that already stand out in Setúbal's city center on the map showing the current situation, there are two more territorial units, located in Setúbal's riverside area and in the inner area of the Sado river estuary, reflecting the increased susceptibility and exposure to the coastal and estuarine flooding processes.

6. Conclusions

The development of the WebGIS tool was based on a simple architecture and open-source components that made it possible to implement a solution that does not require sending data to a WebGIS server or database server. Considering the large number of possible combinations of climate hazards in the definition of multirisk scenarios arising from climate change, a critical aspect considered in the development of the platform was ease of use, requiring no advanced knowledge in geographic information systems. The ability to visualize and explore current multirisk mapping assessment models in climate change scenarios regarding the different risk components, namely, susceptibility, exposure, and vulnerability, allows highlighting areas with lower or higher risk. On the other hand, the ease in quantitatively assessing risk at the census block by creating future multirisk mapping assessment models in climate change scenarios through a structured and systematic process provides opportunities for the evaluation of different strategies and actions to minimize the adverse impacts of any hazardous event.

Analyzing the multihazard risk, quantified using the Multihazard Risk Index for the present-day and the future (for the RCP 8.5 scenario), it shows that the risk at the end of the twenty-first century will increase significantly. The current risk is maximum in the downtown area of Setúbal as a result of flash floods and will continue to be so in the future. However, the risk will increase in the entire area of the Sado estuary, particularly on the waterfront of the city of Setúbal and in the inland surroundings of the Sado estuary, particularly in the parish of Praias do Sado.

Although the WebGIS Riskcoast platform is associated with a European project and has a defined time-bound, it is possible to identify a number of improvements and future developments. Assessing exposure at the end of the twenty-first century is a complex exercise, mainly due to the enormous uncertainty about population numbers and their distribution. One of the developments in the multirisk assessment model is the possibility of including demographic projections, despite the complexity of spatializing the demographic projection at the census block scale. In functional terms, the interactive delimitation of the geographical area for carrying out the risk analysis, the comparative analysis of scenarios, and the implementation of advanced reports is another evolution of the platform that will make it possible to improve the use multirisk mapping assessment models in climate change scenarios.

To conclude, despite the limitations and possibilities for improvement identified for future development of the WebGIS platform, the implementation of the prototype in the context of the research project allowed to demonstrate that a multirisk assessment can be obtained quickly and easily for any combination of climate hazards for a study region particularly susceptible to climate change, providing a high added-value contribution to supporting local adaptation strategies to increase municipalities resilience to climate change.

Acknowledgements

This work is part of the project Riskcoast—Development of tools to prevent and manage geological risks on the coast linked to climate change (SOE3/P4/E0868, Interreg Sudoe).

Conflict of interest

The authors declare no conflict of interest.

Author details

Nelson Mileu* and José Luís Zêzere
Centre of Geographical Studies, Institute of Geography and Spatial Planning, University of Lisbon, Lisbon, Portugal

*Address all correspondence to: nmileu@edu.ulisboa.pt

References

[1] Lee H, Romero J, editors. Climate Change 2023: Synthesis Report. Contribution of Working Groups I, II and III to the Sixth Assessment Report of the Intergovernmental Panel on Climate Change. Geneva, Switzerland: IPCC; 2023. pp. 35-115. DOI: 10.59327/IPCC/AR6-9789291691647

[2] Forzieri G, Feyen L, Russo S, Vousdoukas M, Alfieri L, Outten S, et al. Multi-hazard assessment in Europe under climate change. Climatic Change. 2016;**137**:105-119. DOI: 10.1007/s10584-016-1661-x

[3] Fischer EM, Sippel S, Knutti R. Increasing probability of record-shattering climate extremes. Nature Climate Change. 2021;**11**:689-695. DOI: 10.1038/s41558-021-01092-9

[4] Gallina V, Silvia T, Andrea C, Anna S, Thomas G, Antonio M. A review of multi-risk methodologies for natural hazards: consequences and challenges for a climate change impact assessment. Journal of Environmental Management. 2016;**168**:123-132. ISSN 0301-4797). DOI: 10.1016/j.jenvman.2015.11.011

[5] Vlachogiannis D, Sfetsos A, Markantonis I, Politi N, Karozis S, Gounaris N. Quantifying the occurrence of multi-hazards due to climate change. Applied Sciences. 2022;**12**:1218. DOI: 10.3390/app12031218

[6] Delmonaco G, Margottini C, Spizzichino D. Report on New Methodology for Multi-Risk Assessment and the Harmonisation of Different Natural Risk Maps. Deliverable 3.1, ARMONIA Research Project, Funded under the Sixth EU Framework Programme for Research and Technological Development. 2006

[7] Mileu N, Queirós M. Integrating risk assessment into spatial planning: RiskOTe decision support system. ISPRS International Journal of Geo-Information. 2018;**7**:184. DOI: 10.3390/ijgi7050184

[8] United Nations General Assembly. Report of the Open-Ended Intergovernmental Expert Working Group on Indicators and Terminology Relating to Disaster Risk Reduction. New York, NY, USA: United Nations General Assembly; 2016

[9] Nicholls RJ, Cazenave AA. Sea-level rise and its impact on coastal zones. Science. 2010;**328**:1517-1520

[10] Rocha M, Oliveira A, Freire P, Fortunato AB, Nahon A, Barros JL, et al. Multi-hazard WebGIS platform for coastal regions. Applied Sciences. 2021;**11**:5253. DOI: 10.3390/app11115253

[11] Negulescu C, Smai F, Quique R, Hohmann A, Clain U, Guidez R, et al. VIGIRISKS platform, a web-tool for single and multi-hazard risk assessment. Natural Hazards. 2023;**115**:593-618. DOI: 10.1007/s11069-022-05567-6

[12] Heron M, Hanson VL, Ricketts I. Open source and accessibility: Advantages and limitations. Journal of Interaction Science. 2013;**1**:2. DOI: 10.1186/2194-0827-1-2

[13] Rostislav N, Pohankova T, Bittner O, Urban D. Geospatial analysis in Web Browsers—Comparison study on WebGIS process-based applications. ISPRS International Journal of Geo-Information. 2016;**12**(9):374. DOI: 10.3390/ijgi12090374

[14] Zêzere JL, Garcia R, Oliveira S, Pereira S, Melo R, Reis E, et al. Relatório técnico - risco atual e futuro em cenários de alteração climática Setúbal Test Site [Technical report - current and future risk in climate change scenarios Setúbal Test Site]. Riskcoast - Development of tools to prevent and manage geological risks on the coast linked to climate change (SOE3/P4/EO868, Interreg Sudoe). 2021

[15] Censos 2011 [Internet]. 2011. Available from: https://censos.ine.pt/xportal/xmain?xpgid=censos2011_apresentacao&xpid=CENSOS [Accessed: November 02, 2023]

[16] Santos PP, Ferreira TM. Social vulnerability in the Lisbon metropolitan area. In: Ferreira TM, editor. Multi-Risk Interactions Towards Resilient and Sustainable Cities. Singapore: Springer; 2023. pp. 27-49. DOI: 10.1007/978-981-99-0745-8_3

[17] Garcia RAC, Oliveira SC, Zêzere JL. Assessing population exposure for landslide risk analysis using dasymetric cartography. Natural Hazards and Earth System Sciences. 2016;**16**(12):2769-2782

[18] De Groeve T, Polansek K, Vernaccini L. Index for Risk Management -INFORM. Publicatio. ed. Luxembourg: Publications Office of the European Union, European Union; 2016

[19] Santos PP, Pereira S, Zêzere JL, Tavares A, Reis E, Garcia AC, et al. A comprehensive approach to understanding flood risk drivers at the municipal level. Journal of Environmental Management. 2020;**260**(4):110127. DOI: 10.1016/j.jenvman.2020.110127

[20] Pereira S, Santos PP, Zêzere JL, Tavares AO, Garcia RAC, Oliveira SC. A landslide risk index for municipal land use planning in Portugal. Science of the Total Environment. 2020;**735**:139463

[21] Bergonse R, Oliveira S, Zêzere JL, Moreira F, Ribeiro PF, Leal M, et al. Differentiating fire regimes and their biophysical drivers in Central Portugal. Fire. 2023;**6**(3):112, 1-22. DOI: 10.3390/fire6030112

Chapter 7

Potential Earthquake Proxies from Remote Sensing Data

Badr-Eddine Boudriki Semlali, Carlos Molina, Mireia Carvajal Librado, Hyuk Park and Adriano Camps

Abstract

At present, there is no clear scientific evidence of reliable earthquake precursors that can be used as an early warning system. However, many studies have also reported the existence of faint signatures that appear to be coupled to the occurrence of earthquakes. These anomalies have traditionally been detected using data from in-situ sensors near high-seismicity regions. On the other hand, remote sensors offer the potential of large spatial coverage and frequent revisit time, allowing the observation of remote areas such as deserts, mountains, polar caps, or the ocean. This chapter revises the state-of-the-art of the understanding of lithosphere–atmosphere–ionosphere coupling. It also presents recent studies by the authors' ongoing investigation on short-to-midterm earthquake precursors. The Earth observation variables discussed are (1) surface temperature anomalies from thermal infrared or microwave radiometer measurements, (2) atmospheric signatures, (3) ionospheric total electron density fluctuations or scintillation measured from GNSS signals, and (4) other geophysical variables, including geomagnetic field fluctuations, changes in the Schumann resonance frequency, or low-frequency electromagnetic radiation. However, despite the seismic hazard risk models that exist and the results shown by these studies, it is still very difficult to predict the occurrence of earthquakes.

Keywords: earthquakes, precursors, remote sensing, land surface temperature, GNSS, ionosphere, TEC, scintillation, magnetic field, lithospheric-atmospheric-ionospheric coupling, statistical analysis

1. Introduction

Earthquakes are natural disasters that cause devastating damages and significant human casualties. Over the past 50 years, earthquakes have been the first cause of death from natural hazards [1]. Despite the thousands of earthquakes occurring annually, only a tiny fraction is observable. Between 1998 and 2018, earthquakes caused 846 thousand deaths and approximately US$661 billion in economic losses [2]. While earthquakes are unavoidable, efforts have been made to mitigate and restrict their impact [3].

Forecasting earthquakes, especially the significant ones that cause severe destruction, has garnered widespread attention [4]. Numerous studies have explored potential earthquake precursors, aiming to understand the physical characteristics of seismic events and their interaction with the environment [5]. Despite the continued efforts, earthquake forecasting remains challenging, as reliable indicators of future earthquakes are infrequently apparent in seismic analysis. A clear physical link between the observed anomalies and the earthquakes is needed to make the correlation analysis more robust [6].

Remote Sensing (RS) has been widely employed in earthquake studies due to its capability to provide short revisit times and large coverage [7], continuity, and more consistent data than point measurements [8].

1.1 Remote sensing observables

There are different physical mechanisms involved in potential seismic precursors. First, the frictional forces involved in the preparation and occurrence of an earthquake may increase the surface temperature locally, which can be detected by satellite instruments. Therefore, despite cloud cover masks optical imagery, infrared data has already shown potential results for detecting temperature irregularities over active faults [9, 10].

Additionally, some atmospheric changes can also be observed before the occurrence of an earthquake [11], such as air ionization by radioactive radon emissions, thermal convection, anomalous linear cloud formation [12], or latent heat anomalies [13] are some of the means used as earthquake precursors.

The ionosphere is also a possible proxy of seismic activity. Anomalous TEC changes [14], and ionospheric scintillation (IS) perturbations related to earthquakes have been published in the last decades [15]. Different techniques to measure these parameters are summarized in this chapter. Traditional methods to measure these parameters include ionograms from ionosondes and GNSS satellite constellations to measure the effects produced in the signals when they are in view from a ground station. This chapter will review novel studies using GNSS RS techniques such as GNSS Reflectometry (GNSS-R) [16] and GNSS-Radio Occultations (GNSS-RO) to detect possible ionospheric precursors of earthquakes.

Other effects not easily categorized in previous environments have also shown the potential to serve as earthquake precursors. Those are, for example, geomagnetic perturbations [17], outgoing longwave radiation [18] and extremely low-frequency wave generation [19], or small perturbations in the Schumann resonance frequency [20].

1.2 Methodologies

Past studies have used mathematical models and statistical tools to study the correlation between these effects and the occurrence of earthquakes [21]. Typical statistical parameters used to detect the anomalies include the calculation of the standard deviation (STD), the interquartile range (IQT), and the Z-score. The influence of external phenomena on the studied parameters can hide or fake the correlation between the physical mechanism studied and the occurrence of the earthquake. Therefore, filtering techniques are very important to improve the data quality and reduce false alarms. Some previous observables can be affected by external factors, such as space weather, which is also largely influenced by solar activity and the interplanetary environment. In this sense, variables such as the planetary index (K_p) [22], the disturbance storm time index (Dst) [23], and the solar flux [24] should be considered to flag the affected observables.

1.2.1 Confusion matrix and receiver operating characteristic (ROC) curve

This section explains the methodologies used to compute the anomalies, highlighting explicitly the STD, the IQT, and the Z-Score techniques [17]. The Precursor deviation is determined from (Eq.(1)), wherein the Precursor mean corresponds to the average of the detrended Precursor over the entire period under consideration. The Dobrovolsky or strain radius (SR) estimates the size of the earthquake preparation area [25], and it is related to the Moment M_w through (Eq. (5)).

$$\text{Precursor}_{\text{deviation}} = \text{Precursor} - \text{Precursor}_{\text{mean}} \quad (1)$$

$$\left|\text{Precursor}_{\text{deviation}}\right| \geq \text{C}\cdot\text{STD} \quad (2)$$

$$\left|\text{Precursor}_{\text{deviation}}\right| \geq \text{C}\cdot\text{IQT} \quad (3)$$

$$\text{ZScorePrecursorPrecursormean} / \text{STD} \quad (4)$$

$$SR_{[\text{km}]} = 10^{0.43Mw} \quad (5)$$

The test performance is assessed from six essential parameters used to compute the confusion matrix (CM): condition positive (CP), condition negative (CN), true positives (TP), false negatives (FN), false positives (FP), and true negatives (TN) [26]. The explanation for each of them is as follows:

- CP: count of samples with actual earthquakes.
- CN: count of samples without real earthquakes.
- TP: count of correct links where (Earthquake [+] and Anomaly [+]).
- FN: count of wrong links where (Earthquake [−] and Anomaly [−]).
- FP: count of wrong links where (Earthquake [−] and Anomaly [+]).
- TN: count of correct links where (Earthquake [+] and Anomaly [−]).

Additional metrics can be derived from them, such as the specificity, the precision, the recall, and the accuracy. **Table 1** summarizes the different formulas used to compute these parameters, including the true positive rate (TPR), the false positive rate (FPR), the false negative rate (FNR), the true negative rate (TNR), the likelihood ratio (LR), the positive predictive value (PPV), the false omission rate (FOR), the negative predictive value, (NPV), the false discovery rate (FDR), the area under the curve (AUC), and the diagnostic odds ratio (DOR) [26].

The receiver operating characteristic (ROC) curve is a graphical plot illustrating a binary classification method's performance that depends on a threshold value. By

CP	TP	FN	TPR = TP/P	FNR = FN/P
CN	FP	TN	FPR = FP/N	TNR = TN/N
Prevalence = CP/CP + CN	PPV = TP/PP	FOR = FN/PN	(LR+) = TPR/FPR	(LR−) = FNR/TNR
Accuracy = TP + TN/CP + CN	FDR = FP/PP	NPV = TN/PN	DOR = LR+/LR−	
	F_1 = 2·TP/(TP + FP + FN + TN)		MCC = TP·TN-FP·FN/((TP + FP) (TP + FN) (TN + FP) (TN + FN))	

Table 1.
The details of the confusion matrix and formulas.

sweeping this threshold, values of TPR against FPR are depicted on an x-y graph, leading to a curve which, ideally, goes from (0,0) to (1,1). The AUC parameter mentioned above measures the integrated area under this curve, which ranges from 0 to 1, being larger for better classifiers.

1.2.2 Machine learning tools for earthquake prediction

The confluence of all these elements constitutes what can be considered a Complex System with different sources of perturbations, interacting non-linearly and resulting in tiny observable changes [27]. These features make the problem suitable to be studied using machine-learning techniques. Neural networks can effectively extract the main features of a system like this, helping to understand and categorize the input sources involved in the problem and their relative impact.

Earthquake forecasting is commonly branded into two main types: model-based and precursor-based techniques. Model-based approaches involve hypothesizing a mathematical or a machine-learning model for reliable forecasting, incorporating an analysis of signals indicative of an approaching significant quake. Unfortunately, these signals are often undetectable before the earthquakes.

An innovative machine-learning algorithm was introduced in [28] for real-time earthquake precursor detection, leveraging the Support Vector Machine (SVM). This technique integrated daily GPS-TEC geomagnetic indices data and identified spatio-temporal anomalies linking them to earthquakes. A case study in Italy from January 1, 2014, to September 30, 2016, utilized various training, validation, and testing periods. During the validation period (266 days), the algorithm successfully detected precursors in 17 out of 21 earthquakes, and during the test period (282 days), it identified 22 out of 24 earthquakes, with 7 and 13 false alarms, respectively.

In [29], the authors explored physical and dynamic changes in seismic data collected from ten infrared and hyperspectral measurements from 2006 to 2013. They introduced the Inverse Boosting Pruning Trees (IBPT) machine-learning method, using satellite data related to 1371 earthquakes with $M_w > 6$ for short-term forecasting. Compared to other machine-learning approaches, IBPT outperformed six selected baselines, enhancing earthquake forecasting across diverse databases.

After this introduction, the rest of this book chapter is structured as follows. Section 2 reviews the studies on LST proxies for seismic activity. Sections 3 and 4 analyze the atmospheric and ionospheric perturbations due to earthquakes, respectively. Section 5 summarizes additional observables to study earthquake precursors. Section 6 presents the main conclusions.

2. Land surface temperature proxies

Detecting thermal anomalies before significant earthquakes is crucial to understand and predicting seismic activities. This importance arises from the ability to recognize phenomena related to thermal radiation during seismic preparation phases [30]. Satellite data provides an effective means to monitor seismic regions globally, nearly in real-time. Various data sources have been incorporated into this field in recent decades, advancing progressive anomaly detection techniques. This section reviews the advances and developments in pre-seismic thermal anomaly detection technology over the past decade.

In Ref. [31], the authors explore the association between ground surface warming and earthquakes by analyzing thermal data from Meteosat. The objective is to mitigate the impact of cloud coverage and daily weather fluctuations, enhancing the resolution of thermal maps. Detected thermal anomalies show a potential link between LST anomalies and future seismic events, notably observed in the L'Aquila earthquake on April 6, 2009.

In Ref. [31], the authors evaluate LST anomalies in various earthquake events of $M_w > 6$ in Pakistan from 2000 to 2020. They use MODIS imagery data over the epicenter region to examine thermal anomalies associated with earthquake prediction. The findings reveal pre-seismic LST variations ranging from 30 to 54°C, indicating a promising correlation between thermal anomalies and earthquakes.

In Ref. [32], the study analyzes AQUA/MODIS and TERRA/MODIS LST products before and after the M_w 7.9 earthquake in Nepal on April 25, 2015. It explores the thermal information related to the earthquake, revealing a steady increase of LST before the event, reaching a maximum during the quake, and returning to normal afterward. This temporal pattern could serve as premonitory information associated with the Nepal earthquake.

In Ref. [33], the study focuses on the trend of LST based on MODIS daily LST products from 2000 to 2017. The study applies an annual temperature cycle model to derive parameters such as the mean annual surface temperature (MAST), yearly amplitude of surface temperature, and phase shift. The study offers insights into the impact of earthquakes on the mountain thermal environment and its diverse responses to changes in vegetation cover and climatic conditions.

In Ref. [2], three earthquakes of $M_w > 6$ were studied using AQUA/MODIS and TERRA/MODIS LST data. Anomalies are detected using the IQT and mean ± 2. STD methods, revealing positive daytime anomalies based on mean $\pm 2\sigma$. However, negative (or positive) anomalies are observed at night before earthquakes in Mexico and Bolivia, and occasionally, a negative anomaly is detected during the 10 days following an earthquake.

The study in Ref. [21] employs a novel approach to detect localized spatio-temporal fluctuations in hyper-temporal, geostationary-based LST data. The study covers ten areas worldwide, analyzing 20 large and small land-based earthquakes ($M_w > 5.5$ and depth < 35 km). The comparison of times and locations with and without earthquakes shows no distinct repeated patterns, indicating that earthquakes do not significantly impact detected LST anomalies.

In Ref. [34], a segmented threshold method was anticipated to detect anomalous microwave brightness LST linked with strong earthquakes in Sichuan province, China. The Index of Microwave Radiation Anomaly (IMRA) computed by this method was found to be improved nearly 2 months before the three strong earthquakes in 2008 Wenchuan (M = 7.8), 2013 Lushan (M = 6.6), and 2017 Jiuzhaigou (M = 6.5). The study acknowledges the limitation in quantitatively evaluating IMRA due to the infrequent occurrence of earthquakes.

In Ref. [35], the authors aimed to examine the anomalies in LST from GOES/ABI 16, 17 [26], and AQUA/MODIS, potentially associated with 1350 global earthquakes with a magnitude $M_w \geq 4$ during 2020. In the same direction, the authors tried to detect and find a correlation between earthquakes and LST data in a long time series from Fengyun-2F/VISSRS [36], Himawari-8/AHI, and MSG/SEVIRI. In other studies, data acquired from numerous satellites [37] was ingested [38] and processed [39] to identify unusual anomalies before earthquakes. The precursor value tends to be high near the earthquake's epicenter 1 to 7 days before the seismic event [8]. Traditionally, this outcome has been attributed to the thermal flux emitted from the Earth's crust in regions prone to seismic activity. To improve the data quality, filtering is applied exclusively to nighttime datasets between 00:00 and 06:00 am, thereby mitigating the influence of sunlight and radiation on precursor values. Subsequently, the data undergo fusion (averaging, standard deviation, minimum, and maximum values) for each location and day. The organized data is then chronologically sorted to form a time series. The subsequent phase involves computing the precursor average using the Strain Radius (SR) formula, delineated in Eq. (5). The final stage encompasses calculating anomalies based on prescribed formulas. Once the anomalies are computed, they are correlated and associated with potential earthquakes. A confusion matrix is generated to optimize detection parameters. Finally, the correlation is visually represented through charts and maps.

Most existing works apply RS data from geostationary and polar satellites equipped with thermal or microwave sensors covering temporal windows spanning a few months to years. Similarities in data processing methods, such as LST data ingestion filtering, time series analysis, and anomaly detection, are obvious. However, differences arise in the data processing and LST anomaly detection. Recent contributions stand out in several critical aspects: (1) satellite constellations by using multiple satellite constellations from geostationary satellites, ensuring comprehensive global LST coverage; (2) data filtering using diverse LST data filters, including quality flags, water bodies, and fire masks; (3) extensive time series from 2010 to 2021, facilitating correlation analyses across different seasons and temporal conditions; (4) a large number of land earthquakes, enabling analyses across varied land covers, altitudes, depths, and altitude regions for realistic statistical results; (5) statistical approaches not used before such as the Confusion Matrix and the ROC curves; (6) various thresholds analytically tested to determine the optimal configuration for correlation analysis; and finally (7) the classified correlation was conducted based on diverse parameters, including M_w, depth, land cover, regions.

Figure 1c1–c6 illustrates the processed AQUA/MODIS LST data near Ankara, Turkey, from January 20 to 25, 2020. The cyan star indicates the earthquake's epicenter. A marginal rise in LST was observed 3 days before the earthquake. However, a notable positive anomaly in LST was documented on the day of the earthquake, followed by a decrease in LST the day after.

2.1 Confusion matrix and ROC curve

Figure 2 illustrates the outcome of the CM: the TPR and the FPR decrease as the "C" coefficient increases. Consequently, a small "C", especially at C = 0.7, yields

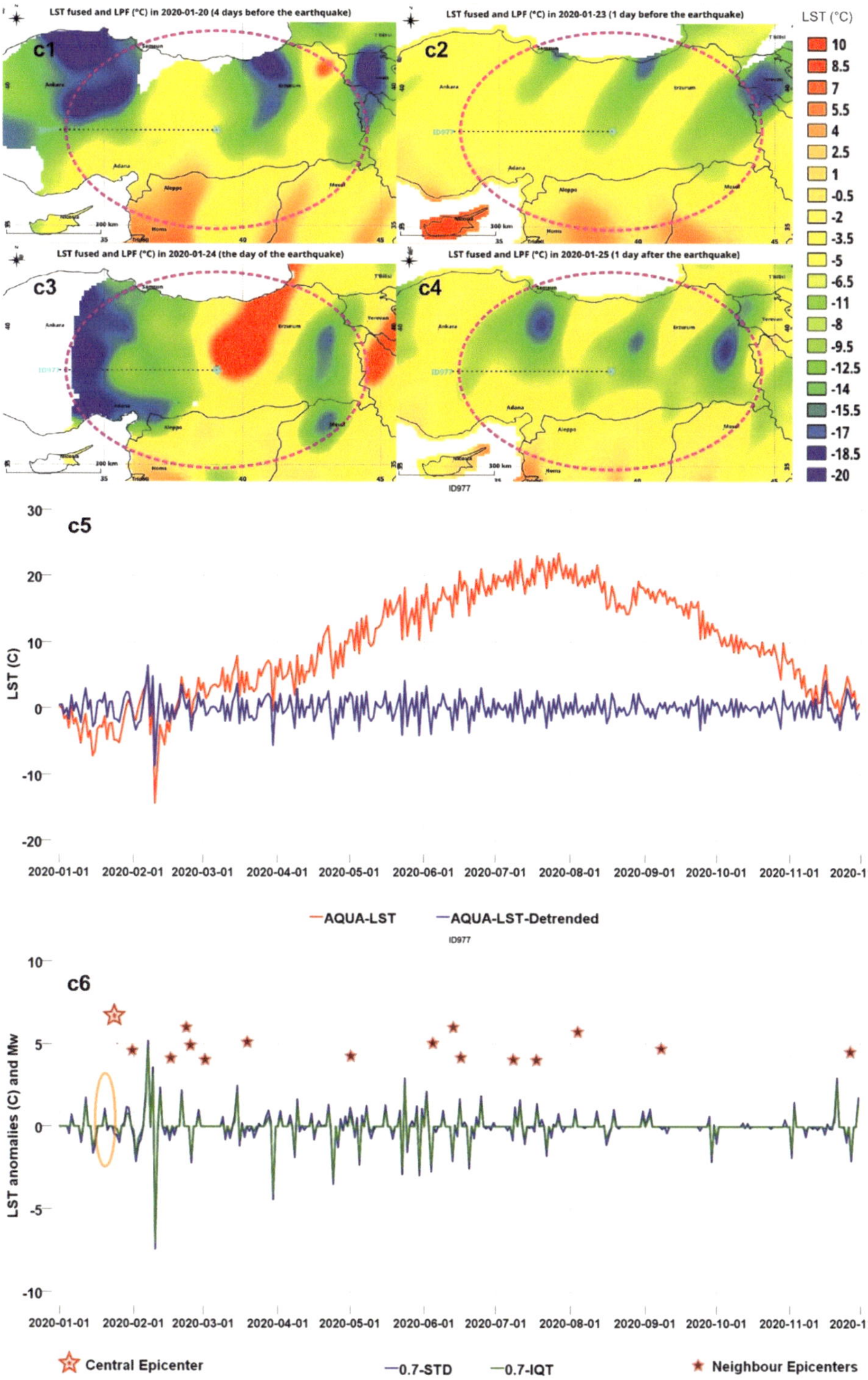

Figure 1.
Sample LST maps, original, detrended, and anomalies time series (°C) measured with ABI/GOES or MODIS/AQUA for epicenters ID977 in 2020 [35].

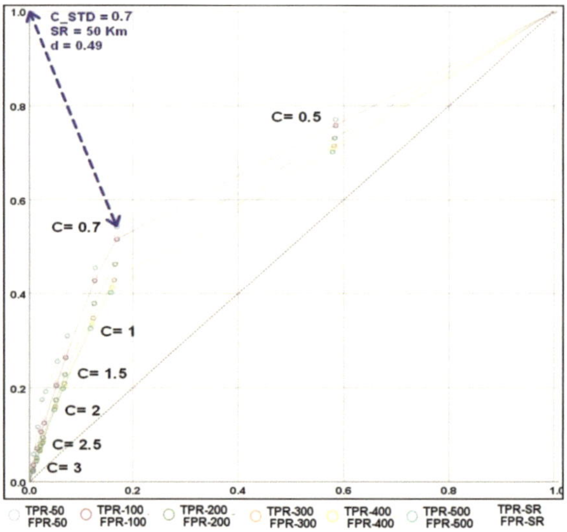

Figure 2.
Optimum ROC curve for C = 0.7 of correlation between LST and earthquake occurrence.

the optimal threshold to detect LST anomalies. However, this comes at the cost of a relatively high rate of false alarms. Conversely, larger "C" coefficients identify fewer LST anomalies. The higher the DOR, the better the test performance. In our study, a DOR of approximately 13.5 corresponds to the optimal values for C = 0.7. The shortest distance (d) from the receiver operating characteristic (ROC) curve to the point (TPR, FPR) (0,1) is approximately 0.49. Therefore, the optimal "C" for earthquake monitoring is 0.7, with an SR of 50 km.

2.2 Impact of Mw, depth, altitude, and land cover on the confusion matrix

Figure 3a shows a consistent TPR, AUC, and DOR increase as M_w increases. The LST anomalies detection is high where the M_w is large. TNR and accuracy remain relatively constant, around 0.8 across different M_w levels. Conversely, FNR and "D" escalate as M_w decreases. The peak TPR, AUC, DOR, minimum FPR, and "D" are observed for deep earthquakes. TNR, FPR, and ACC exhibit uniformity across depths.

From **Figure 3b**, epicenters within 50 to 100 km and over 400 km strongly associate with prior LST anomalies, displaying high confidence and less noise, reaching the highest values of TPR, up to ~0.7.

At low altitudes (1001–2000 m), the highest TPR, AUC, and DOR, coupled with the lowest FPR and "D," are achieved, highlighting numerous correlated LST anomalies. TNR, FPR, and ACC remain consistent across different altitudes, as shown in **Figure 3c**. Consequently, the land cover type and altitude choice are crucial in accurately detecting LST anomalies, as they significantly impact the measurement using RS techniques.

From **Figure 3d**, the maximum TPR, AUC, and DOR are noted for different land covers, while the lowest FPR and "D" occur in mosaic and snow ice covers, respectively. Savannahs, grassland, and cropland record an average detection, whereas forests and shrubs show the worst performance, suggesting that vegetation covers mask LST anomalies. TNR, FPR, and ACC demonstrate similarity across various land

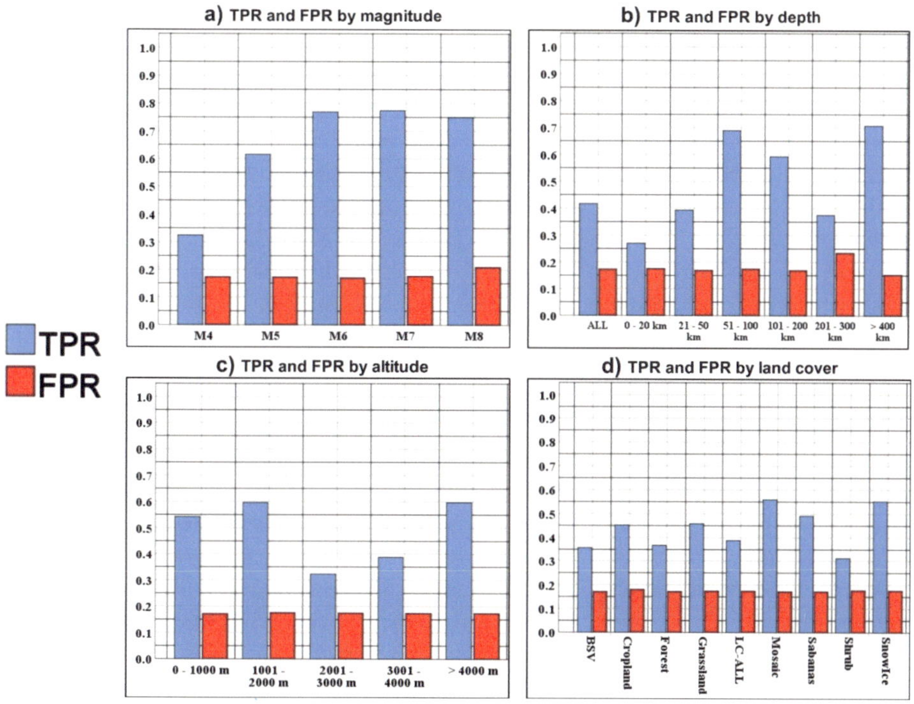

Figure 3.
TPR and FPR separated by magnitude, depth, altitude, and land cover.

covers, emphasizing the significance of land cover type in LST anomaly detection using RS techniques.

Despite various promising approaches to identifying pre-seismic thermal anomalies, the LST anomaly detection method still needs to be more complete and requires higher precision. The detection of pre-seismic thermal anomalies may also be influenced by factors such as topography, soil moisture, vegetation, and other atmospheric conditions.

3. Atmospheric earthquake precursors

The atmosphere is the gaseous layer above Earth's surface. It contains different layers characterized by different composition, density, and temperature. Up to 85% of the total mass of the atmosphere is confined in the lower layer, the troposphere. The overall composition of the Earth's atmosphere is very stable, containing, on average, around 78.08% nitrogen, 20.95% oxygen, 0.93% argon, 0.04% carbon dioxide, and traces of hydrogen, helium, and other noble gases. A variable amount of water vapor is also present, about 1% on average at sea level. Several signatures of seismic activity have been reported within the atmosphere, which can be due to the proximity of these two layers and the energy interchanges between them. Among the affected physical parameters studied, the literature reports on radon emissions in the vicinity of seismic regions, the generation of gravity waves in the atmosphere, the formation of anomalous clouds in the troposphere, the so-called "earthquake clouds," and anomalies in the surface latent heat flux [6].

Radon emissions are one of the most reported seismic precursors within the atmosphere. Radon (Rn) is a noble gas in the sixth period, in which all its isotopes are radioactive. Its most important one (Rn-222) belongs to the radium and uranium-238 decay chains, with a half-life of 3.8 days. Being a radioactive element, it can be easily tracked by the radiation it generates and its sub-products. Additionally, the short half-life limits its detection to short periods after its generation in the atmosphere. It can come from subsurface trapped gas after new faults and cracks appear in preparation for an earthquake.

A very extensive review of the methods using radon emissions, especially as earthquake precursors, is made in [40], where many past studies and techniques are discussed.

An especially large magnitude earthquake was reported in [41], where the sudden increase of the soil-gas Rn-222 was detected 22 days before a 7.1-magnitude earthquake in November 1994 in the Philippines. The peak reached a value of up to six times the background concentration of Rn. In the study, it was discarded that the close pass of typhoon Teresa was the possible origin of the peak because 1 year later, the larger typhoon Angela did not produce the same disturbance in the Rn concentration.

Surface latent heat flux (SLHF) and air and surface temperature anomalies have been reported in [42] for two earthquakes located in a seismic area of Iran in 2010 and 2011, which reached magnitudes 6.5 and 6, respectively. The SLHF is the thermal energy flux transmitted from the Earth's surface to the atmosphere. Therefore, it is highly related to the surface temperature, which may increase during the earthquake preparation.

A study of the SLHF anomalies found before seven earthquakes in India, Taiwan, and Mexico from 1993 to 2003, with magnitudes ranging from 5.4 to 7.8, was presented in [43]. The SLHF anomalies were recorded within 2 to 7 days before the earthquakes, and they were attributed to the increase of infrared thermal temperature in the earthquake preparation region.

In Ref. [44], air temperature anomalies were also found before three earthquakes in Turkey, Canada, and Argentina, with magnitudes of 6.1, 6.2, and 6.4, respectively. This work studied the average and STD values of the atmospheric daily temperature for 5 years before the earthquake date, and the positive or negative anomalies were recorded for the earthquake year. As a result, mostly positive anomalies were found 1 month before the earthquake and some a few days before. Additionally, some anomalies were found after the earthquake.

Another possible atmospheric earthquake precursor reported in the literature is the anomaly observed in the formation and evolution of tropospheric clouds near a fault before the earthquake. In Ref. [15], three case studies are reported using geostationary satellite images in Italy and Iran. The authors claim that the linear stationary shape of clouds, which did not move for about 8 hours, is a clear signature of an earthquake precursor, estimating the magnitude and even the date. According to the study, they only use the cloud shape and size to predict, and the prediction could be improved by using other data, such as radon or groundwater data. Nevertheless, the authors do not describe the methodology used to derive the earthquake occurrence with the shape or size of the anomalies. Additionally, the formation and shape of clouds can be linked to the terrain orography, which needs to be discussed in the report. Two years later, the study was questioned in Ref. [45].

In addition to previous atmospheric signatures, in Ref. [46], atmospheric ozone disturbances were observed before two large earthquakes occurred in Indonesia

(M = 7.3) and Fiji (M = 8.2) in 2019 and 2018, respectively. The study analyzed the ionospheric TEC parameter and the ozone data from AIRS, OMI, and TOMS-like satellites.

4. Ionospheric earthquake precursors

Along with the previously analyzed earthquake precursors, the ionosphere has been reported in many studies to be affected by seismic activity before and after earthquakes. This section discusses recent advances in the precursor signals observed in the ionosphere, particularly in the electron density (Ne) and TEC, as well as the IS phenomena.

4.1 Electron density and TEC anomalies

In Ref. [47], the daily variations of TEC in the ionosphere and the daily energy released near three studied earthquakes were examined and correlated, confirming a minor correlation. It was deduced that more parameters are essential to understand the cause of ionospheric TEC anomalies [48].

In Ref. [49], a study is presented using plasma density, electron temperature, and slant TEC from the ESA Swarm constellation. Two Mw ≥ 6.5 earthquakes were analyzed in this study. The first was a 6.7 magnitude earthquake in Turkey in March, and the second was in the USA in January 2022, with a magnitude of 6.5.

In Ref. [50], positive and negative VTEC anomalies were found before the December 2016 earthquake in Chile. Disturbances occurred 8 to 6 days before the earthquake within the Dobrovosky preparation region. The solar and geomagnetic conditions were also studied, showing that they remained quiet and stable.

In Ref. [51], the Ne was analyzed from the Swarm mission 1 month before and after 12 strong earthquakes of Mw > 6, between 2014 and 2017. The study used the Dobrovosky SR around epicenters to limit the region to correlate with Ne anomalies.

In Ref. [52], an innovative methodology based on TEC measurements was presented to estimate the moment of Co-seismic Ionospheric Disturbances (CID). The study focused on three large earthquakes: the 2015 Illapel, the 2014 Iquique, and the 2011 Sanriku-Oki. Results indicated that CIDs arrived 250–430 seconds after the seismic wave peak or 350–700 seconds after the earthquake onset time.

4.2 Ionospheric scintillation measured from GNSS signals

Ionospheric scintillations (IS) are the rapid fluctuations of the phase and intensity of an electromagnetic wave traversing the ionosphere. It is mainly due to the spatio-temporal changes in the local ionospheric Ne, and it is primarily driven by the solar influence, usually showing a maximum during the post-sunset hours.

Only a small fraction of the studies relating ionospheric perturbations to earthquakes use the IS as a potential proxy. These studies usually take GPS data from ground stations [53] or ground-based ionosondes [54] to measure the S4 intensity scintillation index, and correlate it to regional earthquakes.

This section presents IS signatures as possible earthquake precursors reported in the literature. They are studied using data recorded in ground stations and novel techniques to derive the IS from remotely sensed data from satellites, particularly from GNSS Reflectometry and GNSS radio-occultation.

4.2.1 GNSS ground stations

GNSS ground stations are infrastructures installed worldwide to control and monitor the signals received from GNSS constellations, offering important TEC and IS data. This information is used to make ionospheric corrections for navigation and to evaluate the quality of service. Over the past decades, these data have been used to monitor the state of the ionosphere. However, coverage is limited. GNSS stations are usually strategically placed at fixed ground locations, resulting in sparse installations that only offer data for their immediate local regions. Therefore, ground station data is a good approach when studying a local event or a high-seismicity region.

In Ref. [55], GPS and low-frequency measurements from multiple stations were analyzed to identify ionospheric anomalies related to the South West Banten (Indonesia) earthquake with an Mw of 7.4. The analysis revealed increased IS 5 days before and after the earthquake, with the anomaly being more extensive after the main seismic shock. In [56], the authors examine the impact of the seismic activity associated with the 2021 La Palma (Canary Islands, Spain) volcanic eruption using IS anomalies detected from ground stations near the volcano. The ground station database was pre-processed using the algorithms described in [17]. The S4 data here are obtained using the 1 min averaged signal's intensity received in ground stations from several constellations of GNSS satellites. Only GNSS satellites above 30° of elevation were used to avoid wrong high scintillation values. The results showed a small but detectable correlation between the IS and the earthquake energy released per unit of time. The best correlation between both variables was found to occur 18 h, and 7–8 days after earthquakes. The work also studied IS measured by GNSS-R and GNSS-RO satellites, in this last case where larger correlations were found before the earthquakes. A cross-correlation between the S4 measurements and the Kp and solar flux (F10.7) was also performed, showing smaller and not overlapping in time with the correlations found between earthquakes and S4.

4.2.2 Satellite GNSS-R data

In Ref. [57], the authors demonstrated the efficiency of high-rate GNSS data in detecting low-magnitude anthropogenic earthquakes. This study showed the potential impact of the infrastructure of GNSS ground stations and was successfully integrated into moment tensor calculations, focusing on two earthquakes with magnitudes greater than 3.5.

GNSS-R works as a multi-static radar that uses opportunistic navigation signals. The receiver calculates signal delay and Doppler frequency, enabling simultaneous tracking of multiple transmitters, and their specular reflection points [58]. Applications of GNSS-R include ocean altimetry, surface wind speed, cryosphere studies, and surface soil moisture. NASA's CYGNSS represents the first operational mission based on GNSS-R, featuring eight micro-satellites equipped with SGR-ReSi receiver payloads from SSTL. The SGR-ReSi produces Integrated Delay Doppler Maps (DDMI) by capturing reflected GPS signal power, which depends on delay and Doppler shift [59]. In this study, CYGNSS data at level 1, version 2.0, has been used. The study explores GNSS signals traversing the ionosphere in their down-welling and up-welling paths, providing insights into IS at GNSS frequencies.

In previous contributions [18, 60, 61], the IS data derived from CYGNSS GNSS-R data were scrutinized and correlated to ocean-based earthquakes from 2017 to 2021.

Again, a small, but detectable correlation was identified between S4 anomalies and earthquake occurrences a few days before.

After acquiring CYGNSS data, a nighttime filtering process is employed between 00:00 and 06:00 am local time to eliminate diurnal effects and post-sunset scintillation. While IS may persist beyond midnight on certain occasions, this local time window is chosen to ensure a substantial dataset for processing. The data is then aggregated to determine the mean, STD as shown in **Figure 4a**, IQT, and total count of pixels with identical geolocation (longitude and latitude) and local time. The spatial resolution in the figures is set to 0.5°.

Subsequently, the dataset is converted into a raster format, with a filter applied to exclude pixels with high sea wind speed. This filter is important as a coherent reflection for observing IS in GNSS-R data, which can only occur when wind speeds are below 2–3 m/s. An elevated solar activity flag eliminates potential sources of S4 anomalies unrelated to earthquakes. It is worth noting that the STD, IQR, mean, and median are then computed over 7 years to derive daily S4 anomalies, as illustrated in **Figure 4b**.

In areas near earthquake epicenters, there is a general increase in IS, particularly when the Sea Wind Speed Filter (SWSF) is less than 2 m/s, and the Fixed Search Radius (FSR) is under 50 km. S4 anomalies, ranging from 0.08 to 0.23, can be observed up to 7 days before earthquakes. Despite a visual and numerical correlation, not all earthquakes connect to S4 anomalies, and vice-versa. The forecast of earthquakes remains complex due to the sporadic occurrence of minor S4 anomalies.

For example, in the Solomon Islands earthquake (**Figure 4c**) in February 2018, with an Mw of 4.4 at a depth of 10 km, a positive S4 anomaly of 0.11 was noticed 3 days before the event. Minor positive S4 anomalies, approximately 0.1, were detected 1 and 3 days before the Maldives earthquake (**Figure 4d**) occurred in April 2019.

Figure 5 illustrates that the ROC curves derived from CYGNSS mission data using the STD show remarkable similarity and near-identical patterns. As the parameter "C" increases, the TPR and FPR exhibit distinct trends: TPR rises while FPR decreases. Consequently, detecting S4 anomalies improves for smaller "C" coefficients. However, it is important to note that the false alarm rate is also notably elevated.

For instance, when "C" is set to 0.1, TPR equals 1, and FPR equals 1. Conversely, at "C" = 2.5, TPR is 0, and FPR is 0. The optimal configuration is observed at "C" = 0.5, and a false alarm rate FSR of 50 km. The result is the highest area under the curve (AUC) and a straight distance (d) of 0.62. Therefore, the recommended threshold is "C" = 0.5, paired with a sliding window size factor SWSF of 2 m/s.

4.2.3 Satellite GNSS-RO data

GNSS-RO stands for GNSS Radio-Occultation, and it constitutes a technique employed for atmospheric and ionospheric sounding by using the fluctuations of the received signal from GNSS satellites, but in this case, in a radio-occultation geometry. In this configuration, rays do not reflect on the Earth's surface but cross the ionosphere tangentially when the GNSS satellites fall over or rise above the horizon from another satellite, the GNSS-RO receiver. This technique has the advantage of not relying on surface properties, which may affect the properties of the wave, as was the case in GNSS-R.

In Ref. [62], a novel method was presented for automatically categorizing disturbances in the lower ionosphere. This approach utilized GNSS-RO data from Spire's

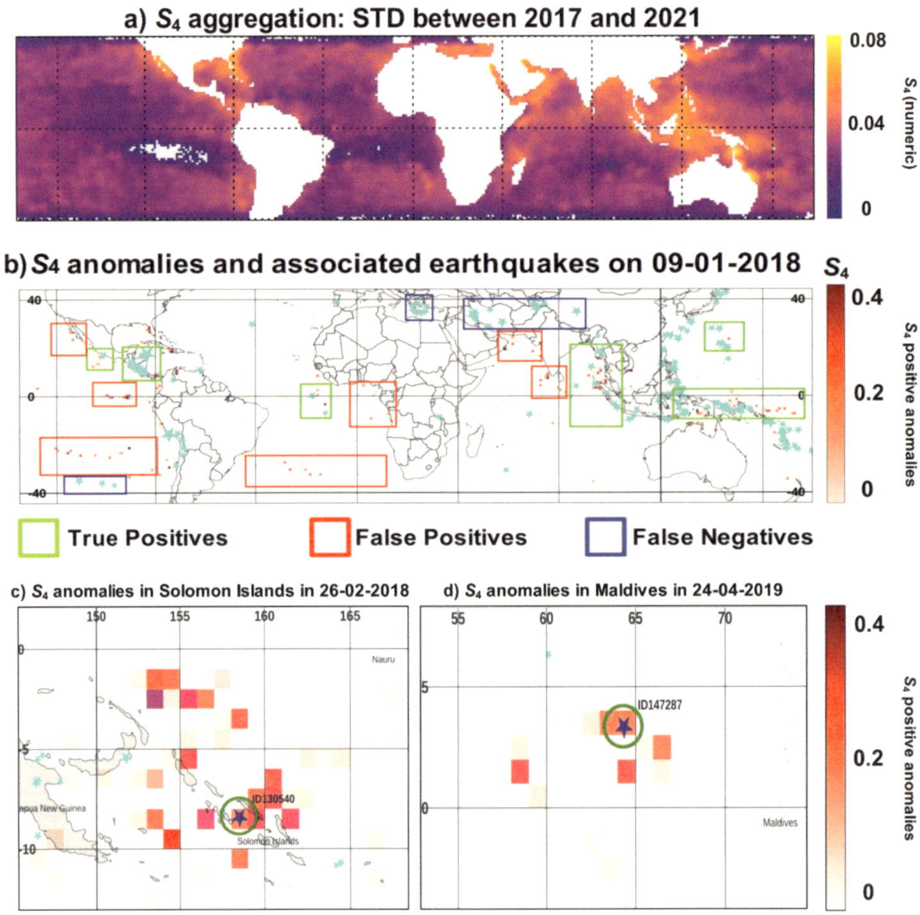

Figure 4.
(a) STD of S_4 with 4 m/s sea wind speed filters between 2017 and 2021, (b) S_4 anomalies classification (TP, FP, and FN), (c) S_4 anomalies associated with the Solomon Islands earthquake, and (d) S_4 anomalies associated with the Maldives earthquake.

CubeSats constellation, integrating signal processing methods and semi-supervised machine learning, employing spectral clustering within a metric space of wavelet spectra. This innovative approach established an automated system for global monitoring of ionospheric disturbances.

As shown in Ref. [63], a similar analysis was conducted, but using GNSS-RO data to examine the S_4 intensity scintillation index throughout 2022 in the Coral Sea region, encompassing Papua New Guinea, Solomon Islands, Vanuatu, and New Caledonia. The selection of this area is based on the substantial occurrence of earthquakes, aimed at enhancing the reliability of the statistical analysis. Within this timeframe and geographical scope, 201 earthquakes with magnitudes $M_w \geq 5$ and 22 earthquakes with $M_w \geq 6$ have been documented.

The IS data in this study was sourced from the GNSS-RO COSMIC-2 constellation [64]. Applying GNSS-RO for IS also enables the investigation of terrestrial and oceanic regions. Earthquake data was obtained from the United States Geological Survey (USGS) database, encompassing earthquake magnitude, date, location, and hypocenter depth [61].

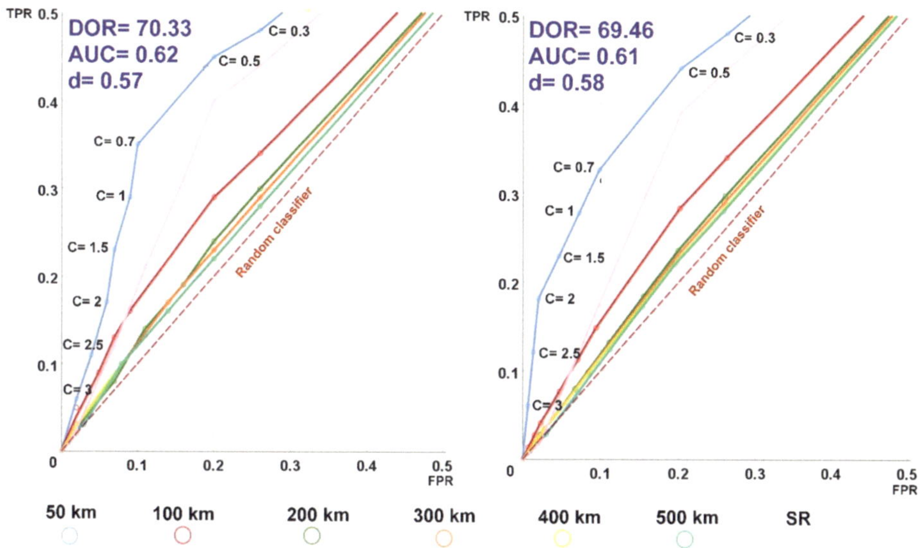

Figure 5.
Confusion matrix for S_4 anomalies using STD and IQT methods with 2 m/s sea wind speed filters.

COSMIC-2 provides GPS and LEO coordinates for each occultation but lacks a reference coordinate for determining the S_4 location. The tangent point is calculated using the GPS and LEO satellite positions from COSMIC-2 using the same methods described in [65], which can be seen in [66]. This is later used to filter data based on the distance around earthquake epicenters or latitude-longitude pairs.

Before data processing, various filters are applied to ensure accurate S_4 estimation. Initially, it is mandated that the pierce points in the profile data of the occultation path account for a Slant TEC value ≥80% of the maximum STEC value in that occultation profile as described in [56] and [63]. This filter enhances the likelihood of isolating points within the ionosphere susceptible to scintillation. Additionally, the elevation must be equal to or lower than 0° to confirm the occurrence of an occultation. Another filter involves a nighttime criterion, as explained in previous sections.

To compare the S_4 scintillation index with seismic activity, the mean and the 95th percentile of S_4 values were computed in 1-day intervals throughout the entire study period. The spatial resolution is set at 1° of latitude and longitude. Throughout the statistical analysis, every latitude-longitude pair within the study area undergoes systematic examination. For each pair, S_4 values within a search radius of 100, 200, and 300 km from the pair's center are gathered, and the mean and 95th percentile of these values were calculated. Epicenter's coordinates are retrieved from the USGS database. To reconcile the spatial resolution disparity between S_4 and earthquake data, the latitudes and longitudes of earthquakes are approximated to their nearest latitude-longitude pair.

This study spans 2022, repeating the analysis for all latitude-longitude pairs within the defined study area. Consequently, there are instances where no earthquakes occur on a specific day and latitude-longitude pair. This aspect of the research aided in evaluating non-detection reliability while simultaneously investigating IS anomalies around earthquake epicenters.

Next, the results of two case studies in this area are shown. These two case studies presented occur 8 days apart, with epicenters positioned approximately 250 km from each other. Consequently, there is a possibility of mutual influence between the earthquakes. For each case study, a temporal analysis and spatial representation are conducted. In the temporal analysis, we examine the daily changes in S_4 values near the epicenter within a temporal window of ±15 days surrounding the event. Concerning the spatial representation, the distribution of S_4 values is illustrated geographically in pixel maps computed daily near the earthquake. The first earthquake under consideration occurred on September 2nd, 2022, with a magnitude of M_w = 6.1 and ID "us7000i4st".

Figure 6 illustrates the daily averages of the S_4 scintillation index within search radii of 100, 200, and 300 km from the epicenter. Notably, ionospheric intensity scintillation anomalies are discernible several days before the earthquake, particularly between 5 and 6 days preceding the event. To quantify this deviation, the median value of the S_4 averages is computed within a 200 km radius, resulting in 0.079. In contrast, values 5 and 6 days before the event are 0.30 and 0.26, respectively—approximately 3 to 4 times larger than the median of the averages within the ±15 days surrounding the event. The same trend is observed if the same analysis using the 95th percentile is represented, as seen in [63].

In addition to the temporal analysis, a spatial representation is examined. This analysis involves 0.2° × 0.2° pixels in latitude-longitude for both case studies, each assigned a distinct color based on S_4 average values. Additionally, a gray circle with a radius of 200 km is delineated, centered on the epicenter of the analyzed earthquake, symbolized by a black star. The spatial representation examines the proximity of high S_4 values to the earthquake's epicenter, providing a visual portrayal of the scene.

For this case study, the spatial representation is presented in **Figure** 7. Various days are depicted, including those with peak S_4 values (**Figure 7a** and **b**), as well as the day of the earthquake occurrence (**Figure 7c**). These figures reveal that the days with peak S_4 values are 5 and 6 days before the earthquake, specifically on the 28 and 27 August, respectively. These days, certain pixels inside the drawn radii exhibit high values of S_4. Conversely, the maximum values are lower on the event day, September 2.

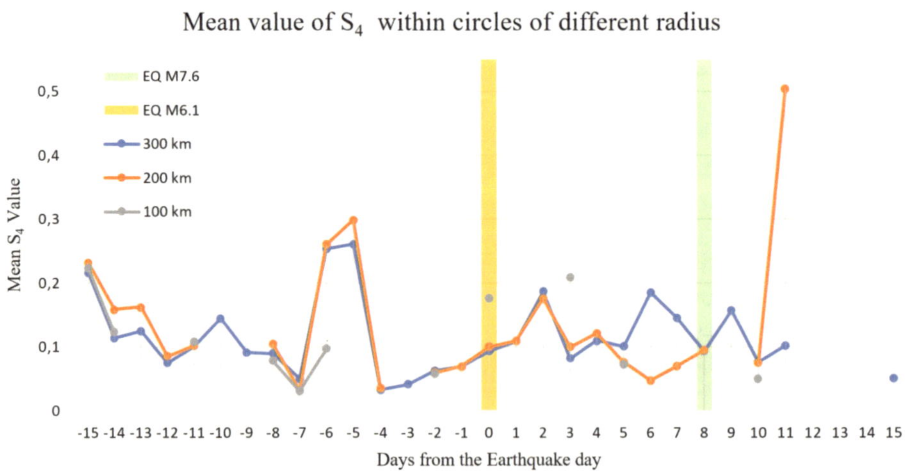

Figure 6.
Average values of the S_4 indexes within circular areas of 100, 200, and 300 km from the earthquake's epicenter "us7000i4st".

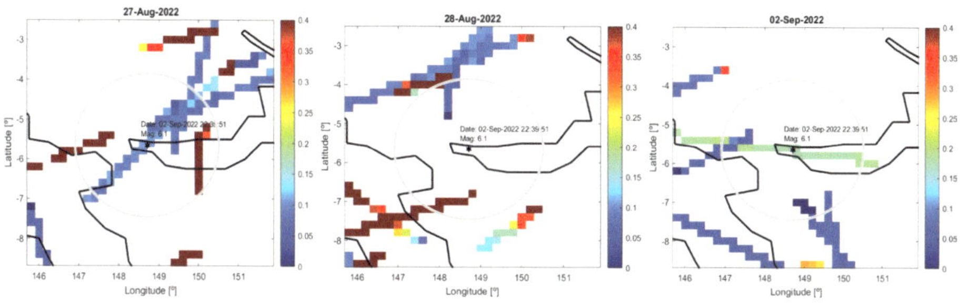

(a) (b) (c)

Figure 7.
Average S_4 maps for selected days centered on the earthquake "us7000i4st".

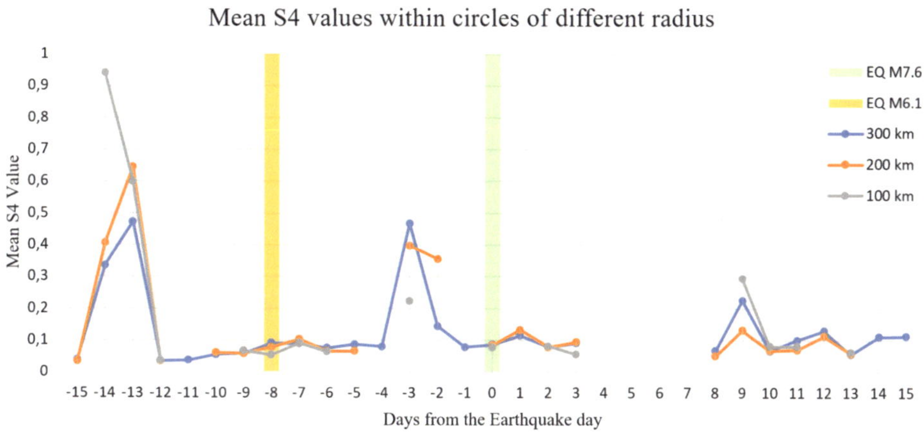

Figure 8.
Average values of the S_4 indexes within circular areas of 100, 200, and 300 km from the earthquake's epicenter "us6000iitd".

The second case study that is shown involves an earthquake on September 10, 2022, with a magnitude of M_w = 7.6 and ID "us6000iitd". **Figure 8** is created using the same method described in **Figure 6**. In this instance, two notable peaks are observed again, occurring 13 and 14 days and 3 and 2 days before the seismic event. The first set of days, corresponding to 5 and 6 days from the "us7000i4st" earthquake (earthquake M6.1 in **Figure 7**), is due to the proximity between the two events. Consequently, the ionosphere is affected by anomalies 3 and 2 days before the earthquake. The S_4 representative values within a search radius of 200 km for the 3 and 2 days before the earthquake are 0.37 and 0.30, respectively. Comparing these values to the dataset's median, which is 0.15, the anomalies are approximately 1.97 to 2.43 times higher than the median.

The spatial representation maps in **Figure 9** correspond to the peaks occurring 2 and 3 days before the earthquake. **Figure 9a** and **b** correspond to the 8th and 7th of October, respectively, and the day of the earthquake itself, **Figure 9c**. It is evident, again, that only in **Figure 9a** and **b** there are values of S_4 around or higher than 0.4. In contrast, there are only values below 0.1 on the day of the earthquake.

An analysis of all earthquakes within the studied region and period is conducted to assess the presence of a positive correlation and to determine the optimal detection

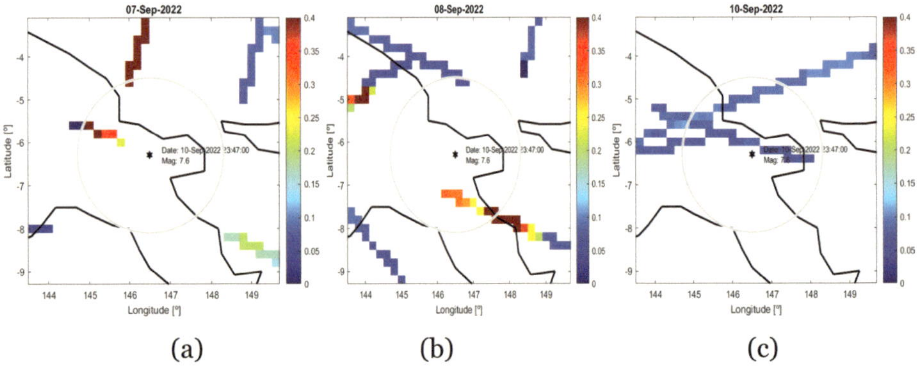

Figure 9.
Average S_4 maps for selected days centered on the earthquake "us6000iitd".

Parameter swept	Values swept
The radius of the search area	100, 200, and 300 km
Type of S_4 representative value	Computed with mean or 95th percentile
Minimum magnitude of earthquakes studied	M_w = 5, 5.5, 6
Depth of the earthquake	[0, 100) km, [100, 200) km, [200, ∞) km
Magnetic field activity	No filter or $K_P \leq 5$
S_4 Anomaly occurring	[−6, −3], [−3, 0], [0,+3] days before/after the earthquake
Detection threshold values	1, 1.25, 1.5, 1.75, 2 ... 4

Table 2.
Description of optimization parameters to generate different confusion matrices.

method through confusion matrices. As detailed in **Table 2**, various parameters are tested to generate diverse confusion matrices and ROC curves. The outcomes of the confusion matrices, along with their associated metrics and ROC curves, contribute to finding the optimal approach.

The investigation of these parameters aims to identify the range from which IS effects can be discerned. This serves as a filtering mechanism, determining the effectiveness of the detection method for earthquakes with specific characteristics. For example, a magnetic field activity filter is implemented to mitigate anomalies from solar or geomagnetic storms. The K_P is utilized to ascertain the presence of a solar or geomagnetic storm, describing the intensity of magnetic field activity.

Ratios are established to determine the likelihood of an S_4 anomaly occurrence numerically. Two types of ratios are defined, leading to two additional iterations to select the optimal detection method. The first ratio ($r_{1,x}$) involves the ratio between a specific day's S_4 mean value and the median of the S_4 representative value within ±15 days from the computed day. The second ratio ($r_{2,x}$) is an IQR-based threshold, calculated as the difference between the median of the S_4 representative values and the daily value of the S_4 representative value, divided by IQR/2. These are also described mathematically in [63], and are computed for all days of the year, not solely around the days of earthquake occurrences. Once both ratios are calculated, they are tested with threshold values ranging from 1 to 4 in increments of 0.25. An anomaly is classified as true or false based on whether $r_{1,x}$, or $r_{2,x}$ exceeds the tested threshold.

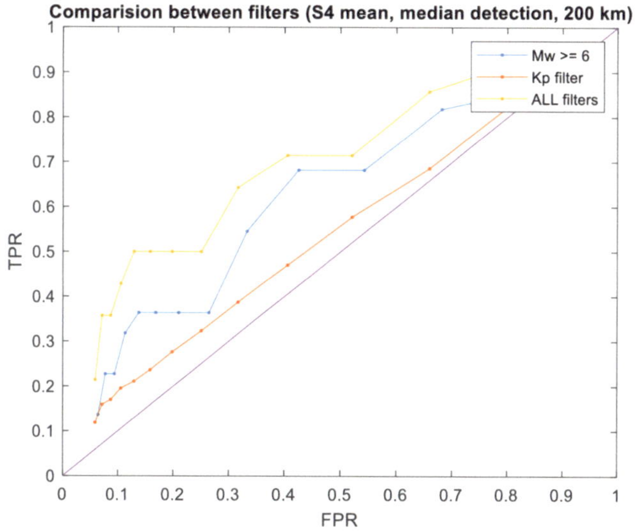

Figure 10.
ROC curve comparison for S_4 average within the searching area of 200 km, ratio $r_{1,x}$, and anomaly occurrence [−6, −3] days with different filters.

After the optimization involving all the parameters, it is shown that the most favorable results are obtained with the window of occurrence day [−6, −3] and the 200 km radius, the S_4 representative value calculated using the mean and the $r_{1,x}$ type of ratio as the method of detection. As for other filters, as shown in **Figure 10**, the best results are obtained when all the filters proposed in **Table 2** are used altogether, obtaining the highest values of AUC.

The final step in this study was to determine the optimal threshold. This involves examining the ROC curve in **Figure 10** and considering metrics such as F_1, MCC, and DOR. After considering those metrics, the best results were obtained using a threshold of 3.75. Out of 14 positive cases, 5 were correctly detected, yielding a TPR of 35.7%. Additionally, an FPR of 7.1% is observed, with 21,027 false positive cases among 295,636 negative cases. The ROC curve suggests a positive correlation, given that the TPR value is approximately five times higher than the FPR, and the AUC is 0.69.

5. Other earthquake precursors

In addition to the previously reported possible earthquake precursors in the LST, the atmosphere, and the ionosphere, this section reviews other related phenomena that are difficult to characterize in any of the previous sections. In this section, geomagnetic field perturbations, electromagnetic emissions, and disturbances in the Schumann resonance as possible earthquake precursors are revised.

5.1 Geomagnetic field perturbations

Geomagnetic disturbances have also been detected in association with earthquakes. Some studies have related post-seismic geomagnetic disturbances. For example, a geomagnetic pulsation lasting for about 3.6 minutes was observed in Thailand 12 minutes after the Sumatra 9.3 magnitude earthquake on December 26, 2004, as reported in [66].

Moreover, in [47], efforts were made to detect perturbations in the geomagnetic field in the ionosphere potentially caused by earthquakes in the lithosphere. A magnetic field perturbation was observed on the orbit track crossing the studied epicenters. The study found 12 satellite tracks with potential co-seismic disturbances related to 10 earthquakes ranging from M_w 5.6 to 6.9, confirming the probability that seismic events produced the detected disturbances.

In Ref. [67], magnetic field data from the Swarm Alpha satellite were analyzed using non-negative matrix factorization (NMF) before the earthquake in Ecuador that occurred in 2016, with $M_W = 7.8$. The study included data collected under quiet and disturbed geomagnetic conditions, minimizing false alarms. Observations suggested a potential association described by one of the Lithosphere–Atmosphere–Ionosphere-Coupling (LAIC) models.

In Ref. [68], authors deployed a magnetic monitoring network in earthquake-prone areas of Sichuan, China, conducting long-term observations and real-time analysis of geomagnetic data. Results showed a high correlation between pre-earthquake geomagnetic anomalies and earthquakes. A new algorithm using geomagnetic anomaly energy was introduced to predict the time of an earthquake's occurrence.

In Ref. [69], the authors inspected magnetic field variations from earthquake-associated satellites, utilizing the China Seismo-Electromagnetic Satellite (CSES) and ESA's three spacecraft Swarm fleet. The study accessed links between specific lithospheric or near-surface sources and ionospheric magnetic field measurements, revealing increased fluctuations in higher-frequency components.

Swarm satellites represent an interesting way to study possible earthquake precursors in the geomagnetic field, as reported in several studies in the last few years.

In Ref. [70], the 2015 Nepal earthquake, with a magnitude of 7.8, is under study. Time series of the magnetic field vector and intensity were analyzed. Only data from night orbits were used to avoid noisy magnetic periods. The time series were derived and detrended to show fast fluctuations. The number of anomalies within the Dobrowolski circular area around the earthquake epicenter was recorded, and it was revealed that they increased in the days before the earthquake. The perturbations are observed mainly in the Y component (East), smaller in the magnetic field intensity and other elements.

5.2 Electromagnetic radiation

Some studies, including some of the previously revised in the chapter, report that earthquakes can generate low to very low-frequency electromagnetic waves, which can travel from the preparation earthquake region. These waves fall into the extremely low frequency (ELF) or very low frequency (VLF) ranges, 3–30 Hz and 3–30 kHz, respectively. Some satellite missions have been particularly designed to try to detect and correlate this electromagnetic emission with earthquakes. The first one was the French CNES Demeter mission, launched in 2004. It could detect changes in the magnetic field and electromagnetic waves; it also had an ion analyzer, a Langmuir probe, and an energetic particle detector.

Another satellite investigating electromagnetic radiation related to earthquake occurrence is the one already mentioned. China Seismo-Electromagnetic Satellite (CSES). This satellite orbits in a 507 km Sun-synchronous orbit with a descending node at 14:00 LT. Its eight instruments include magnetometers, electric field detectors, a GNSS-RO instrument, a plasma analyzer, a Langmuir probe, and high-energy particle detectors. The mission is intended to study the correlation between ionospheric-electromagnetic signatures and seismic activity.

In a study from 1992 [71], seismo-electromagnetic waves were observed by the COSMOS-1809 satellite orbiting above seismic regions in Armenia. The earthquake under study occurred in Spitak, Armenia, in December 1988, with a magnitude of 6.7. Intense EM waves were detected in frequencies under 450 Hz at the L-shells of the earthquake during 12 out of the 13 orbits that passed within 6° in longitude from the epicenter.

In Ref. [72], the VLF frequency band was analyzed as a possible source of earthquake precursors in the shape of seismic radio wave signals. They found pulse-like signals traveling long distances up to 500 km from the epicenter of preparing earthquakes, indicating waveguide mode propagation between the ionosphere and the ground. The great 1994 Kuril Islands Earthquake of magnitude 8.1 occurred East of Hokkaido, and the number of pulses detected in the 5 hours before the earthquake increased about nine times above the background value, with its maximum around 20 minutes before the earthquake.

In Ref. [73], seismic anomalies before two strong earthquakes with $M_w > 7$ in western China were detected using ground-based observation of VLF signals from monitoring stations. A full wave model was used to research possible factors inducing seismic anomalies on the received VLF electric field. Results demonstrated that anomalies could be caused by ascending/descending the ionosphere's bottom height.

In Ref. [74], the research inspected the correlation between earthquakes with $M_w \geq 6$ globally and fluctuations in ionospheric Ne between August 2018 and March 2023. Using data from the Langmuir probe onboard the CSES during nighttime, statistical analysis revealed that larger earthquakes were associated with more anomalous phenomena. Anomalies appeared at specific intervals, including co-seismic effects and possible precursor candidates.

In Ref. [75], Ne disturbances were detected around the Maerkang earthquake ($M_w = 6$) on June 9th, 2022, using multi-data from the global ionospheric map, ionospheric TEC inverted from GPS observations, and the critical frequency of the F2 layer from the ionosonde. Seismo-ionospheric disturbances were observed 7 to 5 days before the earthquake, serving as significant signals of the upcoming main shock.

5.3 Schumann resonance frequency

The Schumann resonance occurs due to the resonant cavity formed between the Earth's surface and the ionosphere, acting as an electric conductor. The global extension of the cavity adjusts some resonance frequencies that fall in the ELF band, with its fundamental frequency at 7.83 Hz, and higher modes at 14.1 Hz and 20.3 Hz. The excitation of the cavity comes from natural sources such as electric discharges from lightning.

A possible link between earthquakes and the Schumann resonance has been recently theorized, and several studies have been conducted on the detection and correlation of these phenomena. A possible physical explanation of these anomalies is that ELF radiation can be generated during the preparation of earthquakes, which can rise from underground and interact with the Schumann resonance [76].

In Ref. [55], two anomalies on the Schumann resonance are discussed before large earthquake occurrences in Taiwan and Japan. The first phenomenon is the enhancement of the Schumann resonance fourth harmonic in Japan, possibly associated with the September 1999 Chi-chi earthquake in Taiwan. This earthquake had an magnitude of 7.6, and the hypocenter was 30 km deep. The second phenomenon was associated with two huge earthquakes in Japan. The resonance frequency was shifted by about 2 Hz from the typical Schumann resonance frequency: ~16 Hz and 18 Hz in two different modes. In this case, the earthquakes producing these anomalies were the 2004 Niigata-chutes earthquake, and the 2007 Noto-Hanto earthquake, with magnitudes of 6.8 and 6.9, respectively. The anomalies started around 20 h before the earthquakes, were maximum around 2 h before, and decreased after the earthquake.

In Ref. [77], a theoretical study is performed on how earthquakes affect the Schumann resonance spectra. The authors utilized a numerical model of the ELF radio wave scattering generated by a localized non-uniformity of the air conductivity profile likely to appear before an earthquake. The results showed an overall elevation of the Schumann resonance spectra, with some dependence from the distance to the earthquake epicenter.

6. Conclusions

This chapter has summarized the potential application of remotely sensed data as proxies for earthquake prediction. Numerous subtle indicators linked with earthquakes have been detected, such as land displacements, accelerations, gas emissions, temperature anomalies, and ionospheric effects. RS data offers benefits over in-situ data collection, such as extensive spatial coverage in remote regions. This chapter highlights investigations into various Earth observation observables, including anomalies in surface temperatures, atmospheric and ionosphere precursors, and additional physical mechanisms like fluctuations in the geomagnetic field. Despite the presence of seismic hazard risk models and the discoveries from these investigations, accurate earthquake forecast remains a significant challenge.

Acknowledgements

This work was supported in part by the grant GENESIS PID2021-126436OB-C21 from the Programa Estatal para Impulsar la Investigación Científico- Técnica y su Transferencia, del Plan Estatal de Investigación Científica, Técnica y de Innovación 2021-2023 (Spain), and in part by the European Social Fund (ESF), Grant RYC-2016-2018 financed by MCIN/AEI /10.13039/501100011033. Badr-Eddine Boudriki Semlali received support from an FI grant: 2021 FI_B 00471 from FI AGAUR 2021.

Data availability statement

The GOES/ABI, MSG/SEVIRI, and Himawari-8/AHI data used in this study comes from the Copernicus global land service https://land.copernicus.eu/global/products/lst and earthquake data comes from the open USGS earthquake database available at https://earthquake.usgs.gov/earthquakes/search/.

CYGNSS GNSS-R data is freely available in the NASA PODAAC database. https://doi.org/10.5067/CYGNS-L1X30.

GNSS-RO data from the COSMIC-2 satellite is freely available in the UCAR/NCAR CDAAC database. https://doi.org/10.5065/t353-c093

Other data are the property of a third-party organization, such as the Spire GNSS-RO data.

Author details

Badr-Eddine Boudriki Semlali[1], Carlos Molina[1,2], Mireia Carvajal Librado[1], Hyuk Park[1,2,3] and Adriano Camps[1,2,4]*

1 CommSensLab—UPC, Department of Signal Theory and Communications, Universitat Politècnica de Catalunya—BarcelonaTech, Barcelona, Spain

2 IEEC—Institut d'Estudis Espacials de Catalunya, Barcelona, Spain

3 Department of Physics, Universitat Politècnica de Catalunya—BarcelonaTech, Castelldefels, Spain

4 ASPIRE Visiting International Professor, College of Engineering, UAE University, Al Ain, United Arab Emirates

*Address all correspondence to: adriano.jose.camps@upc.edu

References

[1] Jiao Z-H, Zhao J, Shan X. Preseismic anomalies from optical satellite observations: A review. Natural Hazards and Earth System Sciences. 2018;**18**(4):1013-1036. DOI: 10.5194/nhess-18-1013-2018

[2] IDDR2018_Economic Losses. 2018. Available from: https://www.unisdr.org/2016/iddr/IDDR2018_Economic%20Losses.pdf

[3] Rasul A, Omar LW. Land surface temperature anomalies detection for the strong earthquakes in 2018. ARO. 2020;**8**(2):15-21. DOI: 10.14500/aro.10591

[4] Mignan A, Ouillon G, Sornette D, Freund F. Global earthquake forecasting system (GEFS): The challenges ahead. European Physical Journal Special Topics. 2021;**230**(1):473-490. DOI: 10.1140/epjst/e2020-000261-8

[5] Jiao Z-H, Shan X. Statistical framework for the evaluation of earthquake forecasting: A case study based on satellite surface temperature anomalies. Journal of Asian Earth Sciences. 2021;**211**:104710. DOI: 10.1016/j.jseaes.2021.104710

[6] Pulinets S, Davidenko D. Ionospheric precursors of earthquakes and global electric circuit. Advances in Space Research. 2014;**53**(5):709-723. DOI: 10.1016/j.asr.2013.12.035

[7] Hassini A, Belbachir AH. Thermal method of remote sensing for prediction and monitoring earthquake. In: 2014 1st International Conference on Information and Communication Technologies for Disaster Management (ICT-DM). Algiers, Algeria: IEEE; 2014. pp. 1-6. DOI: 10.1109/ICT-DM.2014.6917790

[8] Zoran M, Savastru R, Savastru D. Earthquake anomalies recognition through satellite and in-situ monitoring data. European Journal of Remote Sensing. Jan 2016;**49**(1):1011-1032. DOI: 10.5721/EuJRS20164952

[9] Tramutoli V, Corrado R, Filizzola C, Genzano N, Lisi M, Pergola N. From visual comparison to robust satellite techniques: 30 years of thermal infrared satellite data analyses for the study of earthquake preparation phases. Bollettino Di Geofisica Teorica Ed Applicata. 2015;**56**:167-202

[10] Zhong M, Shan X, Zhang X, Qu C, Guo X, Jiao Z. Thermal infrared and ionospheric anomalies of the 2017 Mw6.5 jiuzhaigou earthquake. Remote Sensing. 2020;**12**(17):2843. DOI: 10.3390/rs12172843

[11] Pulinets SA et al. Thermal, atmospheric and ionospheric anomalies around the time of the Colima M7.8 earthquake of January 21st 2003. Annales de Geophysique. 2006;**24**(3):835-849. DOI: 10.5194/angeo-24-835-2006

[12] Guangmeng G, Jie Y. Three attempts of earthquake prediction with satellite cloud images. Natural Hazards and Earth System Sciences. 2013;**13**(1):91-95. DOI: 10.5194/nhess-13-91-2013

[13] Ghosh S et al. Unusual surface latent heat flux variations and their critical dynamics revealed before strong earthquakes. Entropy. 2021;**24**(1):23. DOI: 10.3390/e24010023

[14] Su Y-C, Sha J. A study of possible correlations between Seismo-ionospheric anomalies of GNSS Total electron content and earthquake energy. Remote Sensing. 2022;**14**(5):1155. DOI: 10.3390/rs14051155

[15] Molina C, Boudriki Semlali B-E, Park H, Camps A. A preliminary study

on ionospheric scintillation anomalies detected using GNSS-R Data from NASA CYGNSS Mission as possible earthquake precursors. Remote Sensing. 2022;**14**(11):2555. DOI: 10.3390/rs14112555

[16] Camps A et al. Ionospheric scintillation monitoring using GNSS-R? In: IGARSS 2018-2018 IEEE International Geoscience and Remote Sensing Symposium. Valencia: IEEE; 2018. pp. 3339-3342. DOI: 10.1109/IGARSS.2018.8519088

[17] Stănică DA. Anomalous geomagnetic signal emphasized before the Mw8.2 coastal Alaska earthquake, occurred on July 29th 2021. Entropy. 2022;**24**(2):274. DOI: 10.3390/e24020274

[18] Chakraborty S, Sasmal S, Chakrabarti SK, Bhattacharya A. Observational signatures of unusual outgoing longwave radiation (OLR) and atmospheric gravity waves (AGW) as precursory effects of may 2015 Nepal earthquakes. Journal of Geodynamics. 2018;**113**:43-51. DOI: 10.1016/j.jog.2017.11.009

[19] Hayakawa M, Schekotov A, Izutsu J, Nickolaenko AP. Seismogenic effects in ULF/ELF/VLF electromagnetic waves. IJEAR. 2019;**06**(02):1-86. DOI: 10.33665/IJEAR.2019.v06i02.001

[20] Sierra Figueredo P, Mendoza Ortega B, Pazos M, et al. Schumann Resonance anomalies possibly associated with large earthquakes in Mexico. Indian Journal of Physics. 2021;**95**:1959-1966. DOI: 10.1007/s12648-020-01865-6

[21] Pavlidou E, van der Meijde M, van der Werff H, Hecker C. Time series analysis of land surface temperatures in 20 earthquake cases worldwide. Remote Sensing. 2018;**11**(1):61. DOI: 10.3390/rs11010061

[22] Shi Liu, Guo Liu, You, Wang. "Pre-earthquake and Co-seismic ionosphere disturbances of the mw 6.6 Lushan earthquake on April 20th 2013 monitored by CMONOC". Atmosphere. 2019;10(4):216. DOI: 10.3390/atmos10040216

[23] Khoshgoftar MM, Saradjian MR. Estimation of date and magnitude of four major earthquakes using integration of precursors obtained from remote sensing data. 2021:23

[24] Baral R et al. Spectral features of Forbush decreases during geomagnetic storms. Journal of Atmospheric and Solar-Terrestrial Physics. 2023;**242**:105981. DOI: 10.1016/j.jastp.2022.105981

[25] Dobrovolsky IP, Zubkov SI, Miachkin VI. Estimation of the size of earthquake preparation zones. PAGEOPH. 1979;**117**(5):1025-1044. DOI: 10.1007/BF00876083

[26] Semlali BEB, Molina C, Park H, Camps A. Study of land surface temperature anomalies associated to earthquakes using GOES data. In: IGARSS 2022 - 2022 IEEE International Geoscience and Remote Sensing Symposium, Kuala Lumpur, Malaysia. 2022. pp. 5732-5735. DOI: 10.1109/IGARSS46834.2022.9884887

[27] De Santis A et al. A comprehensive multiparametric and multilayer approach to study the preparation phase of large earthquakes from ground to space: The case study of the June 15th 2019, M7.2 Kermadec Islands (New Zealand) earthquake. Remote Sensing of Environment. 2022;**283**:113325. DOI: 10.1016/j.rse.2022.113325

[28] Xiong P et al. Towards advancing the earthquake forecasting by machine learning of satellite data. Science of the Total Environment. 2021;**771**:145256. DOI: 10.1016/j.scitotenv.2021.145256

[29] Jiao Z-H, Zhao J, Shan X. Pre-seismic thermal anomalies from

satellite observations: AReview. Earthquake Hazards, preprint. Jul 2017. DOI: 10.5194/nhess-2017-211

[30] Piroddi L, Ranieri G. Night thermal gradient: A new potential tool for earthquake precursors studies. An application to the seismic area of L'Aquila (Central Italy). IEEE Journal of Selected Topics in Applied Earth Observations and Remote Sensing. 2012;**5**(1):307-312. DOI: 10.1109/JSTARS.2011.2177962

[31] Fatimah H, Bangash S, Tariq A, Ali Naseem A, Ahmed Z, Ahmad Bangash A. Time series temperature anomalies for earthquake prediction using remote sensing techniques: A case study of five major earthquakes in Pakistan's history. Advances in Space Research. 2023;**71**(12):5236-5255. DOI: 10.1016/j.asr.2023.01.058

[32] Blackett M, Wooster MJ, Malamud BD. Exploring land surface temperature earthquake precursors: A focus on the Gujarat (India) earthquake of 2001. Geophysical Research Letters. 2011;**38**:L15303. DOI: 10.1029/2011GL048282

[33] Zhao W, He J, Yin G, Wen F, Wu H. Spatiotemporal variability in land surface temperature over the mountainous region affected by the 2008 Wenchuan earthquake from 2000 to 2017. Journal of Geophysical Research – Atmospheres. 2019;**124**(4):1975-1991. DOI: 10.1029/2018JD030007

[34] Jing F, Singh RP, Cui Y, Sun K. Microwave brightness temperature characteristics of three strong earthquakes in Sichuan Province, China. IEEE Journal of Selected Topics in Applied Earth Observations and Remote Sensing. 2020;**13**:513-522. DOI: 10.1109/JSTARS.2020.2968568

[35] Boudriki Semlali B-E, Molina C, Park H, Camps A. First results on the systematic search of land surface temperature anomalies as earthquakes precursors. Remote Sensing. 2023;**15**(4):1110. DOI: 10.3390/rs15041110

[36] Boudriki Semlali BE, Molina C, Park H, Camps A. Fengyun-2F/VISSR Land Surface Temperature Anomalies Between 2014 And 2022 and their Potential Correlation with Earthquakes

[37] Boudriki Semlali B-E, El Amrani C. Satellite big data ingestion for environmentally sustainable development. In: Ahmed MB, Mellouli S, Braganca L, Abdelhakim BA, Bernadetta KA, editors. in Advances in Science, Technology & Innovation*Emerging Trends in ICT for Sustainable Development*. Cham: Springer International Publishing; 2021. pp. 269-284. DOI: 10.1007/978-3-030-53440-0_29

[38] Boudriki Semlali B-E, El Amrani C, Ortiz G. SAT-ETL-integrator: An extract-transform-load software for satellite big data ingestion. Journal of Applied Remote Sensing (JARS). 2020;**14**(1):28. DOI: 10.1117/1.JRS.14.018501

[39] Boudriki Semlali B-E, El Amrani C. Towards remote sensing datasets collection and processing. In: Hameurlain A, Wagner R, Dang TK, editors. Transactions on Large-Scale Data- and Knowledge-Centered Systems XLI. Vol. 11390. Berlin, Heidelberg: Springer Berlin Heidelberg; 2019. pp. 286-294. DOI: 10.1007/978-3-030-11196-0_26

[40] Ghosh D, Deb A, Sengupta R. Anomalous radon emission as precursor of earthquake. Journal of Applied Geophysics. 2009;**69**(2):67-81. DOI: 10.1016/j.jappgeo.2009.06.001

[41] Richon P et al. Radon anomaly in the soil of taal volcano, the Philippines:

A likely precursor of the M 7.1 Mindoro earthquake (1994). Geophysical Research Letters. 2003;**30**(9):2003GL016902. DOI: 10.1029/2003GL016902

[42] Alvan HV, Azad FH, Mansor S. Latent heat flux and air temperature anomalies along an active fault zone associated with recent Iran earthquakes. Advances in Space Research. 2013;**52**(9):1678-1687. DOI: 10.1016/j.asr.2013.08.002

[43] Dey S, Singh RP. Surface latent heat flux as an earthquake precursor. Natural Hazards and Earth System Sciences. 2003;**3**(6):749-755. DOI: 10.5194/nhess-3-749-2003

[44] Mahmood I. Anomalous variations of air temperature prior to earthquakes. Geocarto International. 2021;**36**(12):1396-1408. DOI: 10.1080/10106049.2019.1648565

[45] Thomas JN, Masci F, Love JJ. On a report that the 2012 M6.0 earthquake in Italy was predicted after seeing an unusual cloud formation. Natural Hazards and Earth System Sciences. 2015;**15**(5):1061-1068. DOI: 10.5194/nhess-15-1061-2015

[46] Kumar S, Singh PK, Kumar R, Singh AK, Singh RP. Ionospheric and atmospheric perturbations due to two major earthquakes (M >7.0). Journal of Earth System Science. 2021;**130**(3):149. DOI: 10.1007/s12040-021-01650-x

[47] Marchetti D, De Santis A, Jin S, Campuzano SA, Cianchini G, Piscini A. Co-seismic magnetic field perturbations detected by swarm three-satellite constellation. Remote Sensing. 2020;**12**(7):1166. DOI: 10.3390/rs12071166

[48] Smirnov S. Earth electric field negative anomalies as earthquake precursors. E3S Web Conferences. 2020;**196**:01004. DOI: 10.1051/e3sconf/202019601004

[49] Satti MS, Ehsan M, Abbas A, Shah M, De Oliveira-Júnior JF, Naqvi NA. Atmospheric and ionospheric precursors associated with M ≥ 6.5 earthquakes from multiple satellites. Journal of Atmospheric and Solar-Terrestrial Physics. 2022;**227**:105802. DOI: 10.1016/j.jastp.2021.105802

[50] Sotomayor-Beltran C. Positive and negative ionospheric disturbances prior to the 2016 christmas earthquake in Chile. Geomatics, Natural Hazards and Risk. 2019;**10**(1):622-632. DOI: 10.1080/19475705.2018.1536081

[51] De Santis A et al. Magnetic field and electron density data analysis from swarm satellites searching for ionospheric effects by great earthquakes: 12 case studies from 2014 to 2016. Atmosphere. 2019;**10**(7):371. DOI: 10.3390/atmos10070371

[52] Sanchez SA, Kherani EA, Astafyeva E, De Paula ER. Rapid detection of Co-seismic ionospheric disturbances associated with the 2015 Illapel, the 2014 Iquique and the 2011 Sanriku-Oki earthquakes. JGR Space Physics. 2023;**128**(9):e2022JA031231. DOI: 10.1029/2022JA031231

[53] Guha Bose A, Das A, Chowdhury S, Deb A. Studies of scintillations and TEC variations with GPS satellite links together with soil radon anomalies preceding Nepal earthquakes of April–May 2015. Natural Hazards. 2022;**112**(2):1137-1163. DOI: 10.1007/s11069-022-05221-1

[54] Nayak C et al. Peculiar features of the low-latitude and midlatitude ionospheric response to the St. Patrick's day geomagnetic storm of march 17th 2015. JGR Space Physics. 2016;**121**(8):7941-7960. DOI: 10.1002/2016JA022489

[55] Hayakawa M, Hobara Y, Ohta K, Izutsu J, Nickolaenko AP, Sorokin V.

Seismogenic effects in the ELF Schumann resonance band. The Transactions of the Institute of Electrical Engineers of Japan. A. 2011;**131**(9):684-690. DOI: 10.1541/ieejfms.131.684

[56] Molina C, Boudriki Semlali B-E, González-Casado G, Park H, Camps A. The 2021 La Palma volcanic eruption and its impact on ionospheric scintillation as measured from GNSS reference stations, GNSS-R and GNSS-RO. Natural Hazards and Earth System Sciences. 2023;**23**(12):3671-3684. DOI: 10.5194/nhess-23-3671-2023

[57] Kudłacik I, Tymińska A, Lizurek G, Kapłon J, Paziewski J. High-rate GNSS data in seismic moment tensor inversion: Application to anthropogenic earthquakes. Geomatics, Natural Hazards and Risk. 2023;**14**(1):2232084. DOI: 10.1080/19475705.2023.2232084

[58] Molina C, Boudriki-Semlali BE, Park H, Camps A. A Preliminary Study on Ionospheric Scintillation Anomalies Detected Using GNSS-R Data from NASA CYGNSS Mission as Possible Earthquake Precursors. 2022. p. 22. DOI: 10.1109/IGARSS46834.2022.9883701

[59] Camps A, Park H, Foti G, Gommenginger C. Ionospheric effects in GNSS-reflectometry from space. IEEE Journal of Selected Topics in Applied Earth Observations and Remote Sensing. 2016;**9**(12):5851-5861. DOI: 10.1109/JSTARS.2016.2612542

[60] Boudriki Semlali B-E, Molina C, Park H, Camps A. On the correlation between earthquakes and prior ionospheric scintillations over the ocean: A study using GNSS-R data between 2017 and 2021. IEEE Journal of Selected Topics in Applied Earth Observations and Remote Sensing. 2024;**17**:2640-2654. DOI: 10.1109/JSTARS.2023.3346204

[61] USGS Earthquakes. Available from: https://www.usgs.gov/natural-hazards/earthquake-hazards/earthquakes [Accessed: August 8, 2021]

[62] Savastano G, Nordström K, Angling MJ. Semi-supervised classification of lower-ionospheric perturbations using GNSS radio occultation observations from spire Global's Cubesat constellation. Journal of Space Weather and Space Climate. 2022;**12**:14. DOI: 10.1051/swsc/2022009

[63] Librado MC, Molina C, Semlali BEB, Park H, Camps A. Correlation between ionosphere scintillation and earthquakes around Coral Sea in 2022. In: IGARSS 2023-2023 IEEE International Geoscience and Remote Sensing Symposium. Pasadena, CA, USA: IEEE; 2023. pp. 2346-2349. DOI: 10.1109/IGARSS52108.2023.10281834

[64] UCAR COSMIC Program. COSMIC-2 data products. UCAR/NCAR – COSMIC. 2019. DOI: 10.5065/T353-C093

[65] Molina C, Semlali BEB, González-Casado G, Park H, Camps A. Ionospheric scintillation anomalies associated with the 2021 la palma volcanic eruption detected with gnss-r and gnss-ro observations. In: IGARSS 2022 - 2022 IEEE International Geoscience and Remote Sensing Symposium, Kuala Lumpur, Malaysia. 2022. pp. 7445-7448. DOI: 10.1109/IGARSS46834.2022.9883701

[66] Iyemori T et al. Geomagnetic pulsations caused by the Sumatra earthquake on December 26th, 2004. Geophysical Research Letters. 2005;**32**(20):2005GL024083. DOI: 10.1029/2005GL024083

[67] Zhu K et al. Analysis of swarm satellite magnetic field data before the 2016

Ecuador (mw = 7.8) earthquake based on non-negative matrix factorization. Frontiers in Earth Science. 2021;**9**:621976. DOI: 10.3389/feart.2021.621976

[68] Huang Y, Zhu P, Li S. Feasibility study on earthquake prediction based on impending geomagnetic anomalies. Applied Sciences. 2023;**14**(1):263. DOI: 10.3390/app14010263

[69] Schirninger C et al. Satellite measured ionospheric magnetic field variations over natural hazards sites. Remote Sensing. 2021;**13**(12):2360. DOI: 10.3390/rs13122360

[70] De Santis A, Balasis G, Pavón-Carrasco FJ, Cianchini G, Mandea M. Potential earthquake precursory pattern from space: The 2015 Nepal event as seen by magnetic swarm satellites. Earth and Planetary Science Letters. 2017;**461**:119-126. DOI: 10.1016/j.epsl.2016.12.037

[71] Serebryakova ON et al. Electromagnetic ELF radiation from earthquake regions as observed by low-altitude satellites. Geophysical Research Letters. 1992;**19**(2):91-94. DOI: 10.1029/91GL02775

[72] Fujinawa Y, Takahashi K. Electromagnetic radiations associated with major earthquakes. Physics of the Earth and Planetary Interiors. 1998;**105**(3-4):249-259. DOI: 10.1016/S0031-9201(97)00117-9

[73] Zhao S et al. Investigation of precursors in VLF subionospheric signals related to strong earthquakes (M > 7) in Western China and possible explanations. Remote Sensing. 2020;**12**(21):3563. DOI: 10.3390/rs12213563

[74] Han C et al. Study on electron density anomalies possibly related to earthquakes based on CSES observations. Remote Sensing. 2023;**15**(13):3354. DOI: 10.3390/rs15133354

[75] Liu J et al. The Seismo-ionospheric disturbances before the June 9th 2022 Maerkang Ms6.0 earthquake swarm. Atmosphere. 2022;**13**(11):1745. DOI: 10.3390/atmos13111745

[76] Sorokin VM, Pokhotelov OA. Gyrotropic waves in the mid-latitude ionosphere. Journal of Atmospheric and Solar-Terrestrial Physics. 2005;**67**(10):921-930. DOI: 10.1016/j.jastp.2005.02.015

[77] Hayakawa M, Nickolaenko AP, Galuk YP, Kudintseva IG. Manifestations of nearby moderate earthquakes in Schumann resonance spectra. IJEAR. 2020;**7**(1):1-28. DOI: 10.33665/IJEAR.2020.v07i01.001

Section 3

Case Studies and Scenario Analysis

Chapter 8

Forecast of Catastrophic Floods Based on Hydrodynamic Modeling: The 2013 Flood on the Amur River Case Study

Alexander Buber and Mikhail Bolgov

Abstract

In 2013, a catastrophic flood occurred on the Amur River, caused by heavy rainfall throughout almost the entire basin. The floodplains of Blagoveshchensk, Khabarovsk, and Komsomolsk-on-Amur were flooded. Similar hydrological phenomena on the Amur occurred earlier (1954, 1972, and 1984), but they did not lead to catastrophic flooding. In the proposed manuscript, the authors investigated three issues: the causes of the 2013 flood, advanced hydrological forecast, and measures to prevent catastrophic flooding. For research, a hydrodynamic model of the Amur River section from Blagoveshchensk to the mouth was developed. Based on the analysis of model calculations, it was shown that some flood problems arose due to the construction of hydraulic structures in recent years (protective dams, bridges, and embankments) and a medium-term forecast scheme proposed.

Keywords: hydraulic structures, hydrological phenomena, floods, hydrodynamic model, forecast, protective coastal dams

1. Introduction

The rivers of the Amur basin, according to the nature of their water regime, belong to the type that is pronounced rain-fed. The share of rain-fed is 47–85%, snow 2–26%, and underground 9–31%.

The main phase of the water regime of the rivers of the Amur basin is rain floods observed in the warm season. The flood period lasts on average from 140 to 170 days (in the eastern and southern regions) to 110–150 days (in the northern and western regions). Floods are mainly observed from July to September. During the summer to autumn period, there are from 5 to 10 to 15 floods.

The large area of the basin and the diversity of conditions determine the existence of several flood sources of flood formation in the Amur basin. The main foci are the Upper Amur, Zeya-Bureya, Sungari, and Ussuri; see **Figure 1** [1, 2]. Each of the listed sources can be the cause of the formation of a significant flood on the Amur; however, outstanding floods are formed as a result of the coincidence of the

Figure 1.
The map of region under investigation.

activity of two or more sources. For example, the flood of 1957 (35,500 m^3/s) was formed in the Sungari (48%) and Zeya-Bureya (42%) sources. In 1958, the main source was the Zeysko-Bureya (62%), supported by the Upper Amur runoff (19%). In 1984 (32,900 m^3/s), the Sungari (37%) and Zeya-Bureya (28%) sources were of greatest importance.

The cause of the historic flood of 2013 for the entire Lower Amur (46,000 m^3/s) was a high degree of synchronicity in the development and reach of flood waves from almost the entire territory of the Amur basin. The main role in the formation of the maximum discharge near Khabarovsk was played by the Zeysko-Bureya (30%), Ussuri (29%), and Sungari (24%) sources [3, 4].

The interval of maximum runoff modules for small and medium-sized rivers was 28.6–384 l/s km^2, i.e., the difference between the minimum and maximum values is fifteenfold, which confirms the wide variety of conditions for the formation of maximums. There is a group of rivers with very high values of peak flow modules (in the range of 200–400 l/s km^2) located in the Zeya basin and the adjacent part of the Middle Amur.

This means reaching maximum values of approximately 1% probability of exceeding, observed in 1 year over a vast territory.

The main Amur high water, which led to the large-scale flood of 2013, began in July in the western part of the basin, where the main precipitation zones were located over the eastern part of the Zeya reservoir catchment, over the flat part of the Upper and Middle Amur and in the People's Republic of China over the upper reaches of the Nonni River (Sungari basin).

The flow of the Upper Amur in 2013 was not extreme, although the water content of the Shilka and Argun in July and in the first half of August increased. As a result, the maximum levels of the Upper Amur were below dangerous levels.

As it developed, the flood covered the Jewish Autonomous Region and the Khabarovsk Territory. The main Amur flood, moving downstream, accepted large amounts of water from the main southern tributaries – the Sungari (PRC), Ussuri, as well as numerous small tributaries [5]. **Figure 2** shows a comparative description of two floods – the flood of 1984, the last of the observed catastrophic ones, and 2013. It can be seen that in the section of the Middle Amur from the city of Blagoveshchensk to the village, the Ekaterino-Nikolskoye flood developed in 2013, almost coinciding with the 1984 flood. Downstream, Amur levels in 2013 were significantly higher. Small tributaries of the Amur, both from the Russian and Chinese sides, were more abundant in 2013, since precipitation zones with good previous moisture continued to cover the basin when the main flood shifted.

Below the confluence of the Sungari, the level of the Amur was already more than a meter higher than in 1984. Both Sungari and Ussuri were more water-rich in 2013. The maximum discharge of the Sungari near the village of Jiamusi was 13,300 m^3/s on August 31. Prior to this, the high-water content of the Sungari was observed only in 1998 (with a maximum flow rate of 16,200 m^3/s), and the most abundant year in the lower reaches of the Sungari, 1960, was characterized by a flow rate of 18,400 m^3/s. The share of the Sungari runoff in 2013 was about 30% of the Amur runoff near Khabarovsk during the period of highest flows, the share of the Ussuri runoff was

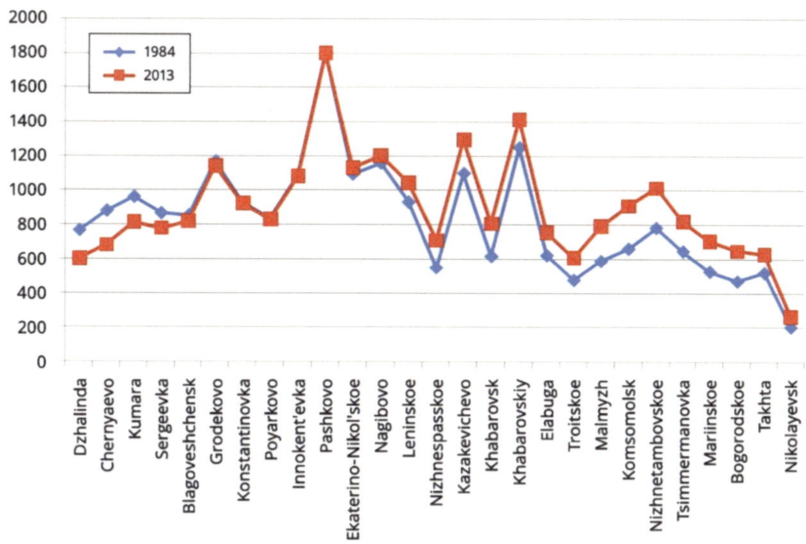

Figure 2.
Comparison of maximum observed levels in 2013 and 1984.

about 16%, and the share of the Bureya runoff was about 6%. In 1984, Sungari gave about 18%, Ussuri – about 9%, Bureya – about 5%.

On a section of more than 1000 km (from the village of Nagibovo in the Jewish Autonomous Region to the village of Takhta in the Khabarovsk Territory), the maximum levels exceeded the historical maximums by 0.40–2.11 m. Moreover, the duration of such high levels (exceeding dangerous levels) amounted to about or more than a month in large cities, and the duration of flooding of the floodplain to depths of 2–4 meters was up to two and, in some places, more than months.

Water level near the city of Nikolaevsk-on-Amur (in the mouth area of the river Amur) was above the critical level in the period from September 10 to October 15, 2013.

2. Materials and methods

To analyze the causes of catastrophic flooding during the passage of a flood wave in August to September 2013, a hydrodynamic model (HDM) of the Amur River in the middle and lower reaches was developed. The model was calibrated for high flow rates (more than 10,000 m^3/s).

The HDM was developed using the MIKE 11 computer program of the Danish Hydraulic Institute, using satellite images, digital elevation model (DEM) of the area, and Geographic Information System (GIS) of the Amur River [6–12].

The modeling area is the Middle and Lower Amur from the village Chernyaevo (454 km above Blagoveshchensk city) to the mouth of the Amur River (Sea of Okhotsk). The model considers all the main tributaries that are significant in terms of runoff volume, including the rivers Zeya, Bureya, Sungari, Big Bira, Ussuri, Tunguska, Gur, and Amgun. **Figure 3** shows the design scheme of the modeling area. The gauging station's position corresponds to its distance from the mouth. For some

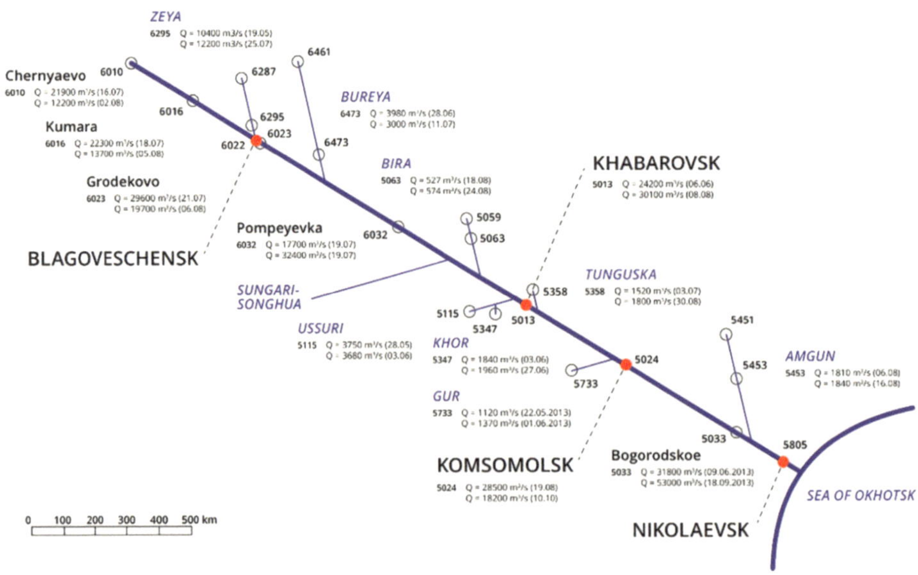

Figure 3.
Design scheme of the modeling area.

gauging stations, the maximum discharges in 1958 and the dates of measurement are shown (except for gauging station 5733 "Snezhny," for which data for 2013 was provided). The position of the largest human settlement areas is marked in red. The names of tributaries are shown in blue.

The modeling used cross-sections obtained from the results of field surveys carried out on morphological sections and from an updated digital terrain model (DEM) of the area (shuttle radar topography mission (SRTM) data) [11, 13, 14]. The digital model was clarified on the basis of cartographic materials of various scales, surveys conducted in previous years in the Blagoveshchensk and Khabarovsk cities area, survey materials, and pilot maps of 2006. Location of cross sections along the Amur River was designated in the locations of morphological sections, at existing (or previously existing) gauging stations, as well as in places requiring a detailed analysis of the hydraulic situation during high floods.

Cross sections formed on the basis of an updated DEM were corrected at the locations of morphological sections, considering the field survey results, which resulted in both systematic DEM errors and local satellite altimetry errors being corrected. **Figure 4** shows an example of a comparison of cross-sections obtained using a DEM and as a result of surveys in 2014 at morphological Section 5 (village Belogorye).

The model accepts two types of boundary conditions, specified in the form of flow (hydrograph) and level time series – external and internal boundary conditions. **Table 1** shows the location of the boundary conditions.

Historical hydrological data for the Amur River and its tributaries were used as initial data for modeling. To calibrate the model, 3 years were chosen – 1958, 1972, and 1984, which were closest in water content to the extreme year of 2013. It should be noted that the maximum flood in these years was formed on the Amur according to different scenarios. So, in 1958, the flood affected the upper and middle sections of the river, in 1972 and 1984 – middle and lower sections.

The main task was to obtain the most representative information for calibrating the hydrodynamic model, for which data on flow rates and levels for seven stations were processed: 6010 (Chernyaevo), 6016 (Kumara), 6023 (Grodekovo), 5012 (Khabarovsk), 5013 (Khabarovsk)), 5024 (Komsomolsk), and 5033 (Bogorodskoye). The most representative data seems to be g/s 5013 (Khabarovsk city station) and g/s 5024 (Komsomolsk) because they have observational data on flow rates and levels for the indicated 4 years.

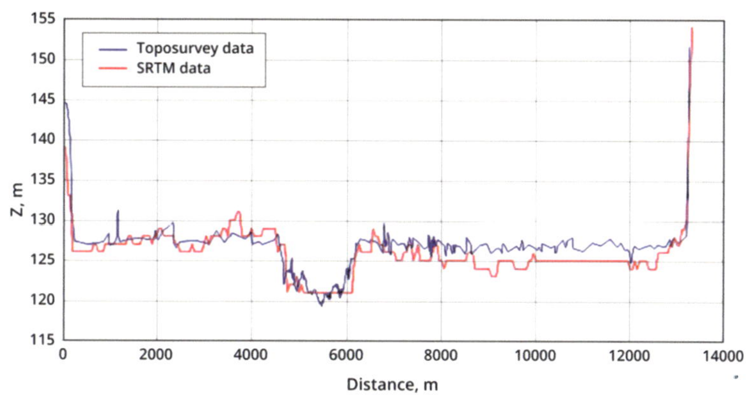

Figure 4.
Comparison of the cross section obtained from the DEM and from the survey results (morphometric point 5).

No.	Distance from source, km	Gauge number	Tributary	Settlement	Type of boundary condition
1	436	6010	Amur	Chernyaevo	External, Q
2	888	6295	Zeya	Belogorye	Inflow, Q
3	1152	6473	Bureya	Malinovka	Inflow, Q
4	2678	5453	Amgun	Guga	Inflow, Q
5	1607		Songhua	Jiamusi	Inflow, Q
6	1816	5115, 5347	Ussuri	Sheremetyevo+Khor	Inflow, Q
7	1696	5063	B. Bira	Birobidzhan	Inflow, Q
8	1874	5358	Tunguska	Arkhangelovka	Inflow, Q
9	2145	5733	Gur	Snezhnoye	Inflow, Q
10	2824		Mouth	Okhotsk Sea	External, Q

Table 1.
Boundary conditions adopted in the hydrodynamic model of the Amur River.

Of the total number of tributaries flowing into the Amur River in the study area, tributaries located on the territory of the Russian Federation and the Sungari River, located in the People's Republic of China, were taken into consideration. Gauging stations on tributaries were selected closest to the place where the tributary flows into the Amur River.

Initial data for all selected gauging stations were presented in the form of daily changes in levels relative to mark "0" (in cm) and absolute flow rates (in m^3/s) for 1958, 1972, 1984 and 2013. Relative changes in levels were converted into absolute level marks.

Based on the recalculated values, graphs of $Q = f(h)$ were constructed for the period from 01.05 to 30.09. Graphs were built for all years for each g/s. **Figure 5** shows $Q = f(h)$ curves for all sections in which the hydrodynamic model was calibrated.

Table 2 shows the observed data of the dependences $Q = f(h)$ for maximum water discharges at the calibration sections of the Amur River HDM.

Figure 6 shows all open modeling editors windows in the MIKE 11: (1) modeling window, (2) river network, (3) boundary conditions, (4) inflow in 1958 to station 6010 chernyaevo, (5) cross-section, (6) hydrodynamic parameters (roughness and sediment conditions).

3. Results and discussion

3.1 Results of hydrodynamic modeling of the 1958 flood wave

Using the developed hydrodynamic model, a calculation of the passage of the 1958 flood was performed [15].

Figure 7 shows diagrams comparing observed data and data obtained as a result of modeling for two gauging stations.

At high flow rates, the observed course of levels is significantly lower than the model one (by more than a meter). This indicates that there have been significant

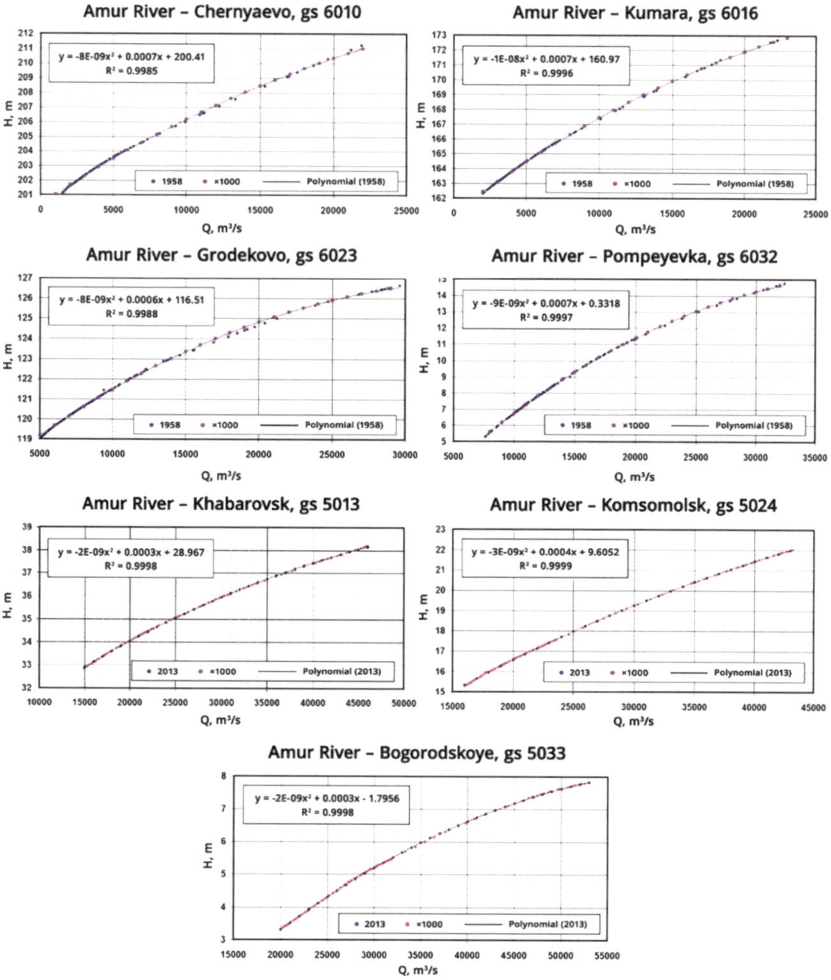

Figure 5.
Graphs of the Q = f(h) dependences at the calibration sections of the Amur River HDM.

Gauge number	Distance from the source, km	Settlement	Maximum flow cubic m/s c	Maximum level, m
6010	436	Chernyaevo	21,900	211,25
6016	656	Kumaras	22,300	172,71
6023	903	Grodekovo	29,600	126,65
6032	1376	Pompeevka	32,400	82,05
5013	1862	Khabarovsk, Gaugest.	46,000	38,23
5024	2210	Komsomolsk	43,200	22,04
5033	2586	Bogorodskoye	53,000	7,84

Table 2.
Dependency data Q = f(h) for calibrating the model for maximum flow rates.

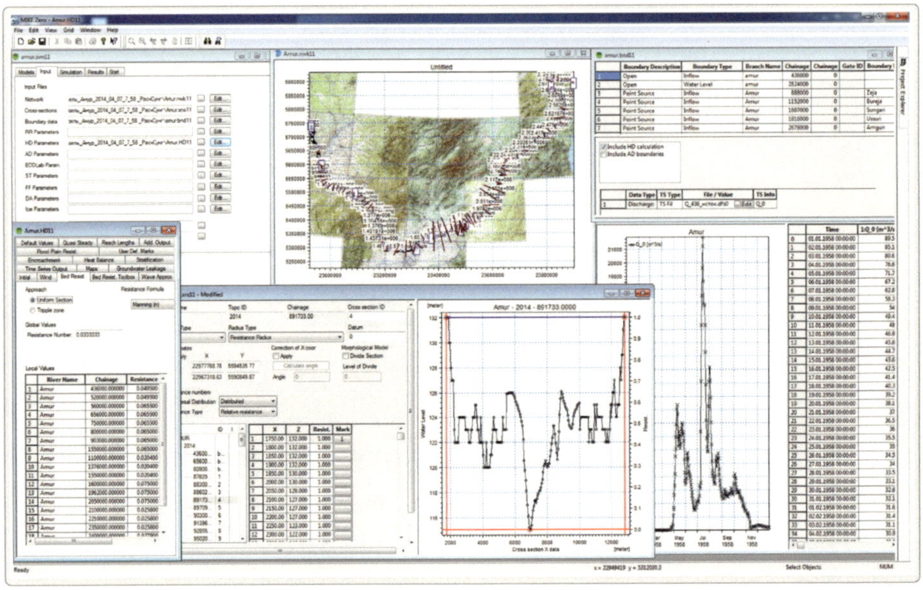

Figure 6.
Working windows of modeling editors in the MIKE 11.

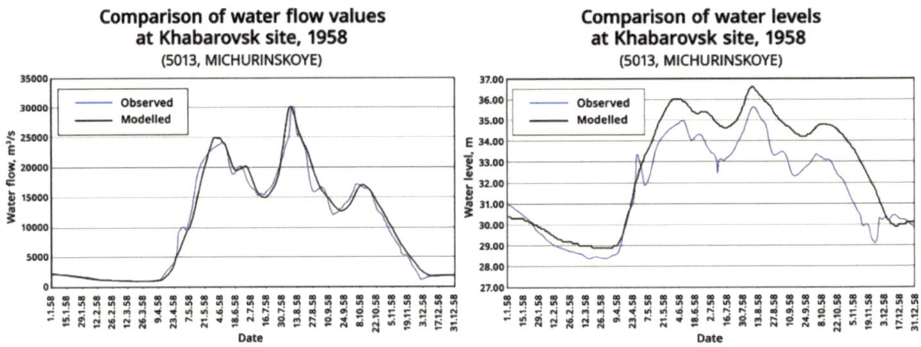

Figure 7.
Comparative diagrams of the passage of the 1958 flood.

changes in the channel and bank morphometry of the Amur River in the Khabarovsk region. The reason for this was the construction of hydraulic structures that create backwater in the Amur River bed: Left bank coastal dams built by the Chinese side, channel gravel dams on the Pemzenskaya and Beshenaya channels, bridge across the Amur River in Khabarovsk. Due to exceeding the design level of coastal dams, flooding of the right-bank floodplain occurred in three regions of China (**Figure 8**).

3.2 Results of hydrodynamic modeling of flood wave movement based on data from the 2013 flood

Using the developed hydrodynamic model, the calculation of the passage of the 2013 flood was performed [16–18]. **Figure 9** shows inflow hydrographs.

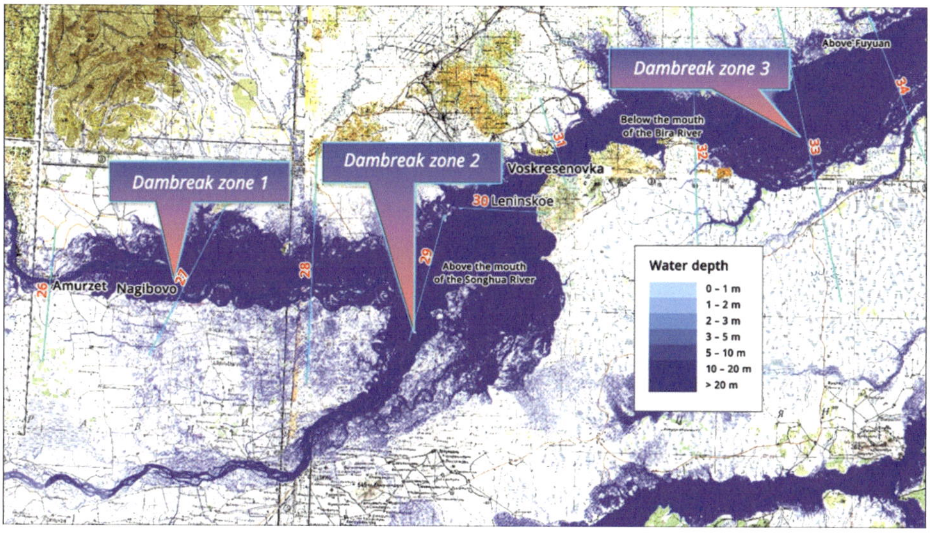

Figure 8.
Dambreak zones during the 2013 flood.

Figure 10 shows the longitudinal profile of the passage of the 2013 flood wave 9 days before the peak of the flood in Khabarovsk (maximum levels are indicated in red).

Figure 11 shows diagrams comparing observed data and data obtained as a result of modeling for two gauging stations. The discharges and levels fluctuations are given for part of the calculated time series during which the flood was observed.

Figure 12 shows the results of calculations of Q/h at control stations g/s 6023 (Grodekovo), 6032 (Pompeevka), 5013 (Khabarovsk, city station), 5024 (Komsomolsk-on-Amur), and 5033 (Bogorodskoye) when passing flood wave 2013 along the Amur River.

Figures 13 and **14** show the level fluctuations in the design sections of two cross sections of Khabarovsk city (before the railway bridge – g/s 5012, 4 km below the railway bridge – g/s 5013). Blue shows the water level line at the current moment, red shows the water line at the maximum level and green shows the water line at the minimum level in 2013. The height difference (difference in maximum values) was 38,9-38,1 = 0,8 m. In 1958, the height difference was 37,2–36,6 = 0,6 m.

3.3 Determination of the main parameters of the passage of high flood waters (time of arrival of the wavefront, wave crest, and time of wave decline to the flood level of 5% probability) in years of high-water content (1958, 1984, 2013)

In accordance with the methodological instructions of the Ministry of Emergency Situations, during the passage of high flood or breakthrough waves, it is necessary to determine the main parameters of wave propagation: the time of arrival of the wavefront, the wave crest, and the time of decline of the wave to a safe flood level of 5% probability. This allows timely implementation of measures to evacuate the population and livestock farms.

Table 3 and **Figure 15** show the model parameters for the passage of the 2013 flood wave.

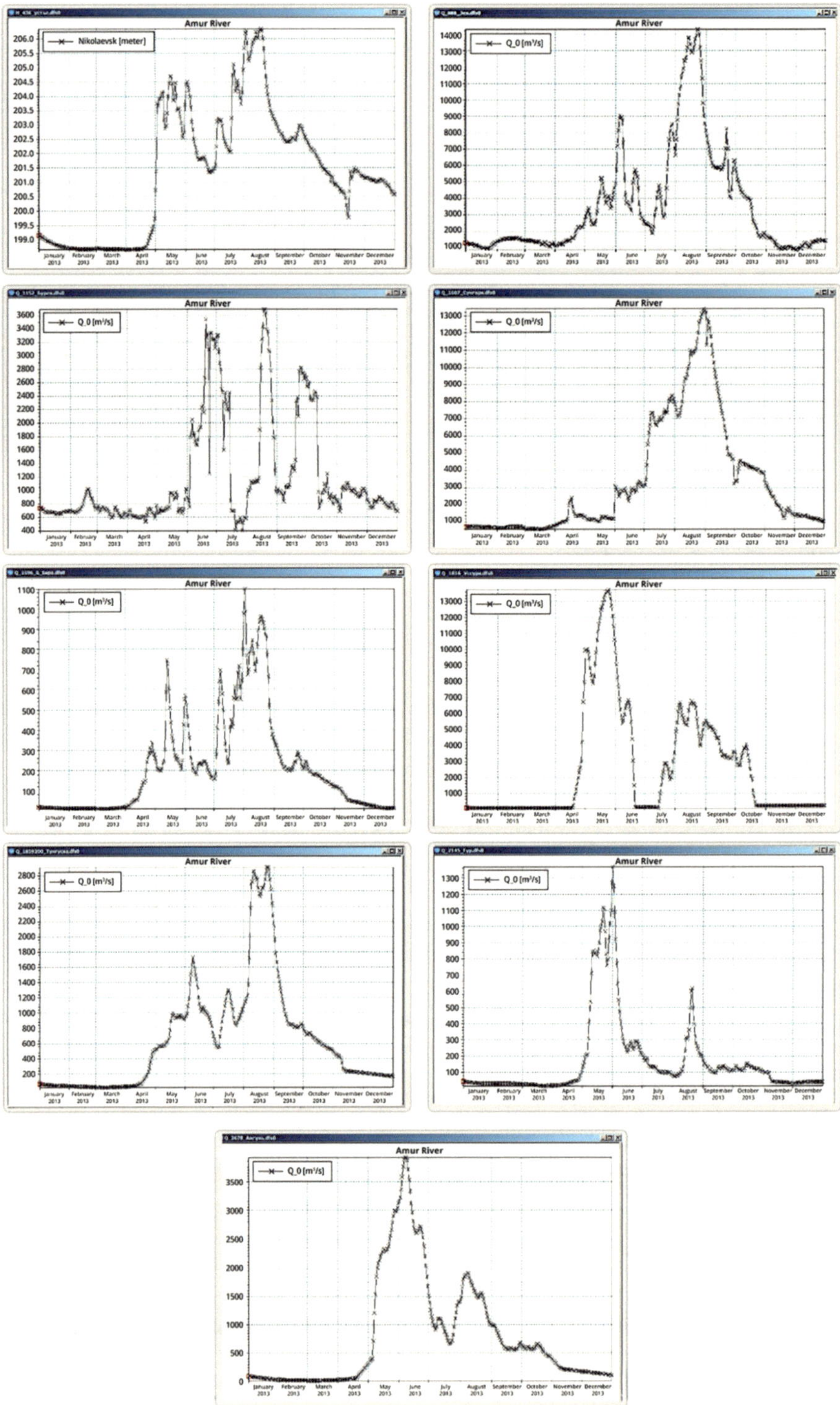

Figure 9.
Hydrographs of inflow of the 2013 flood.

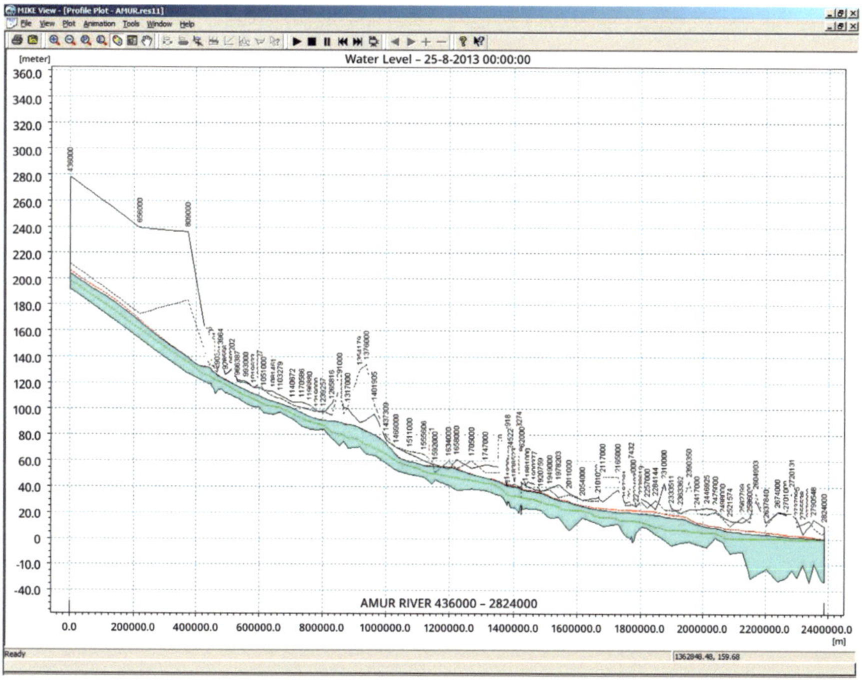

Figure 10.
Longitudinal profile of the passage of the 2013 flood wave.

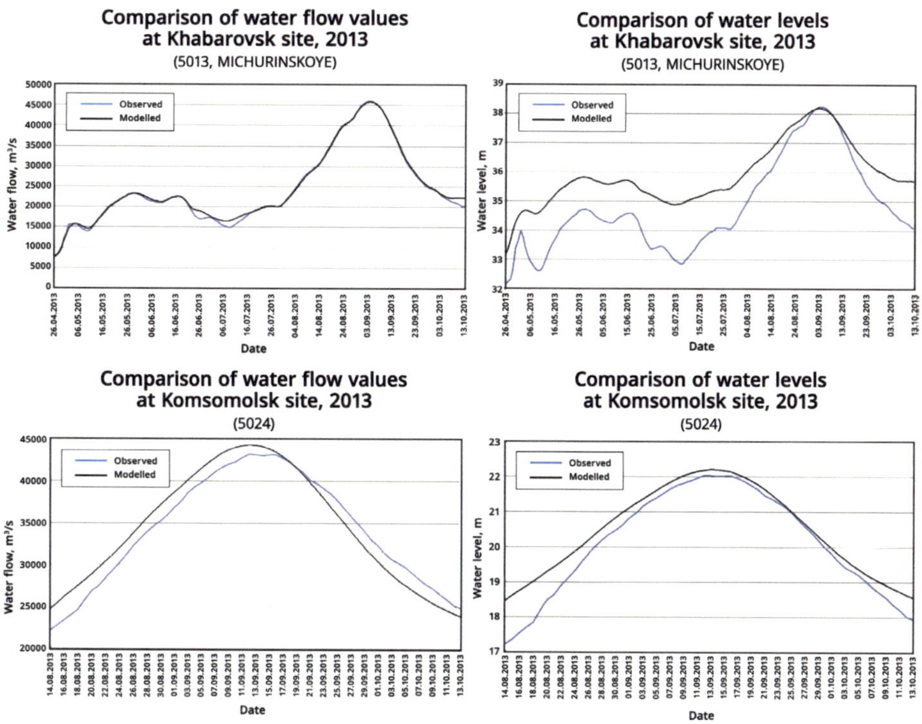

Figure 11.
Comparative diagrams of the passage of the 2013 flood.

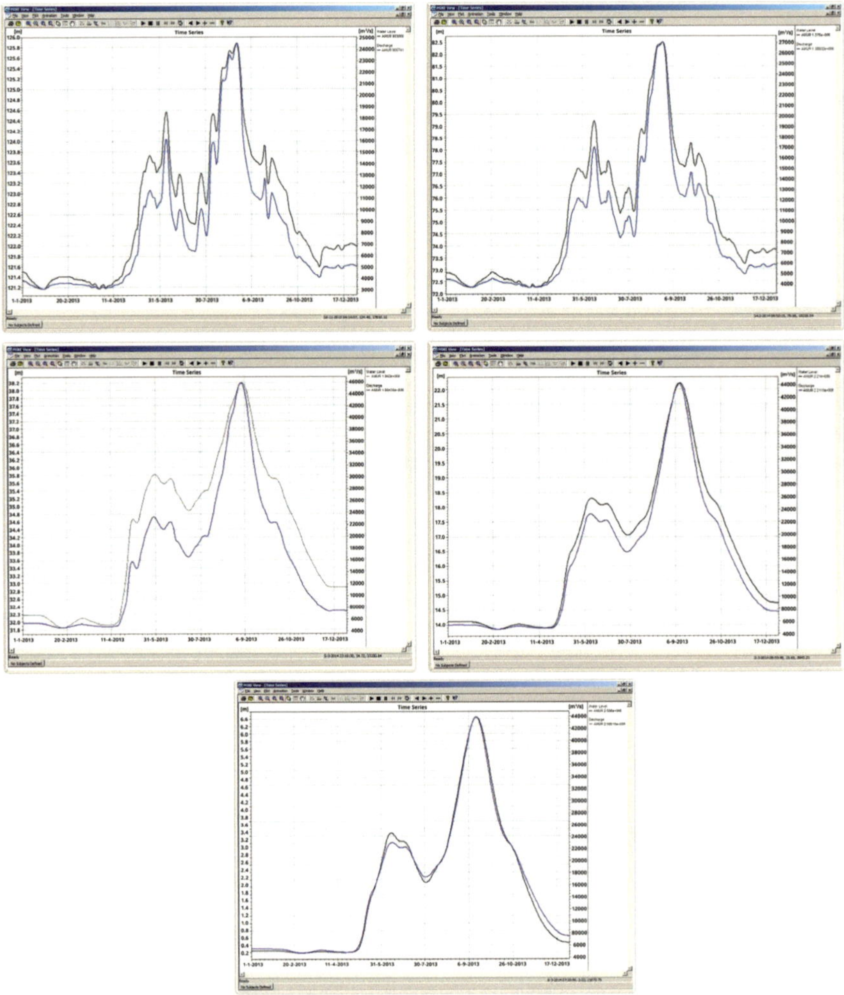

Figure 12.
Results of calculations (flow rates and water levels) based on the hydrodynamic data of the 2013 flood.

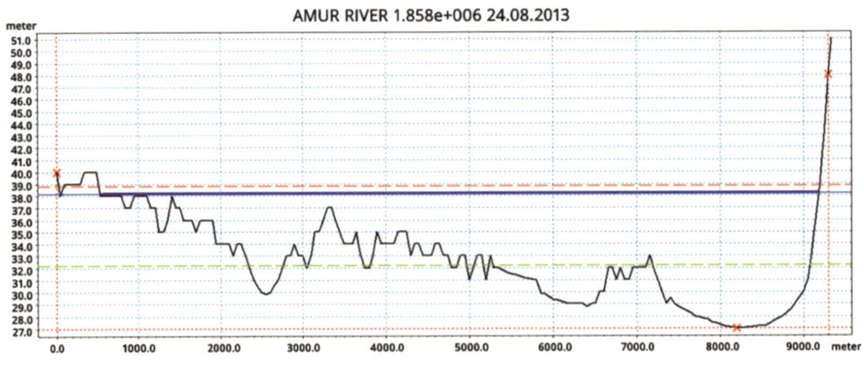

Figure 13.
Range of water levels during the flood period of 2013 (g/s 5012).

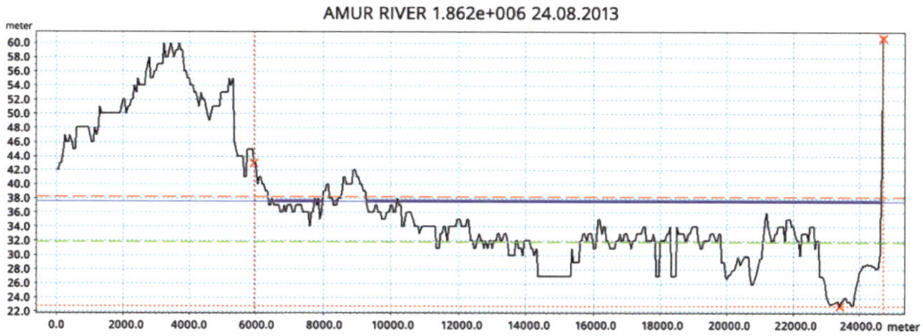

Figure 14.
Range of water levels during the flood period of 2013 (g/s 5013).

No.	Distance from the source, m	Parameters			Wave travel time, days		
		Depth, m	initial level, m	Max. elevation flooding, m	Arrival of front	reaching maximum level	Decline to flood level 25% probability
1	436,000	7.3	199.1	206.4	20	26	28
2	656,000	6.6	161.5	168.1	22	27	30
3	809,000	3.3	136.2	139.5	23	29	32
4	883,000	5.6	124.7	130.4	24	31	33
5	903,000	4.4	121.5	125.9	15	31	36
6	993,000	4.0	110.1	114.1	16	33	38
7	1,051,000	4.4	103.6	108.1	17	33	39
8	1,183,730	4.1	91.2	95.4	18	34	41
9	1,291,000	8.4	81.0	89.4	19	35	42
10	1,376,000	9.7	72.9	82.5	20	35	43
11	1,511,000	3.0	56.2	59.2	21	36	44
12	1,634,000	7.0	48.4	55.4	13	40	63
13	1,858,000	6.2	32.6	38.8	8	44	88
14	1,862,000	6.0	32.2	38.2	0	44	89
15	1,949,000	6.2	26.8	33.0	4	45	92
16	2,054,000	4.2	21.7	25.9	10	49	95
17	2,117,000	6.6	17.1	23.7	11	51	97
18	2,210,000	8.1	14.1	22.2	12	52	99
19	2,310,000	11.3	8.2	19.5	14	52	101
20	2,417,000	7.0	5.7	12.7	14	53	101
21	2,498,000	6.5	2.4	8.8	15	54	102
22	2,586,000	6.4	0.3	6.7	17	56	104
23	2,701,000	3.2	0.1	3.3	9	57	105
24	2,776,000	1.8	0.0	1.9	9	56	106

Table 3.
Parameters of the passage of a high flood wave in 2013.

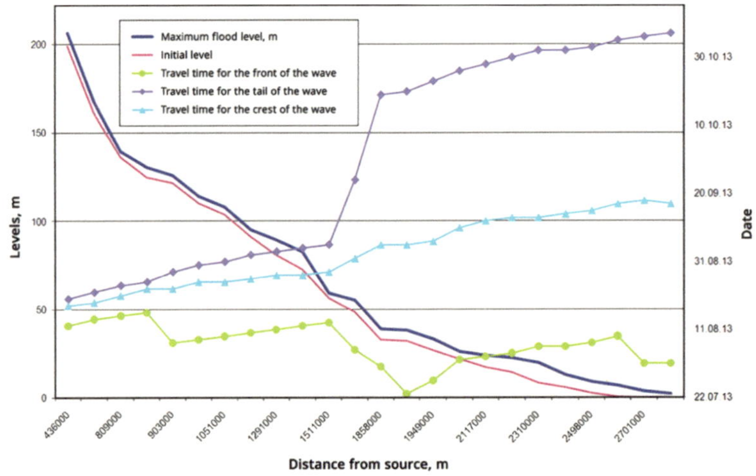

Figure 15.
Parameters of the passage of a high flood wave in 2013.

3.4 Flood zones during the passage of a flood wave during years of high water content (1958, 1984, and 2013)

Based on the results of computer hydraulic modeling, flood maps of the Amur River valley were created for various forecast situations. These maps were constructed on the basis of a differential model obtained as the difference between the water level and relief marks. Positive values in such a model correspond to flood depths, and negative values correspond to the heights of the area above the water level. Negative values are discarded when plotted on a map.

Figure 16 shows a map of the flood depths of territories constructed based on the results of modeling the 2013 flood.

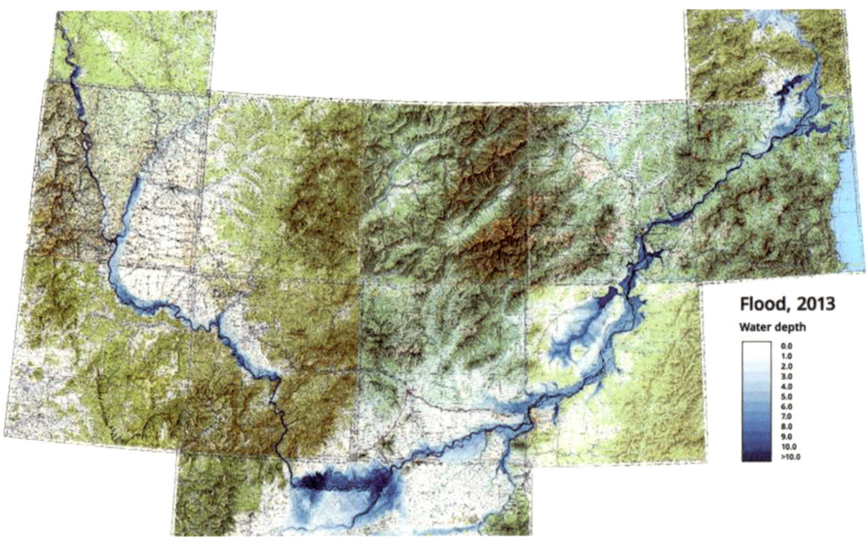

Figure 16.
Map of flood depths, built according to calculated data for 2013.

For the regions of the Khabarovsk and Komsomolsk cities on the Amur, detailed maps of flooding of territories were constructed, obtained from the results of modeling the floods of 1958, 1984, and 2013 (**Figures 17** and **18**).

For selected areas (each with an area of 5932 km^2) in the area of Khabarovsk and Komsomolsk on the Amur, the flooded areas of the territory were analyzed. The areas were calculated for the following flood depths: less than 1 m, from 1 m to 3 m, and more than 3 m. The area occupied by the Amur riverbed and channels during the low-water period for the site in the region of Khabarovsk amounted to 321.5 km^2, for the site in the region of Komsomolsk-on-Amur – 350 km^2. Data on the obtained estimated flood areas are given in **Tables 4** and **5**.

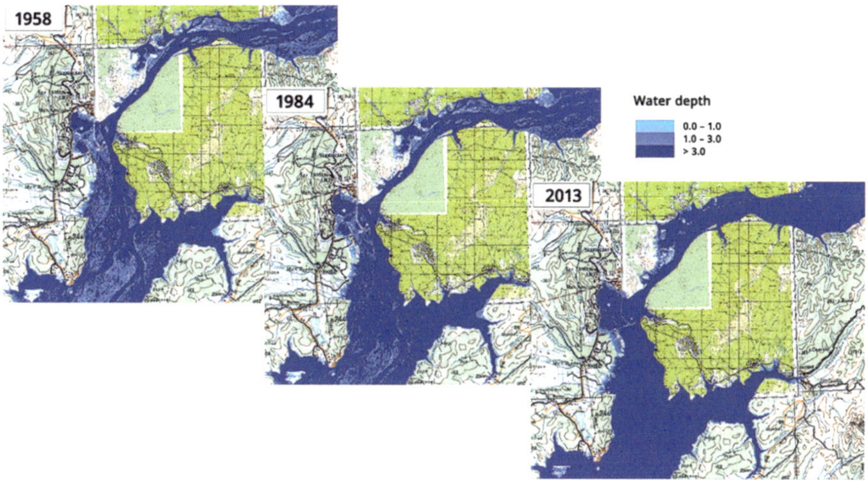

Figure 17.
Maps of flood depths in the Khabarovsk region.

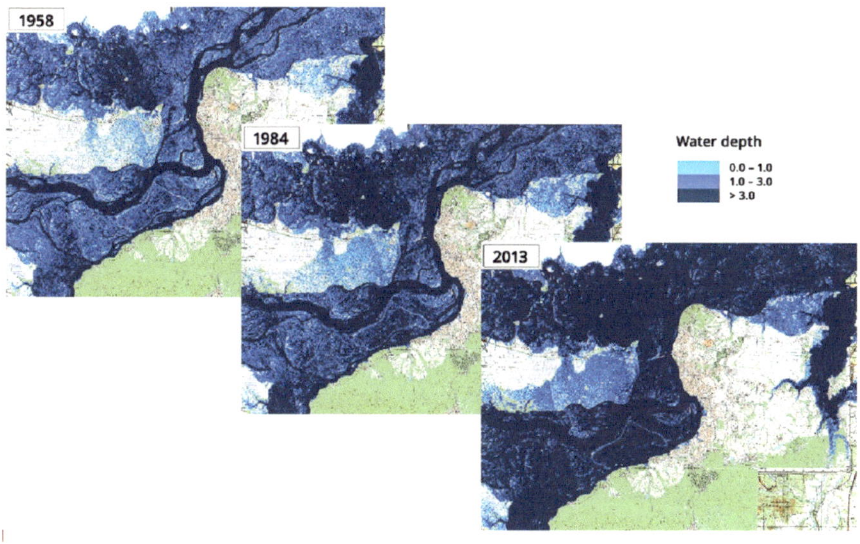

Figure 18.
Maps of flood depths in the area of Komsomolsk-on-Amur.

Flood depths	1958 г.	1984 г.	2013 г.
Up to 1 m	232,5	202,5	137,1
From 1 to 3 m	799,9	599,6	402,5
More than 3 m	1021,3	1437,7	1911,9

Table 4.
Estimation of flood areas in the Khabarovsk city area, km².

Flood depths	1958 г.	1984 г.	2013 г.
Up to 1 m	50,5	37,2	27,6
From 1 to 3 m	244	153,8	61,5
More than 3 m	524,6	661,2	843,4

Table 5.
Estimation of flood areas in the area of Komsomolsk-on-Amur, km².

3.5 Medium-term hydrological forecast based on hydrodynamic modeling

To demonstrate the predictive capabilities of the Amur River hydrodynamic model, model calculations of the 2013 flood were performed at various time intervals. In this case, the inflow forecast was carried out according to the principle: in the following days, the inflow will be the same as on the calculated date (settled hydrological regime).

Figures 19–23 show three curves: observed water flows, obtained on the basis of modeling based on observed data (for the entire period) and made using predicted inflow data (see above). Calculations were performed for the following forecast dates: August 1 (33 days before the peak of the flood), August 8 (26 days before the peak), August 16 (18 days before the peak), August 22 (12 days before the peak), and August 25 (9 days before peak).

Figures 19–21 show how much the forecast curve (green diagram) differs from the observed curve (blue diagram). However, on August 22 (**Figure 22**), the predicted discharge already corresponds to the maximum observed discharge, but the maximum flow occurs at a later date. The final forecast diagram for August 25 (**Figure 22**) almost completely repeats the graph constructed using the observed data (the same result was obtained for water levels). The latter shows that the use of a hydrodynamic model of the Amur River would make it possible to do a fairly accurate forecast of the critical date and values of maximum flow rates and water levels 9 days before implementation (the peak of the catastrophic flood occurred on September 3, 4 and grew very quickly). The Khabarovsk Department of the Hydrometeorological Service provided a real forecast of the flood wave passage only 2 days before the peak, which did not allow the implementation of possible flood protection measures.

Thus, the timely use of the Amur River hydrodynamic modeling survey could reduce the risks of destruction during the catastrophic flood of 2013.

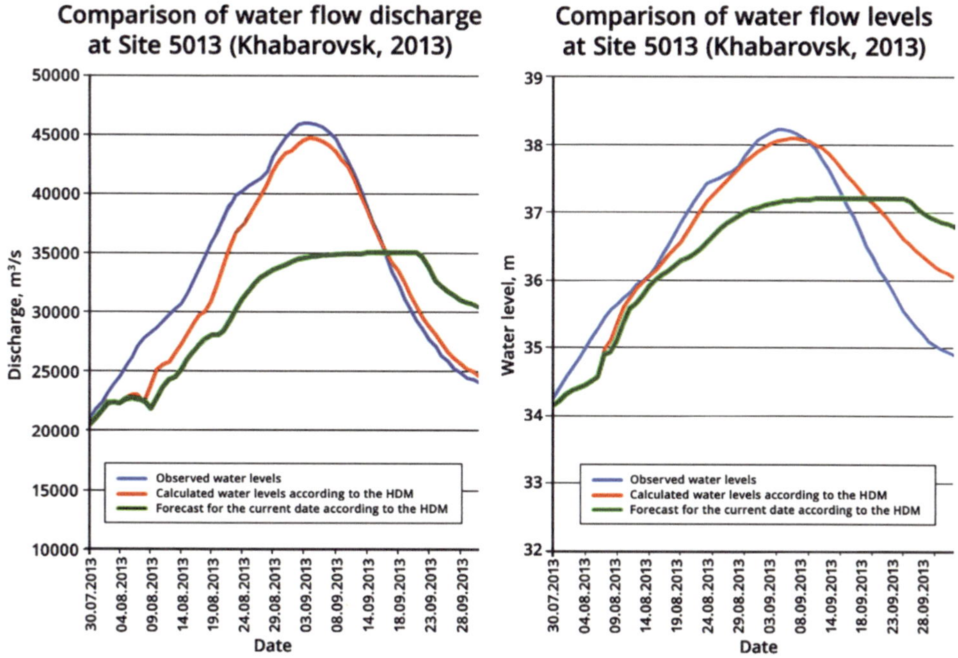

Figure 19.
Inflow forecast on August 1, 2013.

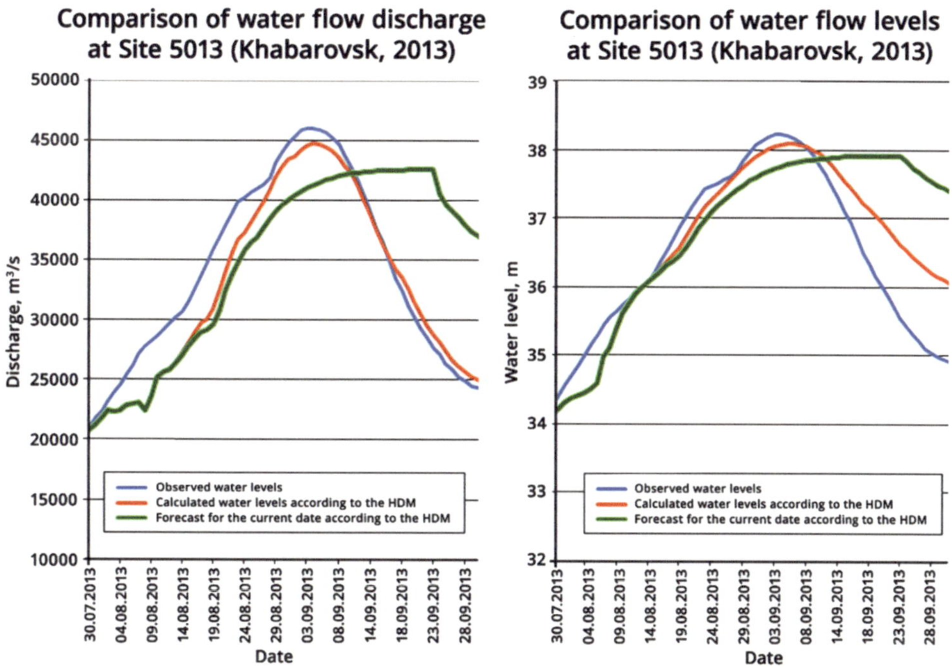

Figure 20.
Inflow forecast on August 10, 2013.

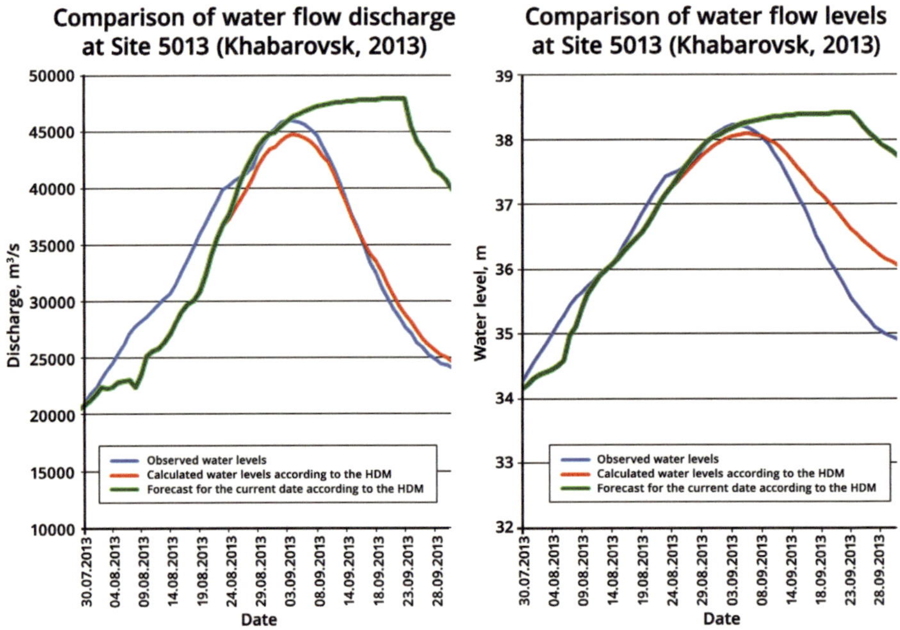

Figure 21.
Inflow forecast on August 16, 2013.

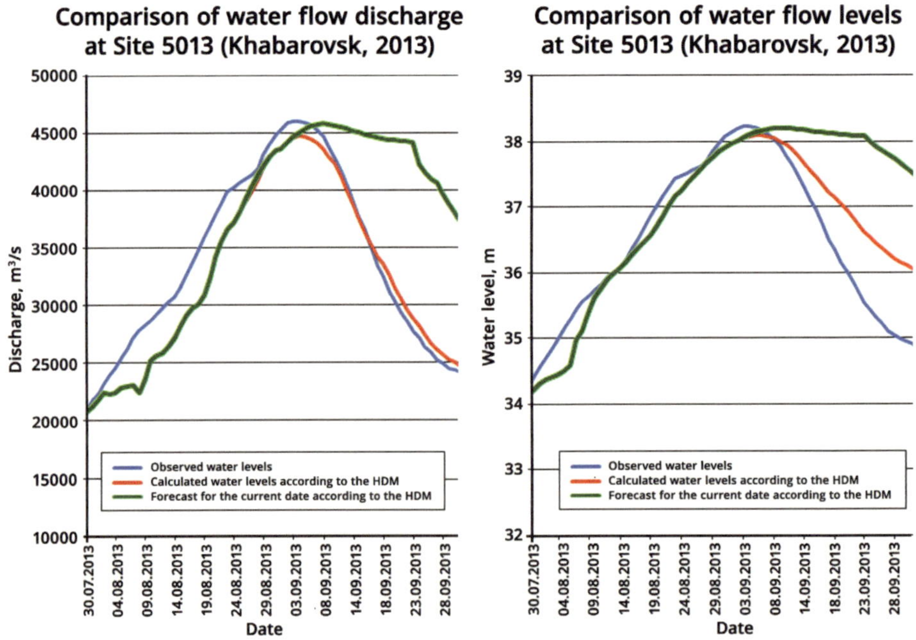

Figure 22.
Inflow forecast on August 22, 2013.

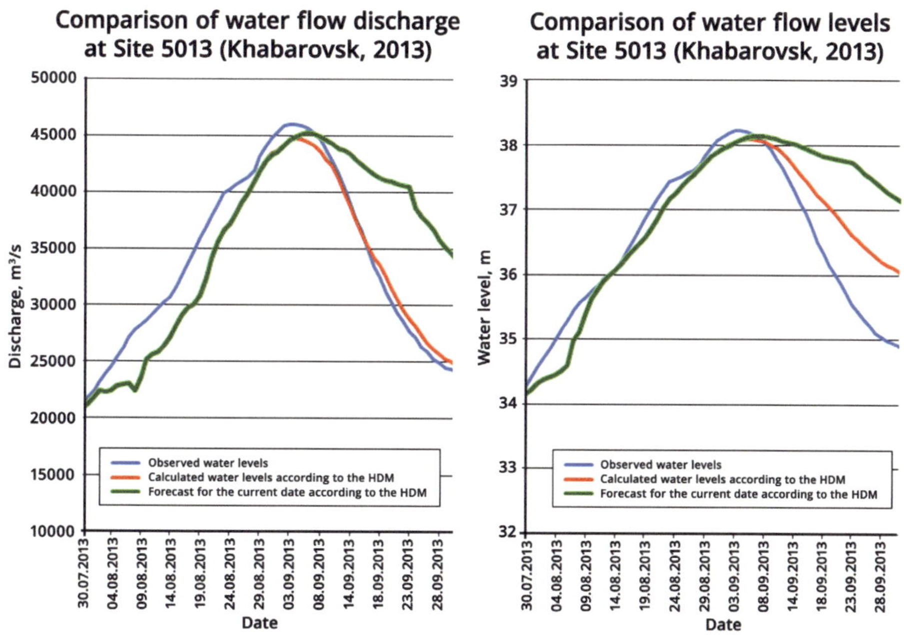

Figure 23.
Inflow forecast on August 25, 2013.

4. Conclusions

1. The hydrodynamic simulation model developed in the MIKE 11 software package adequately reflects the passage of high flood waters of 1958, 1984, and 2013 along the riverbed and floodplain of the Middle and Lower Amur, considering the flow of water through the main tributaries.

2. The results of the performed model calculations correspond to the observed data for water flows comparable to the maximum observed during the period of high floods in 1958, 1984, and 2013.

3. Based on the developed hydrodynamic model, short-term and long-term hydrological forecasts of rain floods in the Amur River basin, calculations can be made of a possible catastrophic hydraulic situation in the main settlements located on the Amur River.

4. The hydrodynamic model of the Middle and Lower Amur guarantees the issuance of the necessary final reports with the calculation of the main hydraulic characteristics of the river Amur bed and floodplain (water levels and flow rates, speeds, time of arrival of the wavefront, wave crest, and time of wave decline to a safe level, etc.). The model allows you to monitor a dangerous flood situation and perform a real-time short-term forecast of the occurrence of such a situation.

5. Using the developed digital terrain model and GIS project tools, possible flood zones can be quickly displayed on maps of various scales, and flood areas can be calculated for different depth ranges.

6. For the integral functioning of the model in a real-time mode, it is necessary to obtain flow characteristics, both for the main tributaries of the Amur River, formed on the territory of the Russian Federation (R. Zeya, R. Bureya, R. Ussuri, R. Amgun), and along the tributary of the River Songhua of the People's Republic of China. It is necessary to restore the operation of gauging stations by measuring water flows, at least during periods of high floods.

7. The modeling results show that the current state of the river bed and floodplain of the Amur reduced its capacity. If the flood of 1958 were repeated under modern conditions, the water level in Khabarovsk would be 1–1, 2 m higher than observed in 1958.

Acknowledgements

This study was carried out under Governmental Order to Water Problems Institute, Russian Academy of Sciences, subject no 122041100222-7, FMWZ-2022-0001 and All-Russian Research Institute of Hydraulic Engineering and Land Reclamation subject no 68.31.00, FGUF-2022-0004.

Author details

Alexander Buber[1*] and Mikhail Bolgov[2]

1 Federal Scientific Center for Hydraulic Engineering and Land Reclamation named after A.N. Kostyakov, Moscow, Russia

2 Water Problems Institute, Russian Academy of Sciences, Moscow, Russia

*Address all correspondence to: buber49@yandex.ru

References

[1] Boikova KG. Floods on the Amur Basin Rivers, Voprosy Geografii Dal'nego Vostoka, Issue 5. Khabarovsk: Khab. Kn. Izd-vo; 1963. pp. 192-259. [in Russian]

[2] Kim VI. The conditions for flood generation within the Amur River Basin. In: Voronov BA, Makhinov AN, editors. Research into Water and Ecological Problems of the Amur Region. Khabarovsk: Dal'nauka; 1999. pp. 66-69. [in Russian]

[3] Dugina IO, Yavkina EN, Ageeva SA, Bol'sheshapova OV, Dunaeva IM, Efremova NF, et al. An Outstanding Flood on the Amur River in 2003 and its Characteristics, Abstract Book of the Plenary Meeting of the Seventh All-Russian Hydrological Congress; November 19-21, 2013, St. Petersburg. Vol. 2013. St. Petersburg: Rosgidromet; 2013. pp. 22-25 [in Russian]

[4] Frolov AV, Georgievskii VY. The extreme flood event of 2013 within the Amur River basin. In: Proc. Joint Meeting on "Extreme Floods within the Amur River Basin: Causes, Forecasts, Recommendations"; January 20, 2014, Moscow. Moscow: Izd-vo NITs "Planeta"; 2014. pp. 5-39 [in Russian]

[5] Bolgov MV, Alekseevskii NI, Gartsman BI, Georgievskii VY, Dugina IO, Makhinov AN, et al. The 2013 extreme flood within the Amur basin: Analysis of flood formation, assessments and recommendations. Geography and Natural Resources. 2015;**36**(3):225-233. DOI: 10.1134/S1875372815030026

[6] MIKE. A modelling system for rivers and channels, reference manual. 2017. Available from: https://manuals.mikepoweredbydhi.help/2017/Water_Resources/MIKE_11_ref.pdf [Accessed: Dec. 18, 2023]

[7] MIKE. A modelling system for rivers and channels, short introduction – Tutorial. 2017. Available from: https://manuals.mikepoweredbydhi.help/2017/Water_Resources/MIKE_11_Short_Introduction-Tutorial.pdf [Accessed: Dec. 18, 2023]

[8] MIKE 11. A modelling system for rivers and channels, user guide. Available from: https://manuals.mikepoweredbydhi.help/2017/Water_Resources/MIKE11_UserManual.pdf [Accessed: Dec. 18, 2023]

[9] MIKE ZERO. The common DHI user interface for project oriented water modelling, user guide. Available from: https://manuals.mikepoweredbydhi.help/latest/MIKE_Zero_General.htm [Accessed: Dec. 18, 2023]

[10] MIKE VIEW. A Results Presentation Tool for MOUSE, MIKE URBAN, and MIKE 11, User Guide. Available from: https://archive.org/details/manualzilla-id-5838695 [Accessed: Dec. 18, 2023]

[11] SRTM. Version 4, available from the CGIAR-CSI SRTM 90m) – Database. 2008. Available from: http://srtm.csi.cgiar.org [Accessed: Dec. 18, 2023]

[12] ASTER. Global digital elevation model (ASTER GDEM) – Database. 2011. Available from: http://gdem.ersdac.jspacesystems.or.jp [Accessed: Dec. 18, 2023]

[13] Botavin DV, Zavadsky AS. Experience in Constructing a Digital Relief Model of the Amur-Zeya Water Node Using Earth Remote Sensing Data. Stability and Dynamics of Erosion-Channel Systems. Moscow: Moscow State University Publishing House; 2012. pp. 94-100

[14] Geoservices and geoportals for distributing remote sensing materials: Google Earth, Worldview alpha. Available from: https://earthdata.nasa.gov/labs/worldview/. http://kosmosnimki.ru/, http://glovis.usgs.gov/ and others. [Accessed: Dec. 18, 2023]

[15] Boykova KG. Floods on the rivers of the Amur basin. In: Questions of Geography of the Far East. Vol. 5. Khabarovsk: Pedagogical University; 1963. pp. 192-236

[16] Danilov-Danilyan VI, Gelfan AN, Motovilov YG, Kalugin AS. Catastrophic flood of 2013 in the Amur River basin: Formation conditions, recurrence assessment, modeling results. Water Resources. 2014;**41**(2):111-122

[17] Gartsman BI, Mezentseva LI, Menovshchikova TS, Popova NY, Sokolov OV. Conditions for the formation of extremely high water content in Primorye rivers in the autumn-winter period of 2012. Meteorology and Hydrology. 2014;**4**:77-92

[18] Makhinov A, Kim V, Shuguang L. Specifiks of the exstreme flood on the Amur river in 2013. In: Resourses, Environment and Regional Sustainable Development in Northeast Asia (Papers and Abstracts). Vol. 2014. Changchun, China: Northeast Institute of Geography and Agroecology, CAS; 2014. pp. 313-318

Chapter 9

Perspective Chapter: The Coastal Migration of the Locations of Tropical Cyclone Rapid Intensification over the North Indian Ocean

Kasturi Singh

Abstract

The migration of maximum intensity poleward is triggering a shift in the rapid intensification (RI) locations of tropical cyclones (TC) towards the coast of ocean-rim countries. The study investigates changes in the distribution of locations of RI during the pre-monsoon and post-monsoon seasons in recent warming climate scenarios over the North Indian Ocean (NIO) basin. Over the Bay of Bengal (BOB), the percentage of annual RI TC frequency exhibits a stable or slightly decreasing trend (20–100%), contrasting with a notable surge (50–100%) over the Arabian Sea (AS) in recent years. The distribution of RI TC location gradient is meridional during the pre-monsoon season and is confined zonally below 15°N during the post-monsoon season over BOB. The corresponding locations over AS are confined between 10°N–15°N and 12°N–17°N latitudinal regions. An inverse relation between the simultaneous rise in SST and RH550 is evident during the pre-monsoon season, while the relation fails during the post-monsoon season over BOB. While sea surface temperature and mid-tropospheric relative humidity play a crucial role in RI, the observed changes in tropospheric vertical wind shear patterns and upper-level divergence alignment in current climate conditions are identified as influential factors shaping the distribution of RI location over BOB and AS.

Keywords: tropical cyclones, rapid intensification, North Indian Ocean, Bay of Bengal, Arabian Sea, warming climate

1. Introduction

The North Indian Ocean (NIO), encompassing the Bay of Bengal (BOB) and the Arabian Sea (AS), associated with hugely populated coastal regions are no stranger to the formidable forces of tropical cyclones (TCs). Unlike other ocean basins, NIO TCs exhibits a bimodal nature in TC genesis, i.e., the primary peak is observed during post-monsoon (October, November, and December) and the secondary peak is seen during pre-monsoon season (March, April, May). These natural phenomena have tremendous potential to wreak havoc on coastal regions. One intriguing aspect of these cyclones is their ability to undergo rapid intensification (RI), a phenomenon characterized by a swift increase in wind speeds $\leq$30knots in 24 hours [1].

Song and Klotzbach [2] reported a significant poleward shift in the annual mean maximum intensity latitude post-1980. Nevertheless, the migration rate is nonuniform on decadal timescales, having a significant increasing trend over the West-North Pacific basin. This observed change is primarily linked to the sea surface temperature (SST) variation over the basin. Studies such as [3–5] indicated that the twenty-first century TCs will most probably form within a broader range of latitudes than those of the past 3 million years as there will be an increasing mid-latitude TC genesis favorability and the average migration rate of TC away from tropics at present is nearly a degree of latitude per decade. The variation in regional Hadley circulation is also found to be impacting the displacement of TC genesis and intensification [6]. Despite several studies focusing on the poleward migration of TC genesis and intensification characteristics, the regional shift in the intensification feature of TC still remains a puzzle, especially the zonal migration of RI locations of TCs.

Studies focusing on the global distribution of TCs, and their annual frequency and intensity have already been carried out by many researchers (e.g., [7–10]). However, studies addressing the propagation of rapid intensification locations are negligible for the NIO region. In the current climate scenario, the weakening of the summer monsoon circulation has resulted in less southward ocean heat transport [11]. The decrease in heat transport has led to the accumulation of heat in the NIO region with an increase in the high-intensity TCs and the consequent RI of these systems. As the majority of the Southeast Asian people live around the NIO coastal region, any long-term change in TC climatology would greatly affect these lives and the economy [12].

The prediction of the maximum intensity of TCs is still a major challenge for most of the operational agencies [13, 14], and the prediction gets worse if the TC suffers RI in its lifetime. As the NIO basin is a comparatively small basin and has a high coastal population, the threat increases by many folds when the TC encounters RI near the coast. Climatologically, the annual TC frequency is decreasing, and the number of highly intensified systems is increasing over the basin. Furthermore, the RI TC frequency is also found to be rising in recent years [15].

Surprisingly, the intense TC activity over AS has drawn the attention of the research community in changing climate scenarios. Though the annual frequency over BOB is higher than AS in post-monsoon season, the rapidly intensifying TC frequency over AS is comparable to that of BOB, confirming that AS is more vulnerable to rapidly intensifying systems in present climate conditions. Numerous studies have shown that the annual TC frequency over AS has been rising in recent years and suggested many possible reasons for the rise such as increased SST, decreased wind shear, and anthropogenic pollutants over the basin [9, 16–18]. The matter of concern is that most of these systems are attaining high-intensity levels in current climate conditions and any shift in their RI location, whether towards the Arabian Peninsula or Indian coast would result in higher socio-economic loss to the countries.

Similarly, the BOB coastal areas experience a significant number of highly intensified landfalling systems every year. India is among the top ten countries in the world in terms of absolute losses from disasters between 1998 and 2017, totalling an estimated 79.5 billion dollars [19, 20]. One unfortunate instance of a landfall concerning rapidly intensifying TCs in the BOB is the 1999 Odisha Super Cyclone. In this case, the system escalated from a severe cyclonic storm (SCS) with a maximum sustained wind (MSW) of at least 48 knots on October 27th to reach an intensity of MSW $\geq$ 120knots by October 28th, by experiencing RI, i.e., from a tropical storm to category 5 storm in Saffir–Simpson hurricane wind scale (SSHWS) in just 24 hrs. The event was classified as a Super Cyclonic Storm (SuCS) according to the cyclone classification by maximum sustained wind speed and pressure deficit adopted by the India Meteorological Department [20, 21]. The system underwent RI when the system was closer to the coast and due to a certain degree of uncertainty in initial track and intensity, the larger size of the storm, and the lack of public awareness back then, the storm caused a huge death toll, enormous destruction and loss of properties. Nonetheless, during TC Phailin (2013) and Hudhud (2014), India was highlighted internationally due to its continuous disaster management improvements [22] and its achievements in saving lives.

The propagation of any of the maximum intensity location or RI location towards the coastal regions is closely linked to the potential destruction by their associated heavy rainfall [23], storm surges, strong wind, and flash floods resulting in inundation of low-lying coastal areas. The coastal propagation of the zone of RI attainment of NIO TCs is a complex interplay of atmospheric conditions, coastal topography, and the dynamics of the storm itself. The NIO basin has encountered an elevated SST, especially during the pre-monsoon and post-monsoon seasons in present climate conditions. Beyond surface temperatures, the NIO exhibits a high ocean heat content, allowing cyclones to draw energy from a deep ocean reservoir. The availability of this extensive heat content contributes significantly to sustaining RI [24]. Vertical wind shear (VWS), the change in wind speed and direction with altitude, plays a pivotal role in cyclone development. Wind shear is one of the important factors as it allows the storm to maintain a well-organized structure [25]. A moist atmosphere is essential for the latent heat release crucial to the intensification process.

Since NIO is an active breeding ground for highly intensified storms, understanding the above-said climatic factors becomes paramount. The RI of North Indian TCs serves as a reminder of nature's immense power and the need for continued efforts in understanding its long-term behavioral changes and adapting to these dynamic forces. Recognizing the implications of this propagation is essential for coastal communities, policymakers, and disaster management authorities to develop resilient infrastructure, implement effective evacuation plans, and mitigate the potential impact of these natural disasters. As climate change continues to influence cyclonic patterns, ongoing research and advancements in the climatology of TCs are essential for improving our ability to predict and respond to these dynamic coastal threats. The current chapter investigates the coastal migration pattern of potential zones for the RI of TCs formed over NIO and subsequent environmental factors influencing the concerned movement of TC locations over the BOB and AS basin by considering the pre-monsoon and post-monsoon seasons separately.

2. Data and methodology

The present study is carried out over the NIO region, i.e., 50°–100°E and 5°–30°N domain (**Figure 1**) for the period 1991–2022. To carry out the present study, the TC position and intensity were considered from the India Meteorological Department (IMD) best track datasets available at www.rmcchennaieatlas.tn.nic.in. The adopted methodology for the generation of the best track data is elaborately discussed in IMD technical note version 2.0/2011 (IMD, 2011). Singh et al. [26] observed that the IMD best track data is reliable enough to be used for studies focusing on TC activity. The tropical systems formed over NIO are categorized into 7 categories, i.e., Depression if

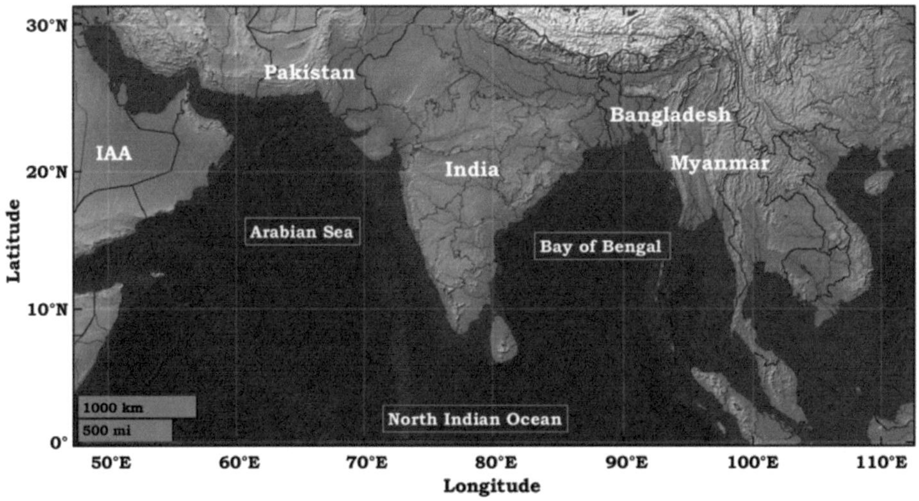

Figure 1.
Schematic representation of the North Indian Ocean (NIO) region and associated coastal countries.

MSW is 17–27knots, deep depression (DD) if the MSW is 28–33knots, cyclonic storm (CS) if MSW is between 34 and 47 knots, severe cyclonic storms (SCS) if MSW is 48–63knots, very severe cyclonic storm (VSCS) if it has MSW between 64–89knots, extremely severe cyclonic storm (ESCS) if MSW is within the range 90–119knots and super cyclonic storm (SuCS) if the MSW is higher than 120knots (www.rsmcnewdelhi.imd.gov.in, [9]). Further, the study considers storms with intensity ≥34 knots only. The RI locations in this chapter are considered to be the latitude/longitude positions of the TCs when the system began to show signatures of the rapid increase in the wind intensity, i.e., thereafter the MSW intensity increased by 30 knots in 24 hours.

The atmospheric parameters such as SST, u and v winds at 1000-200hpa, divergence at 200hpa, and mid-tropospheric relative humidity at 550hPa (RH550) are collected from European Center for Medium Range Weather Forecasts reanalysis 5 (ERA5) hourly data on pressure levels with 0.25° horizontal resolution datasets (cds.climate.copernicus.eu). Malakar et al. [27] analyzed different re-analysis datasets to study the evolution of TCs Over NIO and found that ERA5 gives the best estimate of TC state among all considered reanalysis datasets and hence the study considers ERA5 data for analysis. The deep-layer wind shear (DWS) is computed as the magnitude of the vector wind difference between 200 and 850hPa.

3. Studied gradient of TC activity

3.1 RI TC climatology

The long-term changes in both the annual TC frequencies and rapidly intensifying TCs, as shown in **Figure 2a** and **b**, depict a slightly decreasing trend over BOB and an increasing trend over AS during 1991–2022. Similarly, there is a consistent decline in the percentage of the 5-year running mean of RI TCs that occurred relative to the annual total TC frequencies over the BOB. In contrast, there is a noticeable upward trend in the corresponding percentage for the AS. Here the frequencies include the

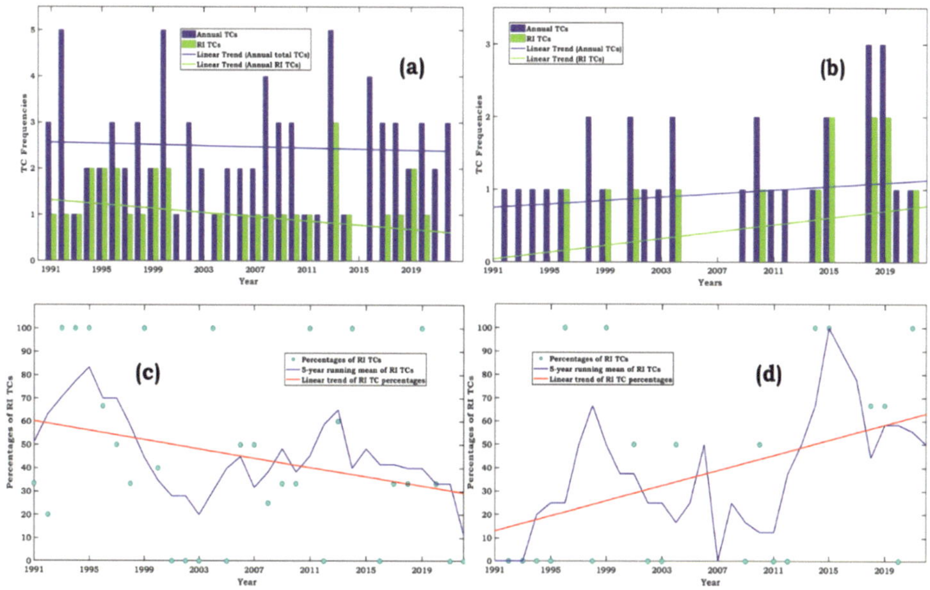

Figure 2.
Time series of annual total TC frequency (blue bar), RI TC frequency (green bar), linear trend for annual frequency (blue lines), and RI TCs (green lines) over (a) BOB and (b) AS. The percentage and 5-year running mean of RI TCs over (c) BOB and (d) AS. Here all the analysis of the annual trends includes pre-monsoon and post-monsoon season systems only.

systems during both the pre-monsoon and post-monsoon seasons. Several studies have shown that the number of global TC suffering RI has increased in recent decades [15, 28]. On the contrary to the global TC trend, the 5-year running mean of the percentage of RI TCs over BOB has decreased from ~80% in 1995 to ~10% during 2022. Over the AS, the corresponding percentage has increased to ~100% during the current decade. Subsequently, the general conception of a rising trend in the RI scenarios over NIO could be attributed to the rise over AS.

3.2 Gradient of RI locations

Regardless of the near decrease (increase) in the overall frequency of TCs (RI TCs) across the NIO, areas with high landfall TC frequencies exhibit significant coastal vulnerability. Additionally, the migration of locations where these systems undergo RI doubles the threat to the coastal areas. **Figure 3** indicates a shift in the RI locations of TCs formed over BOB towards the southern coastal regions of India, i.e., towards Andra Pradesh (15.91°N, 79.74°E) and Tamil Nadu (11.12°N, 78.65°E) coast during both pre-monsoon and post-monsoon season. It is observed that the region of 85°–90°E and 10°–15°N has suffered the greatest number of RI systems during both seasons. During pre-monsoon season, although the frequency of TCs that underwent RI is less (~9 only) during 1991–2022, nearly 70% of systems experienced RI in the latitude region of 10°N to 15°N (**Figure 3**). During the post-monsoon season of the early and mid-90s, the RI positions were nearly parallelly ~5° away from the coast of India. However, with increasing warming conditions over NIO, the corresponding positions are found to be distributed in an arch manner over BOB, mostly confined to regions below 15°N latitudes increasing vulnerability for the lower coastal regions

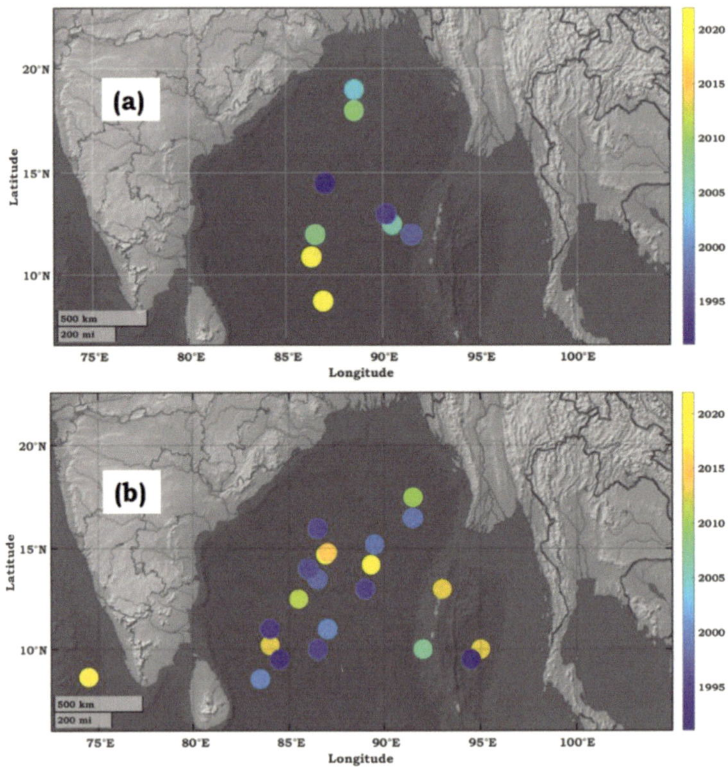

Figure 3.
The gradient of locations of RI during 1991–2022 for (a) pre-monsoon and (b) post-monsoon over BOB. The color bar represents the years of RI TC formation over the basin.

of India as well as the Gulf of Thailand. As it's not only the landfall of these systems with sudden enhanced maximum intensity but also the heavy rainfall and wind of these events that cause destruction to the nearby areas. The location of maximum frequency remaining unchanged as that of pre-monsoon season, the environmental condition over the oceanic region needs to be investigated to understand the reason behind the suitability for RI over this particular grid. It is to be noted that the distribution of the TC RI region is longitudinally distributed during pre-monsoon and nearly latitudinal during the post-monsoon season over BOB (**Figure 3**).

Figure 4 illustrates the shift in the location where tropical systems formed over AS experienced RI. The figure shows that the regions bound between the ~12°N to 17°N are prone to RI throughout the study period. Singh et al. [18] showed that due to a change in wind pattern over the AS basin, the IAA regions or the Arabian Peninsula coast (19.49°N, 47.44°E) and Gujarat (22.67°N, 71.57°E) regions are more prone to TC landfall, i.e., the TCs either travels westwards towards the IAA or recurves to make landfall over Gujarat. The AS coastal regions exhibit similar susceptibility to RI TC cases. During post-monsoon season, the region of RI mostly lies parallel to the coast of IAA and has been migrating towards the coastal areas during the present decade with increase in the frequency, representing the presence of favorable environmental conditions over the region to support the RI of the storm generated over the basin. It enhances the possibility of a continuous rise of threats to both the Indian and Arabian coastal countries.

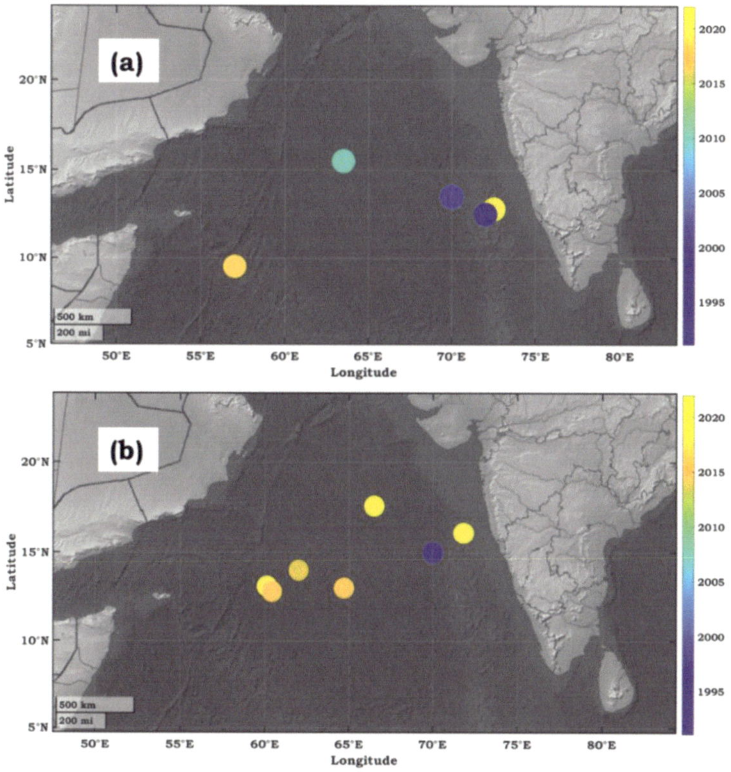

Figure 4.
The gradient of locations of RI during 1991–2022 during (a) pre-monsoon and (b) post-monsoon over AS. The color bar represents the years of RI TC formation over the basin.

The trend for change in latitude and longitude of RI experiencing regions during the pre-monsoon season is not significant due to the small sample size or lesser number of systems during the season (**Figure 5a** and **c**). The quadrennium variation of latitudinal and longitudinal variation over BOB during pre-monsoon season indicates a decrease in latitude (inter-quartile range 8.5°N–11°N) and longitude positions (inter-quartile range between ~86°E and 87°E) of RI over the basin for the period 2019–2022, representing the increase in landfall of RI systems over southern coastal regions in recent years. During post-monsoon season, a broader range of latitude attainment is observed by RI TCs post-2007 (**Figure 5b**). Moreover, the mean longitude is found to be decreasing during the period 2007–2018, from a mean of ~92°E to 81°E with an interquartile range of 75°E–87°E and then an insignificant rise post-2019 (**Figure 5d**). The decreased mean RI latitude (**Figure 5b**), broad range of latitude, decrease in mean RI longitude (**Figure 5d**), and comparatively narrow range of longitude during the quadrennials post-2007 increased the exposure of lower regions of India to TC destruction.

Nevertheless, the increase in RI locations post-2019 represents an increased threat over eastern coastal states of the BOB basin, i.e., Myanmar (21.91°N, 95.95°E) and the Gulf of Thailand (**Figure 5b**, and **d**). It can be seen from **Figure 5** that the zero or lower RI activity during any quadrennium of the pre-monsoon season encourages enhanced RI activity over BOB during the corresponding post-monsoon season and vice versa. The possible explanation for the enhanced RI activity during one season suppressing RI activity during the other for any particular quadrennium could be attributed to the

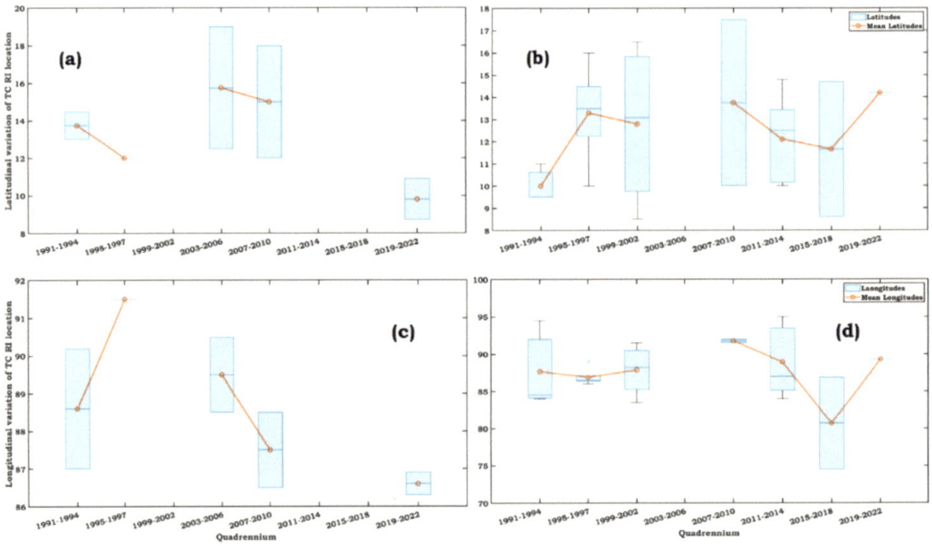

Figure 5.
South-west migration of the latitude (a, b) and longitude (c, d) of RI TCs during pre-monsoon and post-monsoon seasons over BOB respectively. The left panel represents the pre-monsoon season, and the right panel represents the post-monsoon season. The red line represents the mean values of the latitudes and longitudes.

unutilized heat or thermal energy during the pre-monsoon season being harnessed by TCs during the post-monsoon season, leading to RI events and vice-versa. However, the dynamics behind the observed interconnected TC RI activity over BOB needs detailed investigations and is not covered in this chapter. The number of RI TC occurrences over AS is lower, resulting in similar insignificant variations for both seasons (figures not shown for brevity). Nonetheless, an increase in the interquartile range of latitude (~16°N–17.8°N) and longitude (~67°E–72°E) post-2019 shows an increased threat of RI TCs formed over AS to western Indian states during post-monsoon season.

4. Variation of environmental factors

The TC intensification process is related to the internal dynamics of the system and the internal dynamics depend on the environmental factors prevailing throughout the TC activity [29–31]. Under favorable environmental conditions, TCs can continuously intensify until they approach their maximum intensity, and unless any adverse environmental conditions are experienced, the likelihood of TC weakening during the intensification phase is rare. Any long-term change in TC activity is correlated to the change in climatic conditions over the region. Several parameters such as SST, mid-tropospheric relative humidity (RH550), vertical wind shear (VWS), and upper-level divergence are found to modulate the TC intensity [32, 33]. The section below discusses the change in the environmental factors in recent climate conditions and its effects on the propagation of RI locations of TC over the NIO region.

4.1 Sea surface temperature

Emanuel [34] represented TCs as a Carnot heat engine that draws energy from the underlying ocean as entropy flux and loses energy due to surface friction.

TCs intensify when the energy production is greater than the dissipation. Therefore, the SST plays a crucial factor in TC intensification [35]. **Figure 6** represents the area averaged SST (°C) anomaly over 4° latitude and longitude regions over BOB and AS. The tick marks and labels of **Figure 6a** and **b** represent average SST anomaly over 80°–83°E/3–6°N, 80°–83°E/7–10°N, 80°–83°E/11–14°N, 80°–83°E/15–18°N, 80°–83°E/19–21.5°N and then the longitude regions increased to 84–87°E for the latitude ranges defined between 3°N–21.5°N and so on with the same set of latitude ranges. Similar divisions are made for AS for longitudes 50–74°E and latitudes 3–23°N. Since the SST gradient over NIO is not very sharp, averaging the data in chunks helps detect any frequency variability for the analysis period.

It is observed that the average SST over BOB during pre-monsoon ranges between 27.5–31°C, during post-monsoon between 27 and 29.6°C, and over AS the ranges are 23–30.5°C and 24.5–29.7°C respectively. During the analysis, it is observed that TC's rapid intensification is presumably related to the positive SST anomaly for any year over the basin. However, the notably elevated SST anomaly since 2015 does not appear to be conducive for a high rate of RI activity for both seasons in the BOB. For example, the years such as 1998, 2005, 2010, and 2016 during pre-monsoon and years 1997, 2002, 2009, and 2015 during the post-monsoon season where the rise in SST anomaly is abnormally high (**Figure 6a** and **b**), supported zero RI TC activity. As discussed in Section 3.3, for most of the years where high RI TC activity during pre-monsoon suppresses the activity during post-monsoon and vice-versa, the SST anomaly is found to be comparatively higher during that particular season than the other. For example, the quadrennium 1995–1997, and 1999–2002 during pre-monsoon season show zero or negligible RI TC frequency whereas a noticeable rise in activity during post-monsoon season could be seen. Similarly, during 2003–2006, where pre-monsoon season experienced higher RI TCs suppressed the activity during the post-monsoon season over BOB (**Figure 5**).

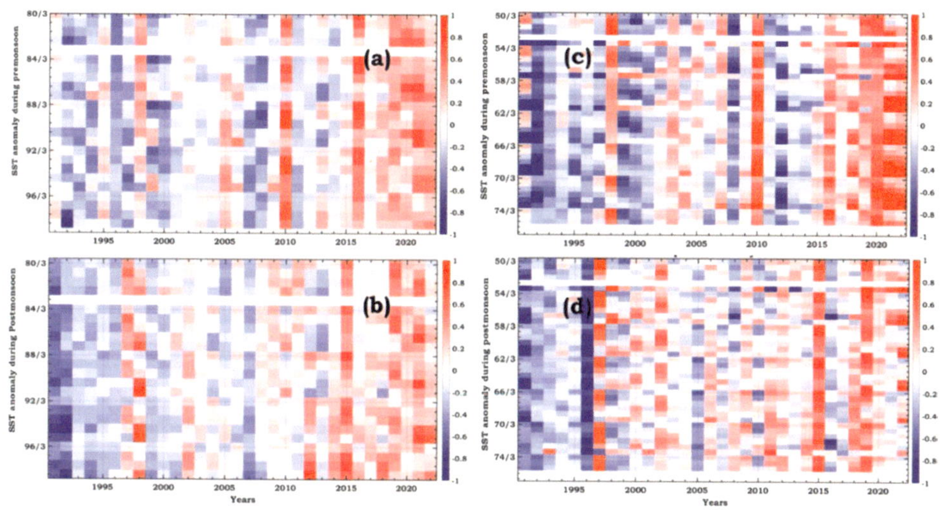

Figure 6.
Average SST anomaly (°C) variation during pre-monsoon (a, c) over BOB and AS and during post-monsoon seasons (b, d) over BOB and AS respectively from 1991 to 2022. Here, the anomalies are normalized by the means for the analysis period.

And, in contrast to the Carnot engine theory, the annual TC frequency over the basin is found to be decreasing despite rising SST over BOB post 2015. The role of SST in intensification depends upon the initial intensity of the storm, size, and other environmental factors to favor further intensification of the system. Similarly, over AS the higher TC activity during pre-monsoon and post-monsoon is supported by higher SST anomaly, contrary to BOB. It is to be noted that the SST over AS was comparatively colder than BOB till 2002 and is rising in the current warming climate supporting a higher number of RI TCs. Most of the AS regions have an average SST range of 27°C to 30.5°C during the pre-monsoon season and regions eastward of 62°E for all considered latitudinal bands have an average SST of 27.5–29.5°C post-2011 during post-monsoon season. Several literatures suggested that the SST range of 27.5–29°C is conducive to TC intensification [9, 36, 37]. Xu and Wang [35] stated that 99.8 and 84.4% of intensifying TC cases occurred over SSTs greater than 26 and 28°C in the WNP, compared with 88.3 and 41.3% of cases in the North Atlantic.

4.2 Mid-tropospheric RH (RH550)

Factors such as RH550 and SST can affect the cloud and precipitation distribution in a TC environment by triggering latent heat release in a system and hence the energy cycle of the system. Consequently, the RI of a TC is often considered to be associated with increased azimuthal and areal coverage of convection [38, 39], motivating further examination of this parameter for their role in the propagation of RI locations over NIO in the present changing climate. Over BOB, during pre-monsoon season the RH550 anomaly is mostly negative till 1998 except for 1994 (**Figure 7a**). During this period, the SST anomaly (**Figure 6a**) and RH550 show a negative relation between both of them over BOB. Similar results have been found by Liu et al. [40], and Sun and Oort [41]. The increase (decrease) in SST weakens (strengthens) the radiative cooling and relative humidity at upper atmosphere [40]. Surprisingly, no significant

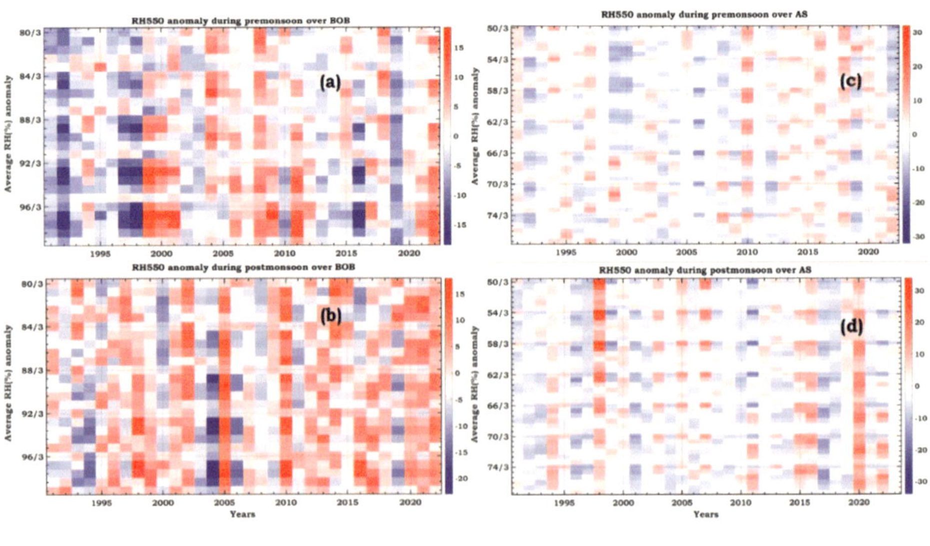

Figure 7.
Average RH550 anomaly (%) variation during pre-monsoon (a, c) over BOB and AS and during post-monsoon seasons (b, d) over BOB and AS respectively from 1991 to 2022. Here, the anomalies are normalized by the means for the analysis period.

RI TC activities were found during these years. Moreover, in the years where remarkable RI TC activities are witnessed during the pre-monsoon season post-2004, the SST anomalies are observed to be negative whereas RH550 is found to have positive anomalies. The average RH550 during pre-monsoon is ~75% and greater than 80% during post-monsoon season over BOB.

For the post-monsoon season, the southern and middle latitude regions of the entire BOB basin have higher RH550, i.e., 60–85%. In the years such as 1995, 1996, 1999, 2000, and 2013, where higher RI TC activities were experienced over BOB, the SST anomaly is negative, and the RH anomaly shows a rise over the basin. However, the RH550 is one of the supporting elements for the intensification of TC over BOB. Surprisingly reduced SST anomaly is found to be favorable for RI of TCs during post-monsoon season. According to Xu and Wang [35] and Park et al. [42], the maximum potential intensification rate (MIPR) of TCs is primarily associated with the regions where the SSTs are higher than 26–28°C over the Atlantic and west-north Pacific (west-north pacific) basins and similar findings can be observed for BOB basin. Furthermore, the RH550 has risen sharply over 84°E to 92°E post-2004 during pre-monsoon season, and over 84°E–100°E for post-monsoon season post-2012 for the latitudinal sections (i.e., 3–6°N, 7–10°N, 11–14°N, 15–18°N) over BOB. The detected rise in pattern over these regions was found to be supportive of the change in the RI locations over the basin.

Over AS, from **Figure 7c**, after 2013 a continuous rise in RH550 during pre-monsoon season could be seen over all longitude regions, though the latitude regions were different for each year. The RI TCs frequencies during 1999, 2001, 2010, 2018, and 2021 coincide with the rise in RH550 by nearly 10–15% over the ocean basin during pre-monsoon season. Analogously, for post-monsoon season, years with higher RI activities such as 2014, 2015, 2018, and 2019 are found to coincide well with the enhanced RH550 years. Until 2007, the elevated RH550 in the western longitudinal regions of the AS basin facilitated increased RI (**Figure 7d**). Subsequently, there is a shift towards the eastern sector of the basin as a rise in RH550 has been experienced in the eastern parts after 2010. The variations are supported by the comparative rise in SST during recent years for both seasons.

4.3 Vertical wind shear (VWS)

The apparent role of the ambient VWS in the intensification process is to encourage the growth of a storm cell by maximizing the strength of vertical motion, low-level inflow and inflow depth, helicity, convective available potential energy, and local shear in the downshear-left quadrant and near the center [43–46]. The sequence of events for RI under a sheared environment as explained by Molinari and Vollaro [44] are; (i) higher vertical wind shear produced persistent downshear rainfall and subsequent downshear vortex redevelopment within the precipitation shield (ii) the tilting of this new vortex by the ambient shear created favorable conditions for a strong cell to develop downshear in the storm core and (iii) this cell dramatically enhanced the heating in the region of large kinetic energy efficiency and contributed to subsequent RI of the storm.

During pre-monsoon season, the decadal average tropospheric VWS is seen to be reducing by ~3.8 m/s over the central BOB region (**Figure 8a**) from the decade 1991–2000 to 2001–2010 (here onwards abbreviated as D1 and D2). The distribution of the difference is zonal, and the RI TCs during D1 is also observed to be nearly zonal (**Figure 3a**). Whereas, during the decade 2011–2022 (here onwards called D3),

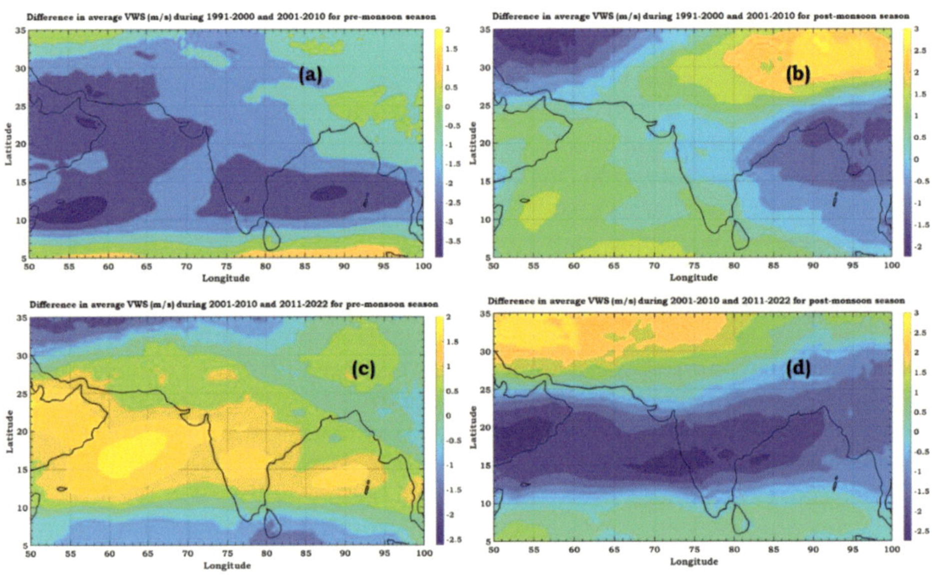

Figure 8.
Differences in average decadal tropospheric VWS (m/s) during 1991–2000 and 2001–2010 for (a) pre-monsoon and (b) post-monsoon season, and the corresponding differences for decades 2001–2010 and 2011–2022 over NIO during (c) pre-monsoon season and (d) post-monsoon seasons respectively.

the gradient of the difference in average VWS between D2 and D3 is observed to be from tropics to central BOB region (**Figure 8c**). This pattern reveals decreased VWS (approximately 1.5 m/s) at lower latitudes transitioning to increased VWS (2 m/s) in the central BOB. The meridional gradient of VWS is found to complement the change in RI locations from being latitudinally aligned during D1 to longitudinal alignment post-2010 over BOB. Over AS, for pre-monsoon season the enhanced tropospheric shear distribution is observed to be NE–SW aligned (**Figure 8c**) during D3. This alignment coincides with the zone of high RI TCs in the current decade, as depicted in **Figure 4a**.

During post-monsoon season, the distribution of the difference between D1 and D2 is meridional with reduced values over the northern bay of BOB to enhanced values towards the equatorial region (**Figure 8b**). The distribution of RI TCs location (**Figure 3b**) before 2005 coincides well with the observed distribution of VWS difference during D2. While experiencing the alteration in VWS pattern in D3 for the post-monsoon season, the most significant VWS gradients are noticed to be limited to the region below 15°N over the BOB with the lowest VWS (~2.5 m/s) near Odisha (20.23°N, 84.27°E) and northern Andhra Pradesh coast (**Figure 8d**). Surprisingly the RI TC location distribution has also shifted from a longitudinal pattern to a latitudinal one, being confined below 15°N after 2010 (**Figure 3b**) under the effect of changed tropospheric shear pattern.

Over AS, the change in the VWS pattern shows a reduction in values of tropospheric shear for the post-monsoon season over the latitudinal belt 13°N–20°N (**Figure 8d**) during D3. The region is seen to be associated with the region of higher RI TC activity locations (**Figure 4b**). The distribution of shear in which TC is embedded defines the track and intensity of the storm. The average tropospheric VWS over the NIO region ranges between 3 and ~20 m/s during pre-monsoon and post-monsoon seasons and 5–12 m/s over the TC formation zones. The MPIR is higher for

SSTs above 28°C over the WNP than over the North Atlantic, which is shown to be related to the weaker (stronger) environmental VWS in regions with higher (lower) SSTs over the WNP than over the North Atlantic. The weaker VWS has been reported to affect the intensification process of TCs over NIO [25], hence although the annual TC frequency has reduced over BOB, the RI TCs distribution is found to be modulated by the prevailing VWS patterns.

4.4 Upper-level divergence

The divergence at higher altitudes leads to a reduction in surface pressure at the storm center. This results in an escalation of the storm's intensity as the speed of converging surface winds rises. The converging wind accumulates enhanced heat and moisture, promoting further warming of the core of the TC. The annual trend of upper-level divergence at 200 hPa (Div200) over BOB is observed to be distributed longitudinally between 85°E and 90°E with a reduction by ~1×10^{-7} s^{-1} per year during the pre-monsoon season and a similar reduction is also observed over AS in the latitudinal region of 10°N–18°N (**Figure 9a**). **Figure 3a** and **4a** show that the distribution of RI TC locations over BOB and AS is determined by DIV200 to a great extent during the pre-monsoon season. The reduced DIV200 over both the basins help in

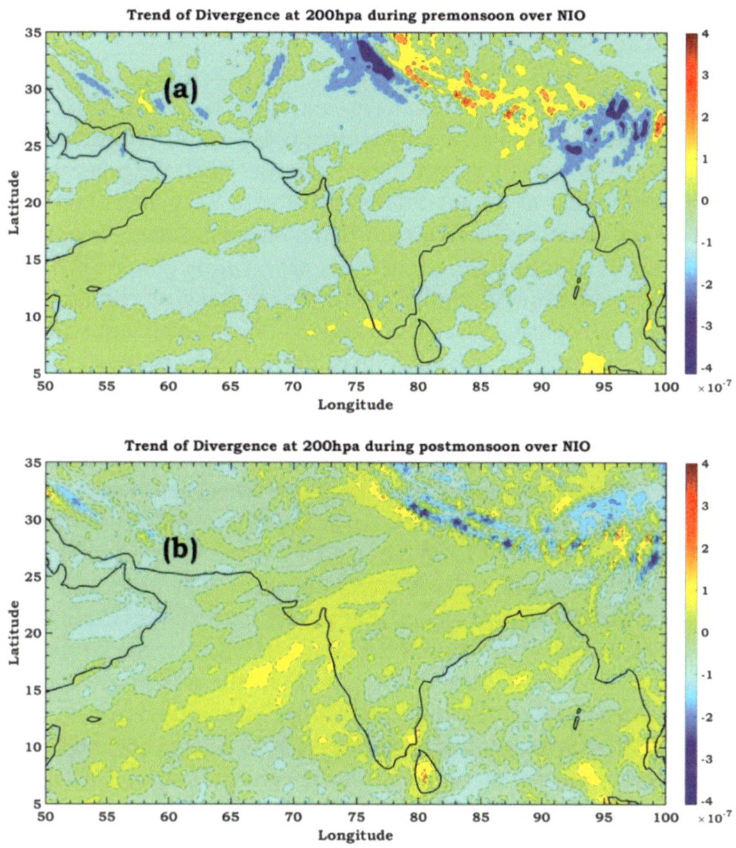

Figure 9.
Spatial distribution of linear trend of annual average upper-level divergence at 200 hPa over NIO region during (a) pre-monsoon season and (b) post-monsoon season.

trapping the lower-level heat flux generated within the TC core environment [47] which helps the system to strengthen the lower-level wind and momentum towards the center of the cyclone and thereby help to attain RI during its lifetime.

During post-monsoon season the regions where the TCs suffered RI, the DIV200 is observed to have an increasing trend. The in-up-out circulation due to enhanced DIV200 in a convective region is associated with low to mid-level convergence and upper-level divergence [48]. The RI location gradient (**Figure 3b**) over BOB during recent decades is analogous to the regions where the increasing annual Div200 trends are spotted (**Figure 9b**). Likewise, the NE–SW aligned increasing DIV200 region from 60°E to 75°E (**Figure 9b**) is found to support the TCs to obtain their RI phases over the AS during post-monsoon season. Interestingly, during the post-monsoon season, the mechanism of intensification is that the increased divergence in the upper level helps in enhancing lower-level convergence by ventilating the latent heat energy from the core of the system to the upper atmosphere, and thereby ensures an undisrupted supply of water vapor, mass, and momentum to increase the lower to middle-level spin up [49] and tangential wind of the TCs to attain RI straightaway. The changing DIV200 distribution is one of the major factors helping RI to migrate towards the coastal regions of NIO. The upper-level divergence ensures the maintenance of the cumulus convection, and thus it acts as a pump for the primary ingredients [50].

5. Conclusions and discussions

The TC frequencies and distribution have clearly changed during recent warming climate scenarios. Studholme et al. [3] reported a poleward migration in latitudes of peak TC intensity at a rate of ~0.5° latitude per decade. With the poleward migration of the TCs, any change in the spatial distribution of TC genesis location or intensification would greatly affect the well-being of coastal lives. The local physics interaction with the TC distribution in the current warming climate is a major topic of discussion among researchers. Under the influence of changing climate, the tropical expansion due to Intertropical convergence zone (ITCZ) movement will be primarily driven by upper atmosphere warming whereas restrained by the surface warming patterns. Consequently, the distribution of genesis and intensification of TCs would be more controlled by the upper atmosphere variation. Similar correlations between RI locations of TCs and atmospheric patterns are observed in the present study too. Kaplan et al. [51] analyzed that VWS and upper-level divergence have the relatively highest weights in predicting the RI of TCs over the Atlantic basin. With a decrease in percentages of RI TCs over BOB and rising RI TCs over AS, presently a shift in the distribution of RI of TCs locations has become apparent. The RI location gradient during the pre-monsoon season is higher over the 85°E–90°E longitude region over BOB. Over AS, the concerned location is found to be high over the latitudinal band 10–17°N. The distribution of RI location over BOB during post-monsoon season lay parallel to the Indian coast during the decade D1 which changed to a curved pattern confined below 15°N with longitude range 75°E to 95°E during D2 and D3, thereby enhancing the vulnerability to the southern Indian coastal states, Andaman Nicobar Islands and gulf of Thailand to the effects of RI TCs. Over AS, the distribution is again observed to be latitudinal (nearly between 12°N–17°N) with increased frequencies, making the Arabian coastal states as equally vulnerable as that of the Indian coastal states. The enhanced SST is found to play a trivial role in the observed shift in the RI locations over BOB and AS during both seasons.

The average SST 27 to 29.6°C over BOB is found to be favorable for intensification of TCs. The overall rising SST over 62°E to 78°E longitudes and below 17°N latitude region during recent decades support the enhanced RI activity over AS basin, however, due to the small data size of the RI TCs the results are insignificant. Moreover, no correlation is observed between the extremely high SST anomaly during any year and the RI TC activity. During pre-monsoon season over BOB, the anomalous rise (decrease) in SST is found to be inversely related to the decrease (rise) in RH550, e.g., years such as 1995, 1998, 2000, 2008, 2010, 2016, 2019, etc. There are three possible reasons explanation behind this behavior (i) the regions with high SST experience strong subsidence, and the relative humidity at the upper level decreases as the air descends (ii) a vice-versa relation of warm SST leading to a more stable lower atmosphere leading to the inhibition of vertical motion and the lifting of moist air to middle atmosphere (iii) increased surface winds in recent climate [9] transporting the available humidity horizontally to nearby regions. The slowdown of the tropical large-scale circulation in a warming climate is linked with a weakening of convective mass flux caused by increased dry static stability and reduced meridional surface temperature gradients in the tropics [52] is another reason for the observed relation between SST and RH550. An enhanced average RH550 (up to 75%) over southern BOB between longitudes 92°E–96°E is also observed during the season. The enhanced VWS pattern over central BOB (85–90°E/10–15°N) during the pre-monsoon season of D3 by 2 m/s than D2 and largest gradient zone over BOB, host the highest number of RI TCs aligned in North-South direction during this season. Bi et al. [53] suggested that the effect of shear on the intensification process depends upon the size of the disturbance with larger TCs affected less by high VWS. The displacement of the upper-level circulation is smaller in larger TCs, indicating that larger TCs are less susceptible to VWS. The decreased DIV200 alignment again agrees well with the observed RI TC distribution over the basin in the pre-monsoon season.

During the post-monsoon season over BOB, unlike the pre-monsoon season, the observed inverse relationship between SST and RH550 is less pronounced. The average RH550 has increased (~85%) in recent years and a prominent rise is seen over southern and middle latitudes. The increasing RH550 and SST support the increasing tendency in RI TCs, despite a decrease in annual TC frequency over the basin. The increase (decrease) in RH550 during pre-monsoon (post-monsoon) over BOB and vice-versa needs further investigation which is out of scope for the present chapter. The gradient of change in decadal average VWS over BOB during post-monsoon season altered from meridional to nearly zonal from D1 to D3, consequently varying the RI TC gradient from NE-SW during early 2000 to confined below 15°N post-2005. A very clear reduction in tropospheric VWS around the coast of Odisha and West Bengal is inhibiting the intensification and landfall over the state in recent decades. High upper-level DIV200 implies large vertical motion, and more air is taken out at the top than is brought in at the bottom promoting the axisymmetric central core of the system favoring RI of the system.

Over AS, for pre-monsoon season the correlation between rise in SST and RH550 is observed for some years after 2010. However, the longitudinal regions that witnessed higher RI TC activity are mostly dominated by a decreased anomaly in RH550 with enhanced SST anomaly. The noticeable VWS gradient in 10°N–15°N facilitates intensification over the basin. The decadal average VWS over the belt is 5–10 m/s. The decreased DIV200 over the region discussed here is found to support the RI TCs. During post-monsoon season, the simultaneous increase in SST and RH550 could be seen for most of the years post-1995, and regimes of higher RI TC activity are associated with increased RH500 and SST anomalies. The rise in RH550 to 60 and 75% during pre-monsoon and post-monsoon over equatorial latitudinal zones supports higher RI activity over the eastern AS basin. The narrowing of the VWS gradient zone over the central AS during the post-monsoon season and the substantial decrease in average VWS by 2.7 m/s over 13°E–20°E favors the rapid growth of the system in the current decade D3. The orientation of low-level mean flow with respect to the vertical shear, and helicity of environmental wind has a relationship with TC intensification and expansion [54–56]. The observed linear enhancement in upper-level divergence is attributed to the observed distribution of RI TC locations which is possibly the key factor for the observed RI locations pattern in recent years. The rising upper-level divergence and reduction in VWS over the near central zone of AS also imply a potential reason behind the increase in the number of TCs over the basin during the post-monsoon period. Apart from the parameters discussed here in this chapter, the pronounced transient zonal and meridional TC migrations are also expected to occur in response to climate warming of the Hadley circulation, jet streams, El Niño–Southern Oscillation (ENSO), and Intertropical Convergence Zone, which will be covered by further extended work of this chapter.

Acknowledgements

The author acknowledges that, 'for the purpose of open access, a Creative Commons Attribution (CC BY) license has been applied to any Author Accepted Manuscript version arising from this submission'. This research was supported by the University of Exeter's Open Access Funds, which provided financial support for the publication fees associated with this manuscript. The author would like to thank the India meteorological department (IMD), New Delhi for the TC best track data available at, rsmcnewdelhi.imd.gov.in and ERA5 (cds.climate.copernicus.eu/) for providing the ocean and atmosphere parameters.

Conflict of interest

The authors declare no conflict of interest.

Author details

Kasturi Singh[1,2]

1 Department of Mathematics and Statistics, University of Exeter, Exeter, UK

2 Met Office, Exeter, UK

*Address all correspondence to: k.singh@exeter.ac.uk; kasturi.env@gmail.com

References

[1] Kaplan J, DeMaria M. Large-scale characteristics of rapidly intensifying tropical cyclones in the North Atlantic basin. Weather and Forecasting. 2003;**18**(6):1093-1108

[2] Song J, Klotzbach PJ. What has controlled the poleward migration of annual averaged location of tropical cyclone lifetime maximum intensity over the western North Pacific since 1961? Geophysical Research Letters. 2018;**45**(2):1148-1156

[3] Studholme J, Fedorov AV, Gulev SK, Emanuel K, Hodges K. Poleward expansion of tropical cyclone latitudes in warming climates. Nature Geoscience. 2022;**15**(1):14-28

[4] Lin II, Camargo SJ, Lien CC, Shi CA, Kossin JP. Poleward migration as global warming's possible self-regulator to restrain future western North Pacific tropical cyclone's intensification. Npj Climate and Atmospheric Science. 2023;**6**(1):34

[5] Kossin JP, Emanuel KA, Vecchi GA. The poleward migration of the location of tropical cyclone maximum intensity. Nature. 2014;**509**(7500):349-352

[6] Sharmila S, Walsh KJ. Recent poleward shift of tropical cyclone formation linked to Hadley cell expansion. Nature Climate Change. 2018;**8**(8):730-736

[7] Murakami H, Delworth TL, Cooke WF, Zhao M, Xiang B, Hsu PC. Detected climatic change in global distribution of tropical cyclones. Proceedings of the National Academy of Sciences of the United States of America. 2020;**117**(20):10706-10714

[8] Deshpande M, Singh VK, Ganadhi MK, Roxy MK, Emmanuel R, Kumar U. Changing status of tropical cyclones over the North Indian Ocean. Climate Dynamics. 2021;**57**:3545-3567

[9] Singh K, Panda J, Sahoo M, Mohapatra M. Variability in tropical cyclone climatology over North Indian Ocean during the period 1891 to 2015. Asia-Pacific Journal of Atmospheric Sciences. 2019;**55**:269-287

[10] Moon IJ, Kim SH, Chan JC. Climate change and tropical cyclone trend. Nature. 2019;**570**(7759):E3-E5

[11] Swapna P, Sreeraj P, Sandeep N, Jyoti J, Krishnan R, Prajeesh AG, et al. Increasing frequency of extremely severe cyclonic storms in the North Indian Ocean by anthropogenic warming and Southwest monsoon weakening. Geophysical Research Letters. 2022;**49**(3):e2021GL094650

[12] Senapati S, Gupta V. Climate change and coastal ecosystem in India: Issues in perspectives. International Journal of Environmental Sciences. 2014;**5**(3):530-543

[13] Courtney JB, Langlade S, Sampson CR, Knaff JA, Birchard T, Barlow S, et al. Operational perspectives on tropical cyclone intensity change part 1: Recent advances in intensity guidance. Tropical Cyclone Research and Review. 2019;**8**:123-133

[14] Munsi A, Kesarkar A, Bhate J, Panchal A, Singh K, Kutty G, et al. Rapidly intensified, long duration North Indian Ocean tropical cyclones: Mesoscale downscaling and validation. Atmospheric Research. 2021;**259**:105678

[15] Manikanta ND, Joseph S, Naidu CV. Recent global increase in multiple rapid intensification of tropical cyclones. Scientific Reports. 2023;**13**(1):15949

[16] Murakami H, Vecchi GA, Underwood S. Increasing frequency of extremely severe cyclonic storms over the Arabian Sea. Nature Climate Change. 2017;**7**(12):885-889

[17] Evan AT, Kossin JP, 'Eddy'Chung C, Ramanathan V. Arabian Sea tropical cyclones intensified by emissions of black carbon and other aerosols. Nature. 2011;**479**(7371):94-97

[18] Singh K, Panda J, Rath SS. Variability in landfalling trends of cyclonic disturbances over North Indian Ocean region during current and pre-warming climate. Theoretical and Applied Climatology. 2019;**137**:417-439

[19] Arrieta MI, Foreman RD, Crook ED, Icenogle ML. Providing continuity of care for chronic diseases in the aftermath of Katrina: From field experience to policy recommendations. Disaster Medicine and Public Health Preparedness. 2009;**3**(3):174-182

[20] Fanchiotti M, Dash J, Tompkins EL, Hutton CW. The 1999 super cyclone in Odisha, India: A systematic review of documented losses. International Journal of Disaster Risk Reduction. 2020;**51**:101790

[21] Aldrich DP. Ties that bond, ties that build: Social capital and governments in post disaster recovery. Studies in Emergent Order. 2011;**4**(December):58-68

[22] Lakshmanan N, Shanmugasundaram J. A model for cyclone damage evaluation. Journal of the Institution of Engineers. India. Civil Engineering Division. 2002;**83**(NOV):173-179

[23] Singh K, Panda J, Kant S. A study on variability in rainfall over India contributed by cyclonic disturbances in warming climate scenario. International Journal of Climatology. 2020;**40**(6):3208-3221

[24] Munsi A, Kesarkar AP, Bhate JN, Singh K, Panchal A, Kutty G, et al. Atmosphere-upper-ocean interactions during three rare cases of rapidly intensified tropical cyclones over North Indian oceans. Journal of Oceanography. 2023;**79**(1):77-89

[25] Mohapatra M, Geetha B, Balachandran S, Rathore LS. On the tropical cyclone activity and associated environmental features over North Indian Ocean in the context of climate change. Journal of Climate Change. 2015;**1**(1-2):1-26

[26] Singh K, Panda J, Mohapatra M. Robustness of best track data and associated cyclone activity over the North Indian Ocean region during and prior to satellite era. Journal of Earth System Science. 2020;**129**:1-20

[27] Malakar P, Kesarkar AP, Bhate JN, Singh V, Deshamukhya A. Comparison of reanalysis data sets to comprehend the evolution of tropical cyclones over North Indian Ocean. Earth and Space Science. 2020;**7**(2):e2019EA000978

[28] Song J, Duan Y, Klotzbach PJ. Increasing trend in rapid intensification magnitude of tropical cyclones over the western North Pacific. Environmental Research Letters. 2020;**15**(8):084043

[29] Wu CC, Cheng HJ. An observational study of environmental influences on the intensity changes of typhoons Flo (1990)

and gene (1990). Monthly Weather Review. 1999;**127**(12):3003-3031

[30] Bosart LF, Bracken WE, Molinari J, Velden CS, Black PG. Environmental influences on the rapid intensification of hurricane opal (1995) over the Gulf of Mexico. Monthly Weather Review. 2000;**128**(2):322-352

[31] Hong X, Chang SW, Raman S, Shay LK, Hodur R. The interaction between hurricane opal (1995) and a warm core ring in the Gulf of Mexico. Monthly Weather Review. 2000;**128**(5):1347-1365

[32] Hong J, Wu Q. Modulation of global sea surface temperature on tropical cyclone rapid intensification frequency. Environmental Research Communications. 2021;**3**(4):041001

[33] Li X, Zhan R, Wang Y, Xu J. Factors controlling tropical cyclone intensification over the marginal seas of China. Frontiers in Earth Science. 2021;**9**:795186

[34] Emanuel KA. The maximum intensity of hurricanes. Journal of the Atmospheric Sciences. 1988;**45**(7):1143-1155

[35] Xu J, Wang Y. Dependence of tropical cyclone intensification rate on sea surface temperature, storm intensity, and size in the western North Pacific. Weather and Forecasting. 2018;**33**(2):523-537

[36] Waliser DE. Formation and limiting mechanisms for very high sea surface temperature: Linking the dynamics and the thermodynamics. Journal of Climate. 1996;**9**(1):161-188

[37] Ting X, Fei L, Guiwen F, Xiaorong G. An analysis of tropical cyclones rapid intensification and the statistical characteristics of sea surface temperature distribution. Haiyang Xueb. 2023;**45**:1-9. DOI: 10.12284/hyxb2023130

[38] Kaplan J, Rozoff CM, DeMaria M, Sampson CR, Kossin JP, Velden CS, et al. Evaluating environmental impacts on tropical cyclone rapid intensification predictability utilizing statistical models. Weather and Forecasting. 2015;**30**(5):1374-1396

[39] Richardson JC, Torn RD, Tang BH. An analog comparison between rapidly and slowly intensifying tropical cyclones. Monthly Weather Review. 2022;**150**(8):2139-2156

[40] Liu L, Wang X, Li Y, Wei W. The effect of sea surface temperature on relative humidity and atmospheric visibility of a winter sea fog event over the yellow-Bohai Sea. Atmosphere. 2022;**13**(10):1718

[41] Sun DZ, Oort AH. Humidity–temperature relationships in the tropical troposphere. Journal of Climate. 1995;**8**(8):1974-1987

[42] Park DS, Seo E, Lee M, Cha DH, Kim D, Ho CH, et al. Sea surface temperature warming to inhibit mitigation of tropical cyclone destructiveness over East Asia in El Niño. Npj Climate and Atmospheric Science. 2024;**7**(1):24

[43] Molinari J, Vollaro D. Extreme helicity and intense convective towers in hurricane Bonnie. Monthly Weather Review. 2008;**136**(11):4355-4372

[44] Molinari J, Vollaro D. Rapid intensification of a sheared tropical storm. Monthly Weather Review. 2010;**138**(10):3869-3885

[45] Eastin MD, Gray WM, Black PG. Buoyancy of convective vertical motions in the inner core of intense hurricanes.

Part II: Case studies. Monthly Weather Review. 2005;**133**(1):209-227

[46] Braun SA, Wu L. A numerical study of hurricane Erin (2001). Part II: Shear and the organization of eyewall vertical motion. Monthly Weather Review. 2007;**135**(4):1179-1194

[47] Chen X, Zhang JA, Marks FD. A thermodynamic pathway leading to rapid intensification of tropical cyclones in shear. Geophysical Research Letters. 2019;**46**(15):9241-9251

[48] Tory KJ, Frank WM. Tropical cyclone formation. In: Chan JCL, Kepert JD, editors. Global Perspectives on Tropical Cyclones: From Science to Mitigation. World Scientific on Asia-Pacific Weather and Climate. Vol. 4. Singapore: World Scientific Publishing Co. Pty. Ltd.; 2010. pp. 5591

[49] Munsi A, Kesarkar A, Bhate J, Singh K, Panchal A, Kutty G, et al. Simulated dynamics and thermodynamics processes leading to the rapid intensification of rare tropical cyclones over the North Indian oceans. Journal of Earth System Science. 2022;**131**(4):211

[50] Gray WM. Global view of the origin of tropical disturbances and storms. Monthly Weather Review. 1968;**96**(10):669-700

[51] Kaplan J, DeMaria M, Knaff JA. A revised tropical cyclone rapid intensification index for the Atlantic and eastern North Pacific basins. Weather and Forecasting. 2010;**25**(1):220-241

[52] Cheng W, MacMartin DG, Kravitz B, Visioni D, Bednarz EM, Xu Y, et al. Changes in Hadley circulation and intertropical convergence zone under strategic stratospheric aerosol geoengineering. Npj Climate and Atmospheric Science. 2022;**5**(1):32

[53] Bi M, Wang R, Li T, Ge X. Effects of vertical shear on intensification of tropical cyclones of different initial sizes. Frontiers in Earth Science. 2023;**11**:1106204

[54] Chen SS, Knaff JA, Marks FD. Effects of vertical wind shear and storm motion on tropical cyclone rainfall asymmetries deduced from TRMM. Monthly Weather Review. 2006;**134**(11):3190-3208

[55] Chen BF, Davis CA, Kuo YH. An idealized numerical study of shear-relative low-level mean flow on tropical cyclone intensity and size. Journal of the Atmospheric Sciences. 2019;**76**(8):2309-2334

[56] Chen BF, Davis CA, Kuo YH. Examination of the combined effect of deep-layer vertical shear direction and lower-tropospheric mean flow on tropical cyclone intensity and size based on the ERA5 reanalysis. Monthly Weather Review. 2021;**149**(12):4057-4076

Section 4

Vulnerability, Resilience, and Policy Engagement

Chapter 10

People Vulnerability before, during and after a Disaster: A Dynamic Taxonomic Approach

Sofia Karma, Stefano Zanut, Monica Crișan, Consuelo Agnesi and Gabriella Duca

Abstract

This chapter provides a taxonomic approach to vulnerability precursors and relevant inhibitors, tackling the entire disaster management cycle, through the analysis of various vulnerability factors "before, during, after" a disaster. Specifically, following the analysis of various existing approaches and models of human vulnerability in disasters, this work proposes an actionable roadmap to identify the criticality points (Critical Vulnerability Indicators - CVIs) under a comprehensive and inclusive approach. In this regard, the following aspects have been elicited: (i) the background; intrinsic weaknesses or possible non-compliance conditions, as well as the social and economic status; (ii) the human-environment interaction; (iii) the governance model; policies and approaches to disaster risk management; (iv) the tools; application of new technologies and modeling tools. This taxonomic approach allows the comprehensive understanding of vulnerability associated with different vulnerable groups and beyond; it aims at a better understanding of the mechanisms that activate and maintain vulnerability, under a dynamic and contextual perspective, highlighting the importance of human variability and interaction with the physical environment. The proposed vulnerability analysis approach could help the policymakers or the relevant authorities to identify the critical points for building inclusive disaster risk reduction and resilience strategies.

Keywords: taxonomy, human variability, physical environment, inclusiveness, resilience

1. Introduction

In a world facing unprecedented changes in climate conditions, as well as considering urbanization, population growth, and existing social and economic inequalities, building disaster resilience against multi-hazards under the lens of inclusiveness is a challenging and urgent issue. The Sendai framework [1] defines strategies aiming at disaster risk prevention and impact reduction; among them is the need to reduce exposure and vulnerability of social groups and individuals. According to the United Nations Disaster Risk Reduction Office (UNDRR), vulnerability is one of the components that define disaster risk; therefore, it is important to develop a taxonomy of it for a better understanding of the factors that can act as precursors or inhibitors of the effects of a disaster, as well as for improving the management of contextual variables during a disaster and for building disaster resilience.

Nevertheless, any type of vulnerability analysis should be seen through an inclusive lens, considering human variability in its broadest sense, in which the specific needs of each person are considered. The need to consider human variability is exacerbated in an emergency in terms of identifying the criticalities related to people's ability to respond in an emergency, or the contribution of the environment, devices, and procedures that could affect emergency response capacities. For example, the difficulties that an obese person or someone with heart problems may encounter in an emergency scenario, as well as his/her capabilities to respond with autonomy to the emergency may be different from a person with cognitive or sensorial impairments or a foreigner. Human interaction with the physical environment relates to all the external surroundings and conditions in which someone lives, or maybe when an event occurs; it influences a person's health or well-being and capability to react and reverse a situation. Therefore, a comprehensive understanding of the overall context is vital for tackling any type of vulnerability driver, namely the environment and its context, among the most critical aspects that lead to greater vulnerability in individuals and social groups. Hence, it is crucial to move beyond the idea of classifying the vulnerability factors associated "only" with individuals or groups, rather than understanding the mechanisms that activate vulnerability, as well as the interdependencies among them, thus introducing a more dynamic and contextual perspective.

In that context, this work strives to provide a proposal for a taxonomy of vulnerability considering the above aspects and make it a tool capable of supporting the strategies to be adopted "before, during, and after" an emergency event. It aims at better understanding and identification of the factors that may act as precursors or inhibitors of vulnerability, contributing to its management. Under this framework, several models relevant to human vulnerability were surveyed to tackle the multidimensional and changeable nature of vulnerability, encompassing social, economic, environmental, and institutional factors. As a step forward, this work identifies a series of criticality points (Critical Vulnerability Indicators - CVIs) under a comprehensive and inclusive approach. Developed within the recent research in the framework of the CORE (sCience& human factOr for Resilient sociEty) H2020 project [2], the proposed taxonomic approach provides a classification of vulnerability associated with different vulnerability patterns, tackling the importance of human variability and the interaction with the physical environment. Though, it goes beyond the classification of vulnerable categories, aiming at a better understanding of the mechanisms that activate and maintain vulnerability, under a dynamic and contextual perspective.

2. Analyzing variables and eliciting mechanisms of vulnerability: state of the play

The last two decades have seen an evolution of vulnerability concepts from intrinsic people's, societal characteristics, or categorizations toward a more dynamic and ecosystem-driven approach. The research presented in this work is in line with this trend and has started from a comparative analysis of the most acknowledged vulnerability models with the aim to deliver a framework for vulnerability interpretation and forecast based on the relation of humans with their socio-physical context [3, 4].

'Vulnerability' can be defined as "the quality or state of being exposed to the possibility of being attacked or harmed, either physically or emotionally" [5]. This is a general definition that sets an overall framework for the concept which can be further modified for specific contexts. In more detail, Sendai Framework terminology on Disaster Risk Reduction defines vulnerability as the characteristics determined by physical, social, economic, and environmental factors or processes that increase the susceptibility of an individual, a community, assets, or systems to the impacts of hazards [6].

Bohle [7] proposes a vulnerability model with two aspects: an external side, which is linked to exposure to risks and influenced by various factors such as political-economic approaches, human ecology perspectives, and the entitlement theory; and an internal side, called "coping", which is related to the capacity to anticipate, cope with, resist, and recover from the impact of a hazard, and is influenced by the crisis and conflict theory, action theory approaches, and models of access to assets.

Pelling [8] depicts a human vulnerability model defined by exposure, resistance, and resilience. In his model, exposure is related to the location and characteristics of the hazard. Resistance is related to the economical, psychological, and physical health, as well as the capacity of individuals or communities to withstand the impact of the event, while resilience is defined as the ability to cope with or adapt to the hazard stress through preparedness and spontaneous adaptations once the event has manifested itself.

When looking at the capability to react to and recover from adverse events, a key reference is represented by the International Classification of Functioning, Disability, and Health (ICF) from WHO [9]. The ICF is a multipurpose classification system designed to serve various disciplines and sectors, such as education, transportation, health, and community services in general. It includes a list of environmental factors and is designed to serve various disciplines and sectors, as well as different countries and cultures, providing a common language and framework for describing the level of functioning of a person within its unique environment, which consequently leads to intrinsic and contextual vulnerability features in relation to specific risks and situations [10]. The ICF aims to provide a scientific basis for understanding and studying health and health-related states, outcomes, determinants, and changes in health status and people's functioning. Additionally, it permits the comparison of data across countries, healthcare disciplines, services, and time and provides a systematic coding scheme for health information systems.

Another comprehensive approach to vulnerability assessment has been developed within the EU FP7 MOVE project [11]. It is based on a comprehensive and multidimensional conceptual framework that focuses on vulnerability and risk to natural hazards. This framework is built on two key concepts: the first one views risk as the outcome of society's exposure to hazards as well as its vulnerability, while the second involves risk management and adaptation, aiming to alter the initial vulnerability conditions or hazards. A previous extensive taxonomy of vulnerability was developed by Birkmann [12], in which vulnerability is conceptualized at a series of increased degrees of complexity and scales, from intrinsic factors to the likelihood to experience harm up to the combination of multiple dimensions encompassing physical, social, economic, environmental, and institutional features.

The framework named "Progression of Vulnerability or Pressure and Release (PAR) [13] has a strong focus on social aspects, interpreting vulnerability as a characteristic of the poorest of the poor in every society, especially those who not only suffer income poverty but are also politically marginalized (no voice in decisions that affect them), spatially marginalized (resident in urban squatter settlements or in remote rural locations), ecologically marginalized (livelihoods based on access to meager natural resources or living in degraded environments), and economically marginalized (poor access to markets).

Institutional studies on the epidemiology of injury and death in disasters in the United States, support a dynamic model, avoiding people's categorization which are conventionally considered "vulnerable" and who are suffering disproportionately—women, elderly people, undocumented immigrants, etc. [14]; it is highlighted that patterns of vulnerability are far too complex and dynamic to support such sweeping generalizations [15].

Cannon's model [16] breaks vulnerability down into five components: livelihood strength and resilience, wellbeing and baseline status, self-protection, social protection, and governance, identifying links, connections, and disconnections among those components, as well as allowing to elicit vulnerability mechanisms. Similarly, Turner and colleagues [17] propose a socioecological model, highlighting human-environment linkages with feedback loops, encompassing multiple scales and dynamics that can be cross-scale, in place, and beyond place.

Forensic Investigations of Disasters (FORIN) approach [18] is grounded in a theory of social construction of disaster risk. It focuses on tackling risk creation factors, aiming at underlying, root causes and dynamic processes in critical events, which are linked to triggering events, exposure of social and environmental elements, social and economic structures of exposed communities, vulnerability, resilience, and institutional and governance elements.

Hufschmidt [19] model represents the flow and causal cascade due to the political, economic, cultural, and environmental settings of a society, the socioeconomic profile of a community, as well as the access to resources. According to the same model, access to resources determines the adaptive capacity of a community; particularly, adaptive activities can be realized before and in the aftermath of a harmful event or disaster.

The models and frameworks surveyed above highlighted the multidimensional and changeable nature of vulnerability, encompassing social, economic, environmental, and institutional factors. They emphasized the importance of understanding exposure, resistance, resilience, coping capacity, access to resources, and the socioecological context in assessing and addressing vulnerability to hazards and disasters. However, it can be observed that interactions among the identified layers and category of variables are only mentioned at a higher level, making it poorly applicable in practice the identification and consequent management of specific factors that influence at small and local scales the level of vulnerability for individuals or small specific groups. From this brief overview, it is evident that comprehensively addressing vulnerability in actual preparedness and response operations is a crucial, yet challenging topic for the inclusive and fair implementation of risk and disaster management. Thus, any improvements will strongly rely on the extent to which vulnerability mechanisms before, during, and after disasters are known, understood, and incorporated in policies and plans for Disaster Risk Reduction and Management.

3. Needs and challenges for vulnerability consideration in policies and plans

Why does vulnerability matter? Vulnerability should be included in the general understanding of disaster risk because in that way it is acknowledged the fact that it does not only depend on the severity of the hazard, or on the number of people and the assets exposed, but also on certain vulnerability precursors and exposure characteristics. The levels of vulnerability and exposure help to explain why particular hazards, which are not extreme, can lead to extended disaster impact, while others do not. Particularly, it is often people's vulnerability that is the greatest factor in determining their risk [20].

Disasters can have multiple impacts – death, injury and health impacts, displacement, damage to homes and goods, deaths of livestock, food insecurity, disrupted livelihoods, and more. Data concerning 2022, compared with the past years, shows how everyone is vulnerable to natural hazards worldwide; the Emergency Event Database EM-DAT recorded 387 natural hazards and disasters globally [21], resulting in the loss of 30,704 lives (three times higher than in 2021) [22] and affecting 185 million individuals. Economic losses totaled around US$ 223.8 billion. For example, the heat waves caused over 16,000 excess deaths in Europe, while droughts affected 88.9 million people in Africa. Hurricane Ian single-handedly caused damage costing US $ 100 billion in the Americas. The human and economic impact of disasters was relatively higher in Africa, e.g., with 16.4% deaths compared to 3.8% in the previous two decades. However, it was relatively lower in Asia despite experiencing some of the most destructive disasters in 2022 [22].

Based on the UNDRR, Global Survey Report on Persons with Disabilities and Disasters in 2023, it came out that persons with disabilities are considered vulnerable to disasters and hence, the world must act on unacceptable failures to protect them [23]. However, it is necessary to outline the difference between persons with disability and disaster vulnerability to better define the scope of this section. According to the United Nations Convention on the Rights of Persons with Disabilities (CRPD) - art. 1, persons with disabilities include those who have long-term physical, mental, intellectual, or sensory impairments that, in interaction with various barriers, may hinder their full and effective participation in society on an equal basis with others [24]. The World Health Organization (WHO) estimated that about 16% of the global population (1.3 billion people) currently experience significant disability, and this number is increasing due to the population aging and the increase of non-communicable diseases (–hidden disabilities) [25]. Almost everyone will temporarily or permanently experience disability at some point in their life due to different factors [25].

However, it should be mentioned that persons with disabilities in poor countries have heavier consequences due to the fact that vulnerability is determined by conditions linked not only to their disability but also other structural risks, such as high levels of poverty, poor sanitation, low levels of education, limited resources of health and social care, and a lack of safety nets [26]. Hazards may transform into disasters for persons with disabilities in low- and middle-income countries, and disasters can have devastating effects on the social capital of persons with disabilities, disrupting or destroying vital social support systems [27]. Another crucial aspect highlighted in 2007 in the World Disaster Report by the International Federation of Red Cross and Red Crescent Societies (IFRC), is that some disasters also create a significant number of new disabilities.

To promote inclusion and accessibility in the urban environment and to make cities resilient, the UNDRR developed an annex for the inclusion of persons with disabilities [28], which was added and included as a complement to the Disaster Resilience Scorecard for Cities. Specifically, it has been recommended to (essential 01) include persons with disabilities in disaster risk governance and (essential 02) in the creation of disaster risk scenarios because persons with disabilities have first-hand experience of the challenges they face and know better what can be done to enhance their wellbeing. Inclusion of persons with disabilities in land use/building codes (essential 04) to promote accessible and resilient urban design and development. To this end, capacities need to be developed or strengthened in the key areas of disaster risk reduction; understanding risk, prevention, mitigation, response, and recovery planning, all with a disability-inclusive approach, engaging persons with disabilities in institutional capacity (essential 06). The response to disaster must include persons with disabilities and their needs, creating contingency plans for populations most at risk - including persons with disabilities. Moreover, it is necessary to conduct regular training drills and exercises covering all aspects of the overall emergency response "system," including community and volunteer considerations with an inclusive lens. There is a need to integrate various professional groups (engineers, contractors, Organizations of Persons with Disabilities (OPDs), health professionals with expertise on disability, social workers, accessibility experts, etc.) into response preparedness in order to involve them efficiently and effectively in preparedness, response, and recovery operations (essential 09) [28].

A further challenge and need to consider vulnerability in policymaking is the climate change threat that is affecting people's lives now; populations around the world are experiencing intensive floods, storms, droughts, heatwaves, and wildfires, which are expected to get even worse in the coming years. At the same time, climate change is a risk amplification factor; rising temperatures [29] are negatively impacting ecosystems [30], sea levels, and storm surges, affect the rainfall patterns, but also amplify the intensity of extreme weather events; as a result, the increasing volatility and unpredictability of the impacts, exacerbate the vulnerabilities of the affected communities. The IFRC estimates that about 108 million people needed international humanitarian assistance in 2018 because of weather or climate-induced disasters, while many more people were affected and needed to rebuild their lives using their own resources - often with support from family, communities, and governments. If the projected increases are coupled with the frequency and intensity of weather extremes with the growing number of people living in poverty, about 200 million people a year are expected to need humanitarian assistance by 2050.

Disasters and conflicts themselves also play a major role in driving vulnerability and exposure to future hazards [31]. Disasters can keep people in, or return people to, poverty and other situations of vulnerability. When hazards combine, they can multiply each other's impact in ways governments, civil society, and the humanitarian sector have not faced before. These include not only rising climate- and weather-related threats, but also other shocks, such as pandemics and epidemics, earthquakes, and financial crashes [32].

According to the above, there are certain challenges to tackle vulnerability issues in disasters that are interconnected with inequalities: (a) lack of information and knowledge of disability issues among governments and relief organizations; (b) exclusion of persons with disabilities from disaster management and relief; (c) inaccessibility of physical environments, preparedness measures, shelters, and relief aid; and (d) stigma and discrimination [22]. Another challenge is the systemic risks; as multiple disasters increase in frequency and intensity there is less time for the communities to recover between them. These multiple shocks act like a domino effect that obstructs populations from acting timely and coping with all the possible risks; for example, the dangers of cyclones, flooding, drought, and heat waves existed while the world was adapting to the COVID-19 pandemic.

4. Advancing actionability of vulnerability dynamics and pattern

Advancing actionability among the involved parties for combating emerging risks in a scene of uncertainties, dynamic conditions, and multiple emergencies, as described in the previous sections, seems to be at the forefront of building disaster risk reduction strategies and making cities and communities resilient agendas [33]. In this direction, tackling vulnerabilities is a fundamental element for advancing people's coping capacity. However, trying to create just an inventory of vulnerabilities is not enough, if there is not a clear understanding of the mechanisms that activate or preserve such vulnerabilities, or if not be seen from a dynamic and contextual perspective. As the first step, different types of vulnerability patterns were reviewed, and it became apparent that vulnerability is a changing situation, challenging the inclusive and fair implementation of risk and disaster management [34]; this statement was also proved in Section 2 of this work.

As a step forward, in this section, there is an effort to provide a taxonomic approach of vulnerability precursors and relevant inhibitors, tackling the entire disaster management cycle; vulnerability "before, during, after" a disaster, addressing different types of vulnerability. This taxonomic approach could be used to assist the policymakers and the relevant authorities in identifying the critical points for building state-of-the-art disaster risk reduction and inclusive resilience strategies.

Specifically, an actionable roadmap is proposed, where critical vulnerability indicators (CVIs) are identified and classified as follows (see **Figure 1**):

i. Background: Intrinsic weaknesses or possible non-compliance conditions, as well as the social and economic status and inequalities.

ii. Human-environment interaction.

iii. Governance model; policies and approaches to disaster risk management.

iv. Means and tools to cope with disasters; application of new technologies and modeling tools.

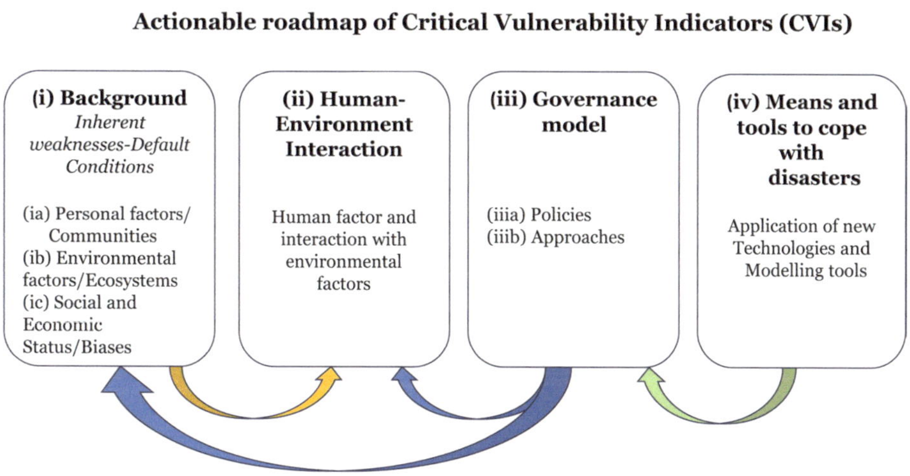

Figure 1.
Actionable roadmap of critical vulnerability indicators (CVIs) and their interdependencies.

The above indicators are to be considered in the context of the phases of the "disaster cycle", to represent what occurs "before, during, and after a disaster event". In this way it is easier to identify the different problems and weaknesses and hence, organize and apply the most efficient methods and available tools of intervention, aiming at reducing the consequences of an emergency event. The idea is that this roadmap is dynamic, meaning that it should be regularly updated based on the specific conditions and the characteristics of the studied region (demographics, social or financial status, geographical area, etc.).

To better understand the usability of the roadmap and the interdependencies among the proposed CVIs illustrated in **Figure 1**, an explanation per indicator is given as follows:

The CVI (i), "Background" describes the inherent weakness and is subdivided into three different categories: (ia)"Personal factors related to Communities"; for example, data gathering by the relevant authorities or the stakeholders, regarding the age, disabilities or impairments of the people that dwell in the area of their responsibility; the limitations and specific needs for assistance of those people before, during and after an emergency should be recorded by the relevant parties in order to prepare the evacuation plans accordingly, (ib) "Environmental factors related to Ecosystems"; this has to do with recording of parameters related to climate crisis and environmental degradation, or of the failures and mistakes of the built environment (e.g., cities built at areas directly exposed to hazards, such as wildfires or floods); non-accessibility standards of buildings or shelters obstructing the safe evacuation for all in case of an emergency; lack of universal design, (ic) Social and economic status; recording of inequalities relevant to gender, nationality, religion, culture; financial, social inequalities and lifestyle, e.g., recording the number of homeless people who are directly exposed to disasters, as well as the undocumented people, the refugees, the migrants, etc.; addressing the educational inequalities and find supportive financial mechanisms by the local government.

The CVI (ii), "Human-environment interaction" is considered critical for vulnerability assessment because it is related to the human factor; human variability, and interaction with various environmental factors. According to ICF [9] there are several environmental factors, meaning the physical, social, and attitudinal environment in which people live and conduct their lives, that can function either as barriers to or facilitators of the person's functioning. In that prospect, it is important to tackle the built environment usability and affordance, specifically for people with impairments; the lack of accessible environment is the reason for the vulnerability of those people, e.g., in case of an emergency evacuation, accessible design of the building affects the self-evacuation capacity of people with impairments. Another issue to tackle under CVI (ii), is people's risk perception in case of an emergency to be self-protected, and hence reduce their vulnerability; this requires a widespread and shared safety culture and familiarization of the population in following the guidelines by authorities, as well as training on how to react in emergencies and evacuate under an inclusive perspective. The relevant stakeholders should organize simulation exercises on a regular basis for inclusive disaster preparedness and response, as well as run campaigns for raising awareness, especially for the populations that dwell in disaster-prone areas; engaging citizens in disaster preparedness and response activities with the participation of local organizations, as well as organizations of persons with disabilities (OPDs) which are actively involved in planning for a potential disaster event in any area of the territory, as well as for responding to such events and for post-disaster reconstruction and rehabilitation.

The CVI (iii), "Governance model" is a critical factor that affects vulnerability, since it outlines the level of involvement of different parties, "before, during, and after a disaster", addressing the possible interdependencies and interactions among the various stakeholders and individuals that cope with disaster emergencies. This factor is subdivided into two different categories, namely (ia) "Policies"; for example, the locally-led action policy in terms of using local, traditional, and indigenous knowledge in disaster risk management; multi-sectoral cooperation in disaster management; horizontal application of accessibility strategies in all sectors, as well as risk-informed and inclusive policy-making, by engaging vulnerable groups of population in disaster planning and response; civil protection guidelines provided in accessible formats, and (ib) "Approaches"; for example, application of a bottom-up combined with a top-down approach in disaster risk management by the governments, under a people-centered framework; promote safety culture at multi-sectoral level; involvement of non-governmental organizations (NGOs) in the disaster management cycle phases, namely "before, during, after a disaster".

Finally, the CVI (iv), "Means and tools to cope with disasters" summarizes all the available capabilities to restrain the vulnerability drivers that can be identified based on CVIs (i) to (iii), explained above. It includes the possible application of new technologies and modeling tools for anticipating risks. In that context, this may include tools for accessible broad-casting of emergency messages (multilingual alarms); inclusive and multi-hazard early warning systems; evacuation and risk models; inclusive training platforms and accessible smart-phone apps for emergencies; accessible crowdsourcing and social networks; data aggregation of disaster risks and big data management using AI; integration of robotic platforms, such as UAVs or UGVs in emergency management, before, during, and after a disaster, considering also the human-machine interface interaction. It is apparent that the list of available means and tools for coping with disasters should be dynamically updated to be state-of-the-art.

In **Table 1**, it is shown how the CVIs described above could be used by the policy-makers and the relevant authorities, firstly, to identify all the possible vulnerability drivers (vulnerability precursors and inhibitors), as well as the interdependencies among them (see **Figure 1**) and secondly, to plan accordingly the prevention and mitigation measures for building resilience strategies, based each time on the available means and approaches, as well as the governance model; this should be done "before, during, and after a disaster", considering the vulnerability dynamics under an inclusive perspective.

Disaster phases	Before	During	After
Vulnerability drivers:	Vulnerability Precursors and Inhibitors	Vulnerability Precursors and Inhibitors	Vulnerability Precursors and Inhibitors
	Background (CVI-i) Human-Environment Interaction (CVI-ii)		
Disaster management cycle:	Preparedness-Prevention	Response	Recovery
Coping capacity/ resilience:	Governance Model(CVI-iii)/Instruments &Means (CVI-iv)		

Table 1.
Interconnection of vulnerability "before-during-after" a disaster with the CVIs presented in **Figure 1***, inside the disaster management cycle [34].*

Vulnerability types	Vulnerable group
1. Vulnerability per age	Sub-groups 1.1. (0–5 years old); Newborn (ages 0–4 weeks); Infant (ages 4 weeks - 1 year); Toddler (ages 1–3 years) Preschooler (ages 3–5 years) 1.2 (6–18 years old); School-aged child (ages 6–13 years); Adolescent (ages 14–18 years) 1.3 (19–29 years old); Young adult 1.4 (30–64 years old); Adult 1.5 (>65 years old); Youngest-old (ages 65–74 years); Middle-old (ages 75–84 years); Oldest-old (ages more than 85 years)
2. Vulnerability per type of Disability/ Impairment or Other diseases	Sub-groups 2.1 People with movement disability, e.g. lower or upper limps impairment and/or body disability; Wheelchair users; Users of walkers or canes 2.2 People with sensorial disability; Deaf, Blind; Deafblind 2.3 People with cognitive disabilities/Neurodevelopmental Disorders; Intellectual Disabilities; Communication Disorders; Autism spectrum disorder; Learning disorder; Down syndrome (Neurogenetic disorder), etc. 2.4 People with Schizophrenia and Other Phychotic Disorders; Depressive disorder; Bipolar disorder; Anxiety disorder; Epilepsy, etc. 2.5 People with Neurodegenerative disorders or neurological problems; Alzheimer disease (dementia); Parkinson's disease; Multiple sclerosis; Fibromialgia, etc. 2.6 Substance-Related and Addictive Disorders; Alcohol use disorder; Drug addiction, etc. 2.7 People with other diseases; COVID-19; Hidden disabilities (i.e. early-stage illness such as dementia, SLA, of chemo-therapy patients); Chronic disease e.g., respiratory problems (asthma), high blood pressure disorders, diabetes, allergies, etc. 2.8 People after surgery or accidents; Temporary disability (movement, sensorial, etc.); Permanent disability (movement, sensorial, etc.)
3. Vulnerability due to Pregnancy	No sub-groups were identified for the respective groups on the left column
4. Vulnerability of caregivers and personal assistants	
5. Vulnerability of tourists, foreign workers/exchange students	
6. Vulnerability of homeless people	
7. Vulnerability of discriminated people	
8. Vulnerability of people without a social safety network/Internally displaced people	
9. Vulnerability of Undocumented people/Illegal Migrants	
10. Vulnerability of Refugees/ Asylum seekers/Migrants	

Table 2.
Vulnerability types and related vulnerable groups and/or sub-groups.

This approach has been recently introduced and elaborated under the CORE H2020 project. Specifically, vulnerability analysis "before, during, and after a disaster" was performed for selected "vulnerable groups", explaining why these people are at high risk in emergencies based on the critical vulnerability indicators presented in **Figure 1**. In that context, ten (10) vulnerability sources were identified, together with specific sub-groups (**Table 2**) [34].

In order to better understand how to analyze vulnerability "before, during, and after", considering the human variability, indicative examples are given in **Table 3**, explaining why these people are at risk in each phase of the disaster cycle [34].

Table 3 of the vulnerability analysis indicates the big picture of the problems and weaknesses, as well as the needs and the interaction with the environment for indicative groups of people in each disaster phase, so that to be considered by the policy makers or the relevant authorities for preparing disaster preparedness and response plans, or other prevention and mitigation measures with inclusive criteria. This analysis, together with the roadmap of the critical vulnerability indicators (CVIs) previously showcased, could be a basis for further elaboration on the direction of building disaster resilience strategies.

5. Conclusions

Based on the findings of this work, it seems that to increase communities' resilience, it is necessary to consolidate a new approach that involves multiple aspects and factors in relation to human variability and behavior within the actual contingency conditions and the real physical environment. Climate change, population growth, or migration can also contribute to the variability of disaster risks, emphasizing the need for inclusive strategies. Human and environmental factors, such as settlement patterns, urbanization, and socio-economic conditions, significantly determine disaster risk, with inequalities exacerbating vulnerability [35]. Addressing these inequalities is essential for effective disaster risk reduction [36]. Reducing risk goes beyond minimizing the damage caused by disasters and involves building a culture of safety and resilience, as well as addressing and underlying the risk factors [28]. Policies and practices for disaster risk management should be based on an understanding of disaster risk in all its dimensions of vulnerability, capacity, and exposure of persons and assets [37].

In this context, our work highlights the importance of developing a taxonomic approach for facilitating vulnerability understanding and assessment, as well as for helping in building strategies to anticipate disaster risks and effectively act "before, during, and after" a disaster with inclusive criteria. This work provides policymakers or the relevant authorities with an actionable conceptual tool that could help them identify critical vulnerability indicators (CVIs), support the understanding of vulnerability mechanisms, and improve the management of contextual variables.

Specifically, the proposed (CVIs) included: (i) the background; intrinsic weaknesses or possible non-compliance conditions, as well as the social and economic status; (ii) the human-environment interaction; (iii) the governance model; policies and approaches to disaster risk management; (iv) the tools; application of new technologies and modeling tools. Indicative examples were given to explain the four suggested (CVIs). Tackling vulnerabilities under the proposed approach facilitates the identification of the possible problems and weaknesses per disaster phase and

Vulnerability type	Before	During	After
Age Group: 0–5 years old	• They are depended on their parents, relatives, babysitters or any other helpers • They are not aware of risks • Limited interaction with environment • Non or inadequate training of their supportive members in emergency preparedness/response	• None or limited understanding of the emergency risk /alarms • None or limited evacuation capability on their own • Inability or difficulty to reach safe places on their own	• Demanding recovery (after-shock) • Regular and on-going support by child psychologists • Medical support in case of injury • Financial support needs • Safe Shelter: food suitable for 0–5 years/water needs
Age Group: >65 years old	• Impairments/Difficulties: • *Mental* • *Physical* • *Sensory* • *Mobility* • *Communication* • *Perception* • Physical decline • Age discrimination • Inadequate services for older people/poverty	• Limited risk perception • Low accessibility or understandability of emergency messages • Lack of knowledge on how to react in emergencies/wrong decision making/no self-protection • Difficulty or inability to evacuate and reach accessible safe places by themselves/possibly need help • Difficulty into following guidelines by authorities/cooperate with the rescuers	• On-going support by psychologists/psychiatrists • Medical support in case of injury/ Difficult health recovery • Financial support needs • Safe Shelter/food /water needs/access to medicine
Persons with disabilities Group: People with cognitive disabilities or Neurodevelopmental Disorders	• Low-risk perception • Limited interaction with environment/communication • Limited understanding of disaster risks • Lack of awareness • Inaccessible disaster preparedness plans • Systemic discrimination	• Difficulty to perceive or comprehend the severity of an emergency situation or the need of evacuation in emergency • Difficulty or inability to understand messages/alarms/signs • Difficulty or inability to evacuate by themselves and reach accessible shelters • Difficulty to follow guidelines by authorities or communicate/interact with the rescuers	• On-going support by psychiatrists/special treatment experts • Medical support in case of injury/ Trained paramedics and doctors are needed based on the specific type of impairment • Financial support needs • Difficulty in mental health recovery • Working place or School adaptation difficulties • Safe Shelter: food/water needs/access to medicine

Vulnerability type	Before	During	After
Homeless people	• Directly exposed to environmental hazards • Poor lifestyle • Lack of safety culture • Limited resources • Social isolation • Lack of access to housing and other material needs • Disabilities, chronic physical conditions, and behavioral health needs	• Lack of home/shelter that can be protected • Low accessibility to information and early warning • Lack of knowledge on how to react in emergencies • Difficulty or inability to evacuate by them-selves and reach safe places • Difficulty to communicate and interact with the rescuers	• On-going support by psychologists/sociologists • Medical support in case of injury • Financial support needs/ difficulty to find a safe place to recover • Legal/therapy support if needed to re-integrate them back to society. • Safe Shelter: food /water needs/access to medicine
Tourists, foreign workers/exchanged students	• Limited understanding of disaster risks in a foreign country • Limited communication capacity due to language • Limited or no understanding of signs, labels, escape plans due to language	• Limited understanding of emergency mes-sages in a foreign country • Limited communication capacity due to language in case of emergency • Limited interaction with the rescuers due to language or cultural specificities • Limited or no understanding of emergency signs, labels, escape plans due to language that obstruct them from reaching a safe refuge area	• On-going support by psychologists/sociologists • Medical support in case of injury • Financial support needs in • cooperation with their country • Safe Shelter: food /water needs/free access to medicine • Need to contact with relatives abroad, or embassies of the home countries

Table 3.
Example of vulnerability dynamics "before-during-after" a disaster for indicative groups: why are they at risk? (adapted from [34]).

the definition of vulnerability drivers. It also helps to understand interactions and interdependencies among the different sectors so that we can timely engage all the necessary parties in disaster management and strengthen the coping capacity of the communities at risk, using each time the proper state-of-the-art prevention and mitigation tools each. It should be highlighted that the proposed vulnerability taxonomic approach complies with the overall UNDRR Disaster Resilience Scorecard for Cities approach [28], and can be used complementarily in a variety of contexts.

As a bottom line, it seems that trying to create just an inventory of vulnerabilities is not enough to strengthen the disaster resilience of communities. Based on the surveyed vulnerability models, it became evident that interactions among the identified layers and categories of variables so far have only been discussed at a higher level. This fact makes it challenging to apply them in practice for identifying and managing specific factors that influence the vulnerability of individuals or small groups on a local scale. On the other hand, it is important to realize that the one solution-fits-all models cannot be applied, especially under an inclusive approach; it is important to take into account human variability and behavior, considering the spectrum of functioning in a broader sense and interaction with the physical environment at multiple scales, and adopting an inclusive dynamic approach for building disaster resilience. The identification of critical points (Critical Vulnerability Indicators - CVIs), as well as the vulnerability types "before, during, and after" a disaster provided in this work could, represent an effective and dynamic approach to manage vulnerabilities with inclusive criteria, in which all social, economic, and political aspects are analyzed, transversally, considering them and their impact on individuals in the whole disaster cycle. This approach allows to address multifaceted issues related to vulnerability, with the aim of reducing the consequences of a critical event and taking an extra step toward an inclusive safety concept, going beyond the idea of classifying vulnerability factors and vulnerable groups into a fixed scheme and allowing the consideration of a wider spectrum of human variability within specific and real contexts. It could also support the advances in the design and use of new methods, tools, and technological solutions as a tangible step toward collective resilience, systematizing the entire supply chain, and engaging each individual actor or community in the broadest sense of disaster resilience.

Acknowledgements

This paper includes research results developed as part of the CORE — Science and Human Factors for Resilient Society project, which has received funding from the European Union's Horizon 2020 Research and Innovation Program under grant agreement No. 101021746.

Conflict of interest

The authors declare no conflict of interest.

Author details

Sofia Karma[1*], Stefano Zanut[2], Monica Crișan[3], Consuelo Agnesi[4] and Gabriella Duca[4]

1 National Technical University of Athens, Greece

2 National Fire and Rescue Service, Pordenone, Italy

3 Red Cross Vicenza Branch, Italy

4 Institute for Sustainable Society and Innovation, Naples, Italy

*Address all correspondence to: sofia.karma@gmail.com

References

[1] Mizutori M. Reflections on the Sendai framework for disaster risk reduction: Five years since its adoption. International Journal of Disaster Risk Science. 2020;**11**(2):147-151

[2] EU, sCience and human factOr for Resilient sociEty [Internet]. 2021. Available from: https://cordis.europa.eu/project/id/101021746 [Accessed: April 15, 2024]

[3] Wisner B. Vulnerability as concept, model, metric, and tool. In: Oxford Research Encyclopedia of Natural Hazard Science. 31 Aug 2016. Available from: https://oxfordre.com/naturalhazardscience/view/10.1093/acrefore/9780199389407.001.0001/acrefore-9780199389407-e-25 [Accessed: April 15, 2024]

[4] de Ruiter MC, van Loon AF. The challenges of dynamic vulnerability and how to assess it. iScience. 2022;**25**(8):104720. DOI: 10.1016/j.isci.2022.104720

[5] Clark B, Preto N. Exploring the concept of vulnerability in health care. Canadian Medical Association Journal. 2018;**190**(11):E308-E309. DOI: 10.1503/cmaj.180242

[6] UNDRR. Sendai Framework terminology on Disaster Risk Reduction. Vulnerability [Internet]. Available from: https://www.undrr.org/terminology/vulnerability [Accessed: April 15, 2024]

[7] Bohle H, Downing T, Watts M. Climate change and social vulnerability. Global Environmental Change. 1994;**4**(1):37-48

[8] Pelling M. Review of global risk index projects: Conclusions for sub-national and local approaches. In: Birkmann J, editor. Measuring Vulnerability to Natural Hazards. 2nd ed. Tokyo: United Nations University Press; 2013. pp. 167-196

[9] World Health Organization, ICF: International Classification of Functioning, Disability and Health [Internet]. 2001. Available from: https://www.who.int/standards/classifications/international-classification-of-functioning-disability-and-health. [Accessed: April 15, 2024]

[10] Bukvic O, Carlsson G, Gefenaite G, Slaug B, Schmidt SM, Ronchi E. A review on the role of functional limitations on evacuation performance using the international classification of functioning, disability and health. Fire Technology. 2021;**57**(2):507-528

[11] Birkmann J, Cardona O, Carreño M, Barbat A, Pelling M, Schneiderbauer S, et al. Framing vulnerability, risk and societal responses: The MOVE framework. Natural Hazards. 2013;**67**(2):193-211

[12] Birkmann J. Measuring vulnerability to promote disaster-resilient societies: Conceptual framework and definitions. In: Birkmann J, editor. Measuring Vulnerability to Natural Hazards. Tokyo: United Nations University Press; 2006. pp. 9-54

[13] Wisner PB. At risk: Natural hazards, people's vulnerability and disasters. In: Wisner PBB, editor. At Risk. London: Routledge; 2003. pp. 50-51

[14] Bourque L, Siegel J, Kano M, Wood M. Morbidity and mortality associated with disasters. In: Rodriguez H, Quarantelli E, Dynes R,

editors. Handbook of Disaster Research. New York: Springer; 2007. pp. 97-112

[15] Frerks G, Bender S. Conclusion: Vulnerability analysis as a means of strengthening policy formulation and policy practice. In: Bankoff G, Frerks G, Hilhorst D, editors. Mapping Vulnerability: Disasters, Development and People. London: Earthscan; 2004. pp. 194-205

[16] Cannon T. Reducing people's Vulnerability to Natural Hazards: Communities and Resilience. Research Paper No. 2008/34. Helsinki, Finland: WIDER; 2008

[17] Turner B, Kasperson R, Matson P, McCarthy R, Corell R, Christensen L, et al. A framework for vulnerability analysis in sustainability science. Proceedings of the National Academy of Sciences. 2003;**100**(14):8074-8079

[18] Oliver-Smith A, Alcántara-Ayala I, Burton I, Lavell AM. Forensic Investigations of Disasters (FORIN): A Conceptual Framework and Guide to Research (IRDR FORIN Publication No. 2). Beijing: Integrated Research on Disaster Risk; 2016. p. 23

[19] Hufschmidt G. A comparative analysis of several vulnerability concepts. Natural Hazards. 2011;**58**:621-643

[20] UNDRR. Vulnerability. 2017. [Internet]. Available from: https://www.preventionweb.net/understanding-disaster-risk/component-risk/vulnerability [Accessed: April 15, 2024]

[21] OCHA. 2022 Disasters in numbers [Internet]. 2023. Available from: https://reliefweb.int/report/world/2022-disasters-numbers [Accessed: April 15, 2024]

[22] IFRC. World Disasters Report 2020 [Internet]. 2021. Available from: https://www.ifrc.org/sites/default/files/2021-09/20201113_WorldDisasters_2.pdf [Accessed: April 15, 2024]

[23] UNDRR. Global Survey Report on Persons with Disabilities and Disaster 2023. [Internet]. Available from: https://www.undrr.org/report/2023-gobal-survey-report-on-persons-with-disabilities-and-disasters [Accessed: April 15, 2024]

[24] UN. Convention on the Rights of Persons with Disabilities. 2006 [Internet]. Available from: https://www.un.org/development/desa/disabilities/convention-on-the-rights-of-persons-with-disabilities/article-1-purpose.html [Accessed: April 15, 2024]

[25] WHO. Disability. [Internet]. Available from: https://www.who.int/health-topics/disability#tab=tab_1 [Accessed: April 15, 2024]

[26] Kett M, Twigg J. Disability and disasters: Towards an inclusive approach. In: IFRC, World Disasters Report. Focus on Discrimination. Geneva: IFRC; 2007. pp. 86-111. Available from: http://www.ifrc.org/Global/Publications/disasters/WDR/WDR2007-English.pdf [Accessed: April 15, 2024]

[27] Priestley M, Hemingway L. Disability and disaster recovery. Journal of Social Work in Disability and Rehabilitation. 2007;**5**(3-4):23-42. DOI: 10.1300/J198v05n03_02

[28] UNDRR. Disaster resilience scorecard for cities. Annex for the inclusion of persons with disabilities V1.0. 2022. [Internet]. Available from: https://mcr2030.undrr.org/sites/default/files/2023-11/undrr_disabilityinclusionscorecard-v1.0_english-dec2022-accessibledesign-rev2.pdf [Accessed: April 15, 2024]

[29] IPCC. Climate change widespread, rapid, and intensifying. 2021. [Internet]. Available from: https://www.ipcc.ch/2021/08/09/ar6-wg1-20210809-pr/ [Accessed: April 15, 2024]

[30] Scholes RJ. Climate change and ecosystem services. Wiley Interdisciplinary Reviews: Climate Change. 2016;**7**(4):537-550

[31] ICRC. When rain turns to dust. 2020 [Internet]. Available from: https://www.icrc.org/en/publication/4487-when-rain-turns-dust [Accessed: April 15, 2024]

[32] Smith F, Jolley E, Schmidt E. Disability and Disasters: The Importance of an Inclusive Approach to Vulnerability and Social Capital. Haywards Heath: Sightsavers; 2012

[33] UNDRR. Disaster Resilience Scorecard for Cities, Making Cities Resilient, MCR2030 [Internet]. 2017. Available from: https://mcr2030.undrr.org/disaster-resilience-scorecard-cities [Accessed: April 15, 2024]

[34] CORE (sCience& human factOr for Resilient sociEty), D3.1 Critical Analysis of Past Disasters [Internet]. 2023. Available from: https://ec.europa.eu/research/participants/documents/downloadPublic?documentIds=080166e5f864d5d0&appId=PPGMS [Accessed: April 15, 2024]

[35] Raikes J, Smith TF, Baldwin C, Henstra D. Linking disaster risk reduction and human development. Climate Risk Management. 2021;**32**:100291

[36] GSDRC. Climate change and social development 2016. [Internet]. Available from: https://gsdrc.org/topic-guides/climate-change-and-social-development/reducing-risk-and-building-adaptive-capacity/disaster-risk-reduction [Accessed: April 15, 2024]

[37] Khalid Z, Meng X, Rana IA, ur Rehman M, Su X. Holistic multidimensional vulnerability assessment: An empirical investigation on rural communities of the Hindu Kush Himalayan region, Northern Pakistan. International Journal of Disaster Risk Reduction. 2021;**62**:102413

Chapter 11

Science-Policy Interfacing and Community-Building for Disaster Risk Reduction in the Area of Natural Hazards

Philippe Quevauviller

Abstract

Disaster Risk Reduction (DRR) is at the heart of international and EU policies, which involve a wide range of actors and cover multiple threats of natural and man-made (accidental or intentional) origins. In the natural hazards sector, the increasing severity and frequency of extreme weather/climate events, as well as a range of geohazards, are triggering a high level of research worldwide. DRR research covers various risk management features supporting policy development and implementation. In this respect, a huge body of knowledge has been developed in successive EU Research and Development Framework Programmes in the last 15 years. In addition, networking initiatives connected scientists, policy-makers, practitioners, industry and civil society representatives in order to boost research uptake, identify gaps and elaborate research programs at the EU level. Research and networking efforts are crystallising into a wide community-building effort that gathers the above-mentioned DRR actors. This chapter will provide a general overview of these initiatives.

Keywords: science-policy interfacing, community-building, disaster risk reduction, natural hazards, EU research

1. Introduction

1.1 General framework

Disaster Risk Reduction (DRR) policies follow an integrated approach to the management of natural and human-made hazards focusing on prevention, preparedness and response, recovery and learning [1]. The 2015–2030 Sendai Framework for Action [2] "Building the resilience of nations and communities to disasters" adopted by 168 UN member states sets a voluntary commitment towards the improvement of disaster resilience and disaster risk reduction as essential ingredients for the achievement of poverty reduction and sustainable development. It closely relates to EU disaster risk management policies, in particular the EU Civil Protection Mechanism [3]. Disaster risk management is also addressed through the EU Internal Security

Strategy [4] and the resulting European Agenda on Security [5] and Consumer Health Protection policies [6]. In addition, climate-related disasters are covered by environmental and climate policies (DG ENV through the Flood Directive [7] and DG CLIMA through the EU strategy on adaptation to climate change (**Figure 1**) [8, 9].

Further development and implementation of these policies represent a complex and ambitious challenge as they involve a wide variety of players. It involves countries with specific national approaches (national action plans) for dealing with crises. These are also differently organised in terms of disaster risk management capabilities to be adapted to the diverse context. The European Union (EU) framework represents a means and a real opportunity to discuss possible ways to improve coordination and knowledge sharing among the various national approaches. It enables the development of a common EU vision strengthened by a joint strategy and its implementation in this field. This requires bringing together key scientific, policy and industry actors (including SMEs and start-ups), practitioners (e.g. first responders, police representatives, fire fighters, civil protection units), NGO's, citizens as well as other stakeholders.

1.2 Civil protection policies

The Union Civil Protection Mechanism (UCPM) aims to facilitate reinforced cooperation and coordination between the European Union and the EU Member States (MS) in the field of civil protection, to improve the effectiveness of systems for preventing, preparing for, and responding to natural and man-made disasters. It supports and complements the efforts of the MS for the protection, primarily of people but also of the environment and property, including cultural heritage, in the event of disasters. Built upon these policy instruments and on principles of solidarity, the UCPM targets the development of an integrated approach to disaster management. The overall mechanism takes due consideration of laws and international commitments and exploits synergies with relevant Union initiatives such as the European

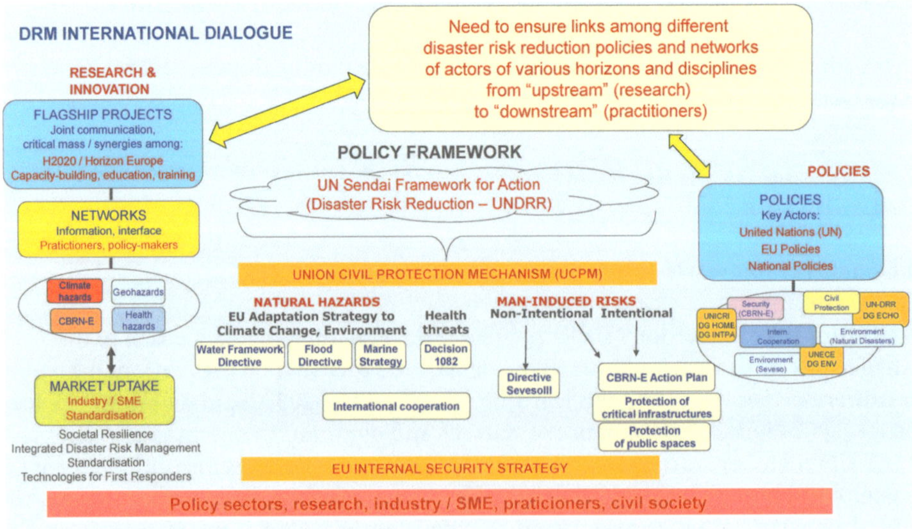

Figure 1.
International disaster risk management landscape (from an EU perspective).

Earth Observation Programmes (Copernicus). The UCPM is operationalised through the Emergency Response Coordination Centre (ERCC) and RescEU in the form of a voluntary pool of pre-committed capacities from the MS and trained experts. It also recognises the role of regional and local authorities in disaster management. Outside the Union, disaster response is coordinated with the United Nations and other relevant international actors. Finally, the use of military means under civilian direction may constitute an important resource to disaster response.

On technical grounds, the UCPM is working towards a general policy framework on disaster risk prevention aimed at achieving a higher level of protection and resilience against disasters by preventing or reducing their effects and by fostering a culture of prevention. From this perspective, it promotes the review of risk assessment, risk management planning conducted at national/regional level, and the development of an integrated approach, linking risk prevention, preparedness and response actions. Among the priorities is the action to improve the knowledge base on disaster risks and facilitate the sharing of knowledge, recommended practices and information, which is one of the core objectives of the recently launched Union Civil Protection Knowledge Network.

The UCPM is also financing actions related to preventing, preparing for and responding to disasters. These include EU Civil protection training programmes, regular large-scale exercises, exchange of experts, prevention and preparedness capacity-building projects (through annual calls for applications), logistical and transport support for response missions, the development of crisis response capacities under RescEU.[1]

At the international level, the UCPM is closely related to the Sendai Framework for Action 2015–2030 [2], which sets out an ambitious set of priorities to place disaster risk reduction as a key element of sustainable development efforts, to define further steps to reduce existing and emerging risks and foster disaster resilience. As stressed in Council Conclusions on this matter, the EU supports a framework that strengthens the contribution of disaster risk management to smart, sustainable and inclusive growth by promoting the use and development of innovative technologies and encouraging a more systematic and reinforced science-policy interface in disaster risk management. These objectives are supported by Intergovernmental Panel on Climate Change (IPCC) recommendations expressed in the special report on extreme events.[2]

1.3 EU strategy on adaptation to climate change

The EU strategy on adaptation to climate change [8] highlights the consequences of climate change and the need for adaptation measures. It gives priority to early, planned and coordinated action rather than reactive adaptation. The Strategy highlights the need for exchanges of best practice on how to best adapt to climate change, also taking into account global impacts that, for example, lead to disruptions in the EU of supply chains, or impaired access to raw materials, energy and food supplies. Its overall aim is to contribute to a more climate-resilient Europe by enhancing the preparedness and capacity to respond to the impacts of climate change at local, regional, national and EU levels, developing a coherent approach and improving coordination.

---

[1] https://ec.europa.eu/echo/what/civil-protection/resceu_en

[2] Managing the Risks of Extreme Events and Disasters to Advance Climate Change Adaptation (SREX) – https://www.ipcc.ch/report/managing-the-risks-of-extreme-events-and-disasters-to-advance-climate-change-adaptation/

The Strategy features a novel international component, aligning with the Paris Agreement and other relevant international policies. A close coordination between climate change adaptation and disaster risk management and policies is also required. Several of the actions proposed, including those related to international aspects of adaptation, are of direct relevance and in full synergy with DRR.

1.4 Water and marine policies

Specific policy instruments are in place in the water sector related to extreme hydrometeorological events such as floods and droughts, in particular the Water Framework Directive [10], which provides a legal tool to address water scarcity,[3] by requiring long-term planning of water resources, economic tools to incentivise efficient use of water resources and specific measures on water abstraction and water efficiency. The strategic approach has been taken forward with the 2007 Commission Communication on water scarcity and droughts [11]. The WFD includes all elements needed to address adaptation to climate change. The definition of water status allows to cater for climate change impacts, and the cyclical approach to regularly update River Basin Management Plans provides the opportunity to constantly re-assess the actual status of water bodies and take the most effective actions to preserve them. The 2019 Fitness Check of EU water law has shown that the Water Framework Directive is flexible enough to address also drought and water scarcity issues, as it requires Member States to achieve the overarching goal of good status of all their waters, and when needed drought management plans should be elaborated and implemented. Finally, the above-mentioned EU strategy on adaptation to climate change calls for the wider use of drought management plans.

Flood prevention and management are tackled by the Flood Directive [7], which requires EU Member States to assess and manage flood risks, with the aim of reducing adverse consequences for human health, the environment, cultural heritage and economic activity associated with floods in Europe. The Floods Directive, which coordinates its objectives with those of the WFD, has no deadlines (in recognition of the ever-changing nature of flooding risk), but it provides a mechanism for assessing and monitoring increased risks of flooding, taking into account the possible impacts of climate change, and for developing appropriate adaptation approaches.

The Marine Strategy Framework Directive [12] is the EU's main tool for marine environmental protection. It looks at all pressures on the seas and ocean and brings them together under one umbrella to tackle the cumulative impacts of human activities in one strategic framework. The Marine Strategy Framework Directive requires EU Member States to set up marine strategies in respect of the marine region of the Member States' marine waters to achieve "Good Environmental Status", one of the criteria of which is the level of chemical pollution in the marine environment. This includes acute pollution events, such as discharges, spills and emissions of hydrocarbons and other harmful chemicals from ships or offshore oil and gas installations, or shipping accidents involving loss of containers. The 2020 report on the implementation of the Directive has shown that a reduction of such events has occurred in the Baltic and Northeast Atlantic regions but more efforts are still needed in the Mediterranean and Black Sea regions.

---

[3] Water scarcity is a demand and supply issue, in which the increasing demand is greater than the available resources.

1.5 Serious cross-border threats to health

The protection of human health is a matter, which has a cross-cutting dimension and is relevant to numerous EU policies and activities. The Commission should ensure, in liaison with the Member States, the coordination and exchange of information between the mechanisms and structures established under Decision 1082/2013/EU [6] on serious cross-border threats to health as well as activities that are relevant to preparedness and response planning, monitoring, early warning of, and combating serious cross-border threats to health. Apart from communicable diseases, a number of other sources of danger to health, in particular, related to other biological or chemical agents or environmental events, which include hazards related to climate change, could by reason of their scale or severity, endanger the health of citizens in the entire Union, lead to the malfunctioning of critical sectors of society and the economy and jeopardise an individual Member State's capacity to react. The legal framework set up under the above Decision should, therefore, be extended to cover other threats and provide for a coordinated wider approach to health security at the Union level. In the context of this Decision, an important role in the coordination of recent crises of Union relevance has been played by an informal group composed of high-level representatives from Member States, referred to as the Health Security Committee, which plays an important role in responding to health threats (notably in terms of crisis preparation, exercises on CBRN events and the listing of pathogens and chemicals which pose a health threat) whilst the European Centre for Disease and Control (ECDC) provides risk assessments for communicable diseases and biological incidents. To address the COVID-19 pandemic, the Commission adopted diverse measures such as guidelines on EU Emergency Assistance on Cross-Border Cooperation in Healthcare related to the COVID-19 crisis and recommendations on a common Union toolbox for the use of technology and data to combat and emerge from the COVID-19 crisis, in particular concerning mobile applications and the use of anonymised mobility data.

1.6 Critical infrastructure protection and critical entities resilience

The EU Security Union Strategy for 2020–2025 stresses the importance of ensuring resilience in the face of various risks. The smooth functioning of the internal market depends on the reliable provision of services that are essential for societal and economic activities in various sectors. Those essential services often are reliant upon one another, thus disruptions in one sector can generate severe and long-lasting effects on the provision of services in others. Those situations have significant consequences for individual EU Member States and across the EU, and the functioning of the Internal Market. It is therefore important that entities providing essential services and the infrastructure they rely on are resilient against natural and man-made threats. The EU Directive on the resilience of critical entities (CER Directive) underpins these services [13] in ten sectors (energy, transport, banking, financial market infrastructures, health, drinking water, wastewater, digital infrastructure, public administration and space) and creates a framework to ensure greater harmonisation of rules and to support Member States in ensuring that critical entities are resilient against disruptive incidents, no matter if they are caused by natural hazards, accidents, terrorism or public health emergencies. The CER Directive introduces an obligation for Member States to have a strategy for ensuring the resilience of critical entities, carry out a national risk assessment and identify critical entities. Those critical entities will be required to carry out their own risk assessments, take appropriate technical and organisational measures in order to enhance their resilience and report disruptive incidents to national authorities.

2. DRR actors/interfacing

Our societies nowadays have to deal with complex and transboundary crises within which a more systemic approach with strict interconnection between risk reduction and sustainable development is needed. Partnerships and collaboration are mandatory to jointly address these challenges, in research, policy development and their implementation [14].

As described in Section 1 of this chapter, risk reduction of any kind of disaster is regulated by a number of international, EU, national and local policies and strategies covering various sectors and features, such as awareness raising, prevention, mitigation, preparedness, monitoring and detection, response and recovery. In this respect, the implementation of international policy initiatives (e.g. the Sendai Framework for Disaster Risk Reduction) requires cross-border and cross-sectoral cooperation as well as enhanced collaboration among different actors and strengthened knowledge covering the whole disaster management cycle, from prevention and preparedness to response and recovery (and learning).

Policy and research development and implementation, require establishing dialogues between a wide range of actors, naturally including policy-makers and scientists, as well as practitioners, civil protection units, industry/SMEs and civil society. While different sectors (e.g. health, civil protection, environment, security) face common features regarding the overall crisis risk management cycle (from prevention to preparedness, detection, response and recovery), they have all specific needs and practices related to their field operations, which condition the proper transfer (and implementation) of research outputs. In other words, this diversity of actors requires that dissemination and communication of project results be tailor-made to different sectors and matching policy needs. The large span of research (and capacity-building) projects often leads to fragmentation of knowledge and sometimes to dispersion of resources (e.g. risks of duplication, insufficient critical mass, etc.). The high level of fragmentation of information often leads to a poor awareness of policy requirements by research and industry communities and a poor transfer of research results to policy and stakeholder communities. There is hence a strong need to establish a mechanism enabling better information exchanges with regular updates for all possibly interested organisations and effective interactions among projects and between different communities. The transfer of knowledge and synergy-building needs are manyfold, they concern interactions among policies (covering various sectors), scientific disciplines, and interfacing about science, policy and "users" (e.g. practitioners). This transfer also requires efficient relays from the international/EU to the national levels (e.g. through advisory groups/committees linked to policy entities (UN, EU, Ministries/Agencies), as well as from national to regional/local levels where operational actions take place in case of disasters.

Bringing together multiple actors from various sectors and disciplines has a very strong impact on various aspects of science-policy interactions. From the policy viewpoint, access to research and innovation at different levels in the policy cycle (design, development, implementation, review) relies on various research dynamics (programming to uptake) in which all DRR actors should in principle be involved (**Figure 2**).

This interfacing mechanism has been used in the water policy sector for more than 15 years [15, 16] and is operational in many policy sectors including DRR. It contributes to enhanced cross-sectoral collaboration, cross-border interactions and cross-discipline dialogue and networking between the scientific community, research institutions and programmes, fostering a faster transfer of results from science into practice. It has a close link with research and capacity-building programmes and networking initiatives referred to in Section 4. This long learning process is expected to be strengthened in the forthcoming policy/research programming cycles.

Complementing this interfacing mechanism, the fact that various DRR actors are now used to work together in research and/or capacity-building initiatives naturally tends to create strong synergies among projects from different origins (i.e. funding by different financial instruments). Besides the direct impact on improved societal resilience and strengthened capacities of first responders in all operational phases related to any kind of natural and man-made) disasters, this move to project clustering has a high potential in terms of building up a critical mass of resources and issuing stronger recommendations to DRR policies actors. This has an impact on cross-border coordination of the disaster risk management cycle and governance from international to local levels. This trend is now embedded into community-building developments (see Section 4) and contributes to enhanced exploitation of the latest scientific results and integrated technologies into enhanced understanding of high-impact hazards and improved prevention, preparedness for mitigation, response and recovery tools. Another important lesson learned is related to networking involving as many DRR actors as possible to exchange information and experiences, thus strengthening the interfacing depicted in **Figure 2**. What remains to be developed is a mechanism to build-up an international network of (international/national/regional) networks. Such entities readily exist but the situation remains fragmented, which is largely due to the fact that civil protection policies are of regalian nature, which may limit

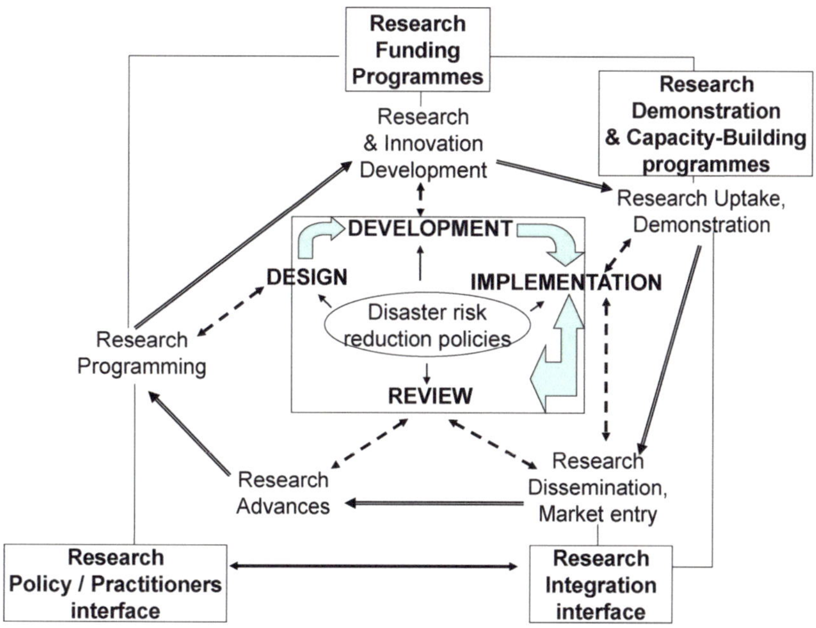

Figure 2.
Science-policy interfacing.

larger networking. In this respect, UNDRR might have a role to play to encourage and implement such network of networks. Coming closer to the regional/local levels would have a strong impact on enhanced understanding and improved knowledge and situational awareness of disaster-related risks by citizens, empowered to act and consider innovative solutions, thus raising disaster resilience.

3. DRR research trends

3.1 Introduction

Within the EU Horizon Europe Framework Programme (2021–2027), the Cluster 3 "Secure Societies" funded and is currently funding research actions that support the implementation of EU policy priorities relating to security, including disaster risk reduction and resilience and improving cross-sectoral aspects of crises in various sectors. The Secure Societies programme includes four thematic areas one of which is called "A Disaster-Resilient Society for Europe" (referred to as DRS in this chapter). Within this thematic area, there is an established culture of comprehensive collaboration at the international level, taking due account of the transnational dimension of different natural and man-made hazards and their drivers (such as climate change). Cooperation includes developing tools and technologies, sharing knowledge, experiences, expertise and mutual learning on disaster-risk management, with synergies sought with the Union Civil Protection Mechanism (UCPM). With an annual budget of ±30M€, the Secure Societies programme's DRS thematic area funds research projects related to societal resilience and disaster risk governance, technologies for first and second responders, and support to interoperability.

The development of multi-hazard disaster preparedness and response plans is required, together with strategies to ensure their implementation on national, regional and local levels, be it through regulation or incentives. Examples of solution-oriented research developed within the Disaster-Resilient Societies area (DRS) are manyfold, the following section gives some examples of current trends that are taken from the Work Programme of the Horizon Europe Secure Societies Programme.

3.2 Current trends

3.2.1 Societal resilience

Societal resilience-related research focuses on various types of areas, one of which is on improved understanding of risk exposure and its public awareness in areas exposed to multi-hazards. The awareness of multiple hazards (natural and man-made) and the understanding and assessment of risks and their consequences is a critical and fundamental step towards the development of local, national and international policies and strategies within all phases of the disaster risk management cycle, in particular preparedness. The availability of reliable scientific data and information (including historical occurrences and climate projections) to anticipate future disaster events or crisis situations, considering uncertainties inherent to natural systems characterisation, and effectively supporting decision-making processes at all levels represents a global challenge for both the research community and governance institutions. Actions at national/local and global/regional levels rely on knowledge of risks in all their dimension and changeable nature. A strengthened understanding of risks by the population (and decision-makers) is therefore needed, based on both records of past events and forecasts and projections (with quantified uncertainties) that reflect consideration of evolving trends and dynamics over time and space. This is particularly acute in the case of multi-hazard risks, that is, occurrences of several disasters either in cascade or at once. Moreover, the work needs to be complemented with improved knowledge on how risk awareness and actions are influenced and shaped by diverse aspects such as past events, cultures and traditions. The understanding of multiple disaster risks (and related awareness) relies on knowledge gained about historical data and information about past events and related lessons learned as well as the ability to forecast and assess future risks under uncertainty (including impacts of pandemics, as well as global change, including climate trends and earth system and environment dynamics). These complex interactions between human decisions and multiple hazards require novel risk assessment approaches such as agent-based modelling and systems dynamics methods. This will result in improved preparedness actions built upon these analyses (e.g. defining evacuation routes, responsiveness of health services, etc.). Social media also plays a role in disaster analytics. For example, an increasing number of location-based social network services can provide time-stamped, geo-located data that opens new opportunities and solutions to a wide range of challenges by analysing the extracted public behaviour responses from social media before, during and after disaster events. When using social media data, the design for data collection and analysis has to respect fundamental rights, privacy and data protection and analyses have to take related societal effects in online and offline environments into account as well as possible disinformation and fake news. Also, risk awareness, understanding and preparedness are unequally distributed along a wide range of variables (socio-economic, cultural, regional, etc.) that may generate drawbacks and conflicting issues with respect to groups' vulnerability.

Improving societal resilience to disasters or crises relies on various features related to the preparedness of citizens, communities, education systems, public administrations, business companies and practitioners. These concern, in particular, ways to react and informed decisions to take in case of an event. Individual, public and multilevel actions are needed in disaster risk management and they have huge implications on potentially reducing losses and increasing the operational capacity of responders, along with significant impacts on the emergency planning and management phases and its relation to continuous operations and existing safety management. In particular, the level of awareness of EU citizens of the risks in their region is an indicator for measuring progress in increasing public awareness and preparedness for disasters and in the implementation of the Union Civil Protection Mechanism legislation. Besides the required risk understanding, research is needed in several domains. With regard to public administrations, it is relevant to conceptualise how to increase risk awareness by building processes capable of fostering a long-lasting coalition with citizens around the objective of reducing vulnerability. This implies the definition of action protocols and models of responsibility that mobilise the intervention of individual employees of public administrations. With regard to business companies and practitioners, it is relevant to integrate their emergency activities in the local context. With regard to citizens and communities, it is necessary to design preparedness actions enabling the empowerment of citizens (including particularly vulnerable groups), their communities and NGOs through bottom-up participatory and learning processes. This includes school/university curricula and professional training and trust building among local actors, integrating relevant traditional knowledge, incorporating a gender perspective where relevant, best practices, guidelines and possible changes of regulations, to allow participatory actions. Difficulties in communication to the public in preparedness (and response) phases require the consideration of legal aspects, along with investigations into innovative practices, forms and tools that will enable the more effective sharing of information, taking into account possible risks of disinformation and fake news. These will support citizens in acting efficiently by themselves, through enhanced collaboration and communication and improving information exchanges between local authorities (including first and second responders), vulnerable populations (e.g. socio-economic groups, ethnic groups, people with illnesses or disabilities, children, elderly, hospital patients), and the private sector. Moreover, recent crises have shown that there is a large sense of solidarity among the population during a disaster or crisis situation. Many citizens who were not involved in disaster relief organisations before the crisis want to offer support to their fellow citizens and the broader community in times of crises. These initiatives of "spontaneous volunteers" are however most efficient if they are informed and trained and if their valuable contributions are coordinated with the authorities and first and second responders on the ground. Preparedness plans, tests and continued adaption on how best to manage spontaneous volunteers and integrate those into the response are needed.

3.2.2 Tools for integrated disaster risk reduction of natural hazards

In contemporary society, the capacity of communities and governments to manage expected and/or unexpected extreme climate events depends heavily on effective governance throughout the entire Disaster Risk Management cycle. This covers operational mechanisms ranging from short-term actions (e.g. early warning and forecast-based actions) to long-term adaptation strategies and resilience building, including nature-based solutions. A coherent integration between Disaster Risk Reduction, Climate Adaptation policies and Sustainable Development Goals as fostered by the European Green Deal and major UN initiatives should result in a comprehensive resilience framework while improving synergies and coherence among the institutions and international agencies involved. The effective implementation of global and European risk governance and policies to enable integrated disaster risk reduction for extreme climate events requires a collaborative involvement in risk assessment and information sharing across involved institutions, including the civil and private sectors and the population. Cross-regional, cross-border and cross-sector agreements covering all phases of Disaster Risk Management can improve the knowledge about extreme climate events, such as forest fires, droughts, floods, heatwaves, storms and storm surges. In addition, improving effective prevention, preparedness and response rely upon specific national or local expertise and experience. It is important to overcome silos between technical and political authorities at all levels and advocate integration among involved actors. Multi-risk governance frameworks related to climate extremes, shifting from single to multi-risk thinking in governmental agencies, represent the key challenge for the future, considering how measures to improve the resilience of the built environment and communities may provide effective solutions to strengthen adaptation measures. Creating an overview of existing knowledge, integrating tools and developing new ones for resilience and emergency management should include careful planning for interoperability among many actors. It is important that solutions pay attention to societal side-effects of integrating data about emergencies, for instance, Apps, where persons concerned tend to share more willingly, but do not reflect the consequences of that. Thus, the development of data management tools for emergencies needs to respect fundamental rights, data protection and avoid function creep.

Another axis of research looks at enhanced assessment of disaster risks, adaptive capabilities and scenario building based on available historical data and projections. The assessment of disaster risks requires different types of actions ranging from soft measures to technologies. Simulation-based risk and impact assessments represent an effective approach to make science understandable to decision-makers and streamline national to local mitigation/adaptation actions. This is especially the case if they are integrated with evaluation tools for cost–benefit/effectiveness and multi-criteria analyses, data-farming experiments, serious games, and are tailored to meet end-user's needs, to assess the effectiveness of alternative options in different phases of the Disaster Risk Management cycle. Specific risk assessments should be decision- or demand-driven and informed by scientific evidence, and there is a clear need to translate the results to ensure they are relevant, usable, legitimate and credible from the perspectives of the users. Co-design, co-development, co-dissemination and co-evaluation engaging the intended end users represent in this sense key features of improved risk, resilience and impact assessments. In the first place, the acquisition of data is an essential feature and this requires innovative solutions for faster risk assessment and reduction. This includes the identification of precursors for different types of threats, supporting the design or improvement of risk-targeted monitoring programmes. In addition, risk assessments themselves are primarily designed to predict the likelihood of a specific event, whereas what is of primary concern is the impact of that event on society, infrastructure, governance, etc. Numerous experiences gathered in the natural hazards area showed that an enhanced assessment of risks and scenario building may be improved by taking into account reliable data (both quantitative and qualitative) and historical occurrences, when available, including disaster loss data (studies of past events in particular low-probability/long-time recurrence events). This includes for example a higher completeness of the historical-geological records of volcanic eruptions, major earthquakes, tsunamis, etc. In the case of extreme climate events such as storms and related storm surges, or health crises (outbreaks, pandemics) the analysis should draw on the outputs of state-of-the-art climate projections, including by taking into account the uncertainties brought on by climate change and our state of knowledge of the key processes underpinning the functioning of the Earth system. In cases where there are not enough historical data and a high level of uncertainty, assessments and decision-making will have to rely on qualitative data. The action should take into account disaster loss databases and risk data repositories in Member States and relevant hubs.

4. International networking and community-building

Understanding and exploiting the existing linkages and synergies among policy initiatives such as the Paris Agreement [17], the EU strategy on adaptation to climate change [8], the EU Green Deal [18], the Sendai Framework [2] and the Sustainable Development Goals (SDGs) [19, 20] represents a global priority for future research and innovation actions in the field of natural hazards and man-made disasters. For the response side, international cooperation on research and innovation with key partners has the potential to identify common solutions and increase the relevance of outcomes. Enhanced cooperation and involvement of different sectors and actors are essential, including policy-makers, scientists from various disciplines, industry/Small and Medium Enterprises (SMEs), public administration (both at national and regional/local levels), credit/financial institutions, practitioners, Non-Governmental Organisations (NGOs) and Civil-Society Organisations (CSOs), notwithstanding the citizen dimension. Cross-regional, cross-border and cross-sector agreements covering all phases of Disaster Risk Management can improve the knowledge about natural hazards, such as forest fires, droughts, floods, heatwaves, storms, as well as geohazards [21]. In addition, improving effective prevention, preparedness and response rely upon specific national or local expertise and experience. Like in research, it is important to overcome silos between technical and political authorities at all levels and advocate integration among involved actors.

Research project outputs in various sectors covering natural hazards materialise in many different ways, such as relevant scientific findings, the maturation of promising technology areas, the operational validation of innovative concepts or the support to policy implementation. Over the years, a strong security research community, driven forward by highly committed stakeholders, has been consolidated at different levels (international, European and national). Policy-makers, practitioners, industry, scientists and citizens (civil society, municipalities, etc.) are the pillars on which sound research agendas are built in. The dialogue among different actors guarantees not only that research addresses real needs, but also that the investment in research will deliver tangible results. One example of a wide community-building aiming to facilitate interactions within the research community and users in various areas of security research is the Community of Users for Safe, Secure and Resilient Societies (CoU) launched by the European Commission in 2014. This informal platform enabled to gather policy-makers, scientists, practitioners, industry/SMEs and civil society organisations at international and regional levels, creating dialogues around research in various thematic areas, including natural hazards, and building "bridges" among different sectors (areas, disciplines and actors), which was an essential step forward. Dialogues and events had a clear effect on enhancing the participation of practitioners in research projects, in particular by promoting research results that are relevant to them, including the most promising tools that might have the potential to be taken up by them, and ensuring that their expertise is made available to policy-makers. Synergies were also stimulated between research and capacity-building projects. In light of the positive impacts of this community-building, the initiative has grown up since 2019 into a more ambitious network of networks in support of Horizon Europe: the so-called Community for European Research and Innovation for Security (CERIS). This forum has a large scope with various thematic areas, with the following objectives:

- *Raising awareness* on major updates in relevant policy sectors and on results achieved by related research and non-research initiatives, analyse impacts and provide policy recommendations.

- *Analysing identified capability* needs and gaps in the corresponding thematic areas (within Thematic Working Groups and other networks) and prioritisation of related research orientations based, at least, on criticality and urgency, in order to produce recommendations for a civil security research agenda.

DOI: http://dx.doi.org/10.5772/intechopen.1004640

- *Identifying solutions* available to address the gaps, differentiating state-of-the-art technologies (off-the-shelf and Development & Integration products) and security research trends. It will also take into account other considerations, such as technological maturity, operational relevance, societal acceptance, cost-effectiveness, etc.

- *Translating capability gaps and potential solutions into research needs* (including scenarios linking research needs to capabilities and societal appropriation, Technology Readiness Levels, development roadmaps, research action types, perspectives of research uptake, etc.) and getting feedback from practitioners about prioritisation of the needs, inputs to research programming and involvement in research activities.

- *Identifying funding opportunities and synergies* between different funding instruments and propose measures to facilitate them.

- *Identifying standardisation needs* through existing networks/platforms and prioritise them in close consultations with policy-makers and practitioners.

- *Integrating the views of citizens* so as to promote responsible research and innovation which respects ethical considerations and civil liberties.

5. Conclusions

Enhanced risks related to natural hazards and their impacts call for enhanced capacities in risk and resilience management and governance, including instruments for better prevention and preparedness, technologies for operational practitioners, and where relevant for citizens, and overall for societal resilience. The increasing severity and frequency of geohazards (earthquakes, volcanic eruptions, tsunamis, landslides), of extreme weather events (e.g. floods, heat and cold waves, storms, forest fires) combined with health-related crises such as the COVID-19 pandemic have demonstrated how societies have become more exposed and vulnerable to disaster risks as existing global inequalities often exacerbate both the exposure and vulnerability of communities, infrastructures, health systems, economies and nature. To enhance our resilience, research will obviously play a key role but the community work, gathering many different actors will need to be strengthened to guarantee that research outputs reach the users and find their way into innovation and market uptake pathways. This is certainly one of the greatest challenges ahead, which requires to work together and limit the fragmentation of sectors, disciplines, expertise, etc. as much as possible.

Regarding concrete impacts of research outputs on policy implementation, different situations may occur according to the type of regulations, either of regalian character (prone to Member States sovereignty) or linked to EU policies (binding character). In the first case, sectors concerned and related policy implementation are under the sole responsibility of National authorities, while the European Commission may provide support (e.g. through research funding) and facilitate coordination among Member States (e.g. through Advisory Boards, expert meetings, etc.); this is typically the case of security, civil protection and health sectors. In the second case,

EU policies are adopted by the European Council (hence by the EU Member States) and Parliament, and transposed into National laws; examples of this are water policies (described in Section 1.4) and other EU policies of binding character. We may wonder what are the (positive or negative) effects of these different policy settings regarding the implementation of research outputs?

In the first instance (sectors of regalian character), research is developed as in other fields with the difference that some deliverables may be of a classified nature (EU-restricted or -confidential) and research outputs may be disseminated to National authorities via respective Advisory Groups or Committees, thus possibly supporting national policies. The uptake of research is hence very much pending upon matching needs and decisions by the National authorities. The fact that there are no EU-binding rules may represent a drawback in that research outputs from EU-funded projects might not match directly national needs. On the positive side, many projects are working on tools/technologies that may directly support national capacities, whatever the absence of EU-binding character, and are interoperable, hence usable at the international level (e.g. technologies supporting first responders' operations in the case of a disaster). This is a direct effect of effective coordination on a voluntary basis, an example of which is the Union Civil Protection Mechanism. On the negative side, classified information, albeit shared with National authorities, will likely not have an interoperable character and respond to specific national needs with a less effective dissemination potential at the EU level (e.g. security-related methodologies). In this case, it is delicate to set metrics to check the effectiveness of research support to policy implementation because analyses must be done in consideration of each national policies and practices. This also applies to possible criteria comparing policy implementation costs versus research results.

In the second case (sectors prone to EU regulations), research is designed to respond to well-defined policy requirements that are embedded into identified EU policies. In other words, research outputs will support policy implementation needs that are common to all Member States. Under this circumstance, research uptake will be all the more facilitated that it will respond to national needs regarding the implementation of the (transposed) EU regulations. A typical case concerns the management of flood risks, the framework of which being set by the EU Flood Directive. While the national implementation is carried out at the river basin management level, common rules are applied at the EU level and research outcomes have thus more chances to find their way to concretely support policy implementation. In this case, metrics to check effectiveness of research support to policy implementation are easier to set because they deal with common policy rules at EU level and regular exchanges among research projects and policy implementers (in the case of water issues, from EU to river basin levels). Economic analyses to assess the effectiveness of research results on policy support are in such case facilitated because of the EU character.

Author note

Dr. Philippe Quevauviller is a research programming and policy officer at the Directorate General 'Migration and Home Affairs'. The views expressed in this paper are his own and do not formally represent the views of the European Commission.

Author details

Philippe Quevauviller
European Commission, DG Migration and Home Affairs, Brussels, Belgium

*Address all correspondence to: philippe.quevauviller@ec.europa.eu

References

[1] European Commission. Overview of natural and man-made disaster risks the European Union may face. Edition 2020. 2020. Available from: https://tinyurl.com/p2du9ktc

[2] UNDRR. Sendai Framework for Disaster Risk Reduction 2015-2030. 2015. p. 32.

[3] European Commission. Decision 1082/2013. 2013

[4] European Commission. Internal Security Strategy for the European Union: Towards a European Security Model, 5842/2/2010. 2010

[5] European Commission. The European Agenda on Security, COM(2015) 185 final. 2015

[6] European Commission. Art.5.1(a), council decision No. 1313/2013/EU. Official Journal of the European Union. 2013a;**L347**

[7] European Commission. Directive 2007/60/EC. 2007a

[8] European Commission. Directive 2012/18/EU. 2012

[9] European Commission. COM (2013) 216 Final and COM(2021) 82 Final. 2013b

[10] European Commission. Directive 2000/60/EC. 2000

[11] European Commission. COM(2007) 414 final. 2007b

[12] European Commission. EU Marine Strategy Framework Directive, 2008/56/EC. 2008

[13] European Commission. COM(2017) 610 final. 2017. Available from: https://tinyurl.com/49s6ywkt

[14] United Nations. SDG 17; Partnerships: Why They Matter. 2016. p. 2

[15] European Commission. Forging a climate-resilient Europe - the new EU Strategy on Adaptation to Climate Change. 2021a. p. 23

[16] Quevauviller P, editor. Water Systems Science and Policy Interfacing. Cambridge: The Royal Society of Chemistry; 2010. p. 430

[17] United Nations. Paris Agreement. 2015a. p. 27

[18] European Commission. The European Green Deal. 2019. 24

[19] United Nations. Transforming our world: The 2030 agenda for. Sustainable Development. 2015b;**35**

[20] United Nations and Department of Economic and Social Affairs. Goal 17: Strengthen the means of Implementation and Revitalize the Global Partnership for Sustainable Development. 2021

[21] European Commission. Horizon Europe Work Programme 2021-2022: 6. Civil Security for Society. 2021b. p. 214

Chapter 12

Understanding and Integrating Systemic Risk (SR) into Disaster Risk Reduction (DRR) and Risk Informed Development (RID) for Long-Term Resilient Development

Johanes A. Belle, Carolina Velásquez, Marcus Oxley and Ketevan Getiashvili

Abstract

In the late twentieth century, there was a remarkable paradigm shift from overreliance on disaster response to disaster risk reduction (DRR) (though practically this is not very evident), and from early twenty-first century, the international community made another paradigm shift from single hazard or multiple hazard analysis to evaluating risk from a systematic, cascading, interconnected, and compounding approach. Another pattern is emerging from pure DRR to risk-informed development (RID). RID is supported by many international protocols such as the Sendai Framework for DRR 2015–2030, SDGs or Agenda 2030, Paris Agreement on Climate Change, New Urban Agenda (NUA), Agenda for Humanity, and Addis Ababa Action Agenda. There is increasing evidence that disasters are increasing in frequency and intensity with exponential rise in economic, human, infrastructural, and environmental damages. There is also a growing trend in the dominance of climate-related disasters especially hydrometeorological disasters. Despite many global initiatives, we have also not succeeded in significantly reducing vulnerabilities and exposure of people, their assets, livelihoods, and the environment to growing natural and human-made hazards. Risk to these hazards is therefore growing globally but most especially among the poorest of the poor in developed and developing countries. There is therefore need for innovative approaches to analyzing risks and drafting strategic plans at national and local levels if we hope and intend to build resilient communities and systems. To address the identified lacunas as discussed above, this chapter focuses on expanding the understanding of systematic risks to reveal their interconnectedness and cascading and compounding effects and critically discuss the principles, practice, and relevance of RID alongside the enabling environment of RID in the context of the current and foreseeable changing global environment. Lastly, to use three case studies (Lesotho, Colombia, and Georgia) to highlight the importance of using SR, RID, DRR, and CCA as foundation blocks for resilience and sustainable development. The unique contribution of this chapter is that it integrates SR, RID, DRR, and CCA into resilience building to promote sustainable development.

Keywords: Climate Change Adaptation, disaster risk reduction, enabling environment framework, resilience, risk-informed development, sustainable development, systemic risk

1. Introduction

Risk-informed development (RID) is an approach to development that takes account of multiple threats and complex risks. RID can be achieved by mainstreaming both disaster and climate change risks and their management into everyday decision-making around development [1–3]. Risk is a normal and inseparable part of economic activities and development, which therefore entails proper planning and integration.

RID is supported by many international protocols such as the Sendai Framework for DRR 2015–2030, SDGs or Agenda 2030, Paris Agreement on Climate Change, New Urban Agenda (NUA), Agenda for Humanity, Addis Ababa Action agenda, and so forth (**Figure 1**). Regional frameworks in Africa like New Partnership for Africa Development (NEPAD), Southern Africa Development Community (SADC) DRR

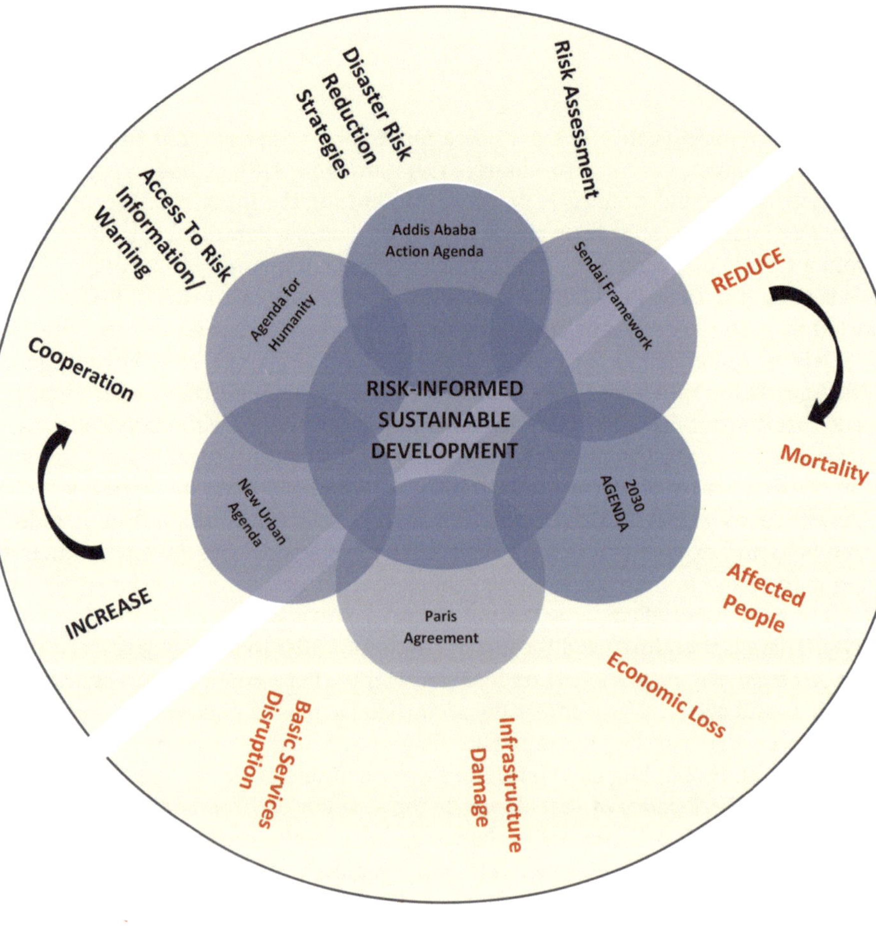

Figure 1.
Components of risk-informed sustainable development (Source: FSD & DCF [4]).

Unit, Africa Strategy for DRR, and so on all support RID [2, 4–7]. However, disaster risks are often not viewed as systemic with interlinkages, connectivity, and cascading and compounding effects. The local and international communities often turn to rely more on managing disasters instead of managing risks. The principles and practice of RID therefore aims to bridge the gap between managing disasters and managing risks.

It is no hidden secret that disasters are becoming more frequent, intense, and with greater impacts. For example, the decade from 2000 to 2019 recorded 7348 major disaster events that affected 4.2 billion people, claimed 1.23 million lives, and resulted in approximately USD 2.97 trillion in global economic losses [8]. Most of the disasters were climate-related with dire effects on human health, livelihoods, food security, water supply, human security, and economic growth [8–10]. Most affected were the world's poorest, whose vulnerability is rising due to socioeconomic and political factors, including urbanization and competition for scarce resources [2, 5, 8]. Besides climate risks, there are other compounding risks such as those linked to fragility, conflict, climate security, and epidemics/pandemics, like COVID-19. Risks are becoming systemic in nature with interlinkages and cascading and compounding effects [1–3]. The top three countries with the highest disaster risk worldwide are the Philippines (WRI 46.86), Indonesia (WRI 43.50), and India (WRI 41.52). The continent with the highest disaster risk is the Americas, followed by Asia, then Africa, Oceania, and Europe [7]. Africa has the highest vulnerability, and the top country with the highest risk index in Africa is Mozambique followed by Somalia [7]. The WRI uses 100 indicators from scientifically proven sources that are comparable and reproducible [7]. The indexes are therefore reliable. Risk is normally a product of the hazard, exposure, and vulnerability, counteracted by the coping and adaptive capacities of the community or the available quality and quantity of the community's assets/community capitals.

In order to reduce community vulnerability and exposure to increasing hazards, a more proactive and sustainable approach is to ensure that all development initiatives are risk informed.

Disaster risks are not adequately taken into account in development planning and programming. Besides, the current international risk management approaches do not consider multiple, interconnected, interlined risks with compounding and cascading effects. Though DRR and DRM are globally preached, the reality is that international, regional, and even national agencies still focus and allocate more resources on managing disasters rather than managing risks. For every USD 10 spent on humanitarian aid, only USD 1 is spent on DRR. There is therefore a great need to integrate disaster risk considerations into development policies, planning, and management, including financial instruments. RID therefore aims to bridge the gap between disaster management and risk management in order to build resilience and promote sustainable development or the UN Agenda 2030 [3, 6, 11].

2. Systematic risks: interconnectedness, cascading, and compounding effects

Systemic risk (SR) and system thinking are interlinked and recommended in risk analysis. Disaster risk is a product of hazard (s), vulnerability, and exposure cautioned by the coping capacity of the community or society (**Figure 2**).

Systemic risk is the interaction between climate change (CC) and natural hazards with complex, interdependent, and interconnected networks of social, technological,

$$\text{Disaster Risk (R)} = \frac{\text{Hazard (H)} \times \text{Exposure (E)} \times \text{Vulnerability (V)}}{\text{Coping Capacity (C)}}$$

Figure 2.
The risk equation (Source: GIZ/GIDRM & RS [12]).

environmental, and economic systems. SR represents a paradigm shift from hazard-to-hazard risk assessment to assessing risk as a complex and interwoven system. The behavior of the network determines exposure and vulnerability at all scales. Systemic risks produce cascading and compounding impacts that spread within and across systems and sectors (e.g., ecosystems, health, and food) [6, 13]. It is therefore imperative to analyze risk not by studying single or multi-hazards but by equally looking at the interconnectedness and compounding and cascading impacts. Some researchers will call this chain impact analysis. For example, a climate risk like reduction in rainfall may end up in violent public protests. **Figure 3** below presents an example of cascading risk with chain impacts.

In order to fully understand the systemic behaviors of disaster risks with their cascading and interconnected impacts, it is vital that the researcher adopts a system thinking approach. The multidimensional nature of risk necessitates a *systems approach* that develops shared understanding across traditional boundaries and builds broad-based constituencies and coalitions for joint actions in alternative development pathways [12]. System thinking requires a whole-of-government or whole-of-society approach involving complementary activities across different dimensions of the enabling environment, including multi-level actions that strengthen linkages with local realities, with regional, national, and international frameworks and processes [12].

The current and emerging risk landscape undermining sustainable development is increasingly multidimensional, with high levels of interdependency, feedback loops, and uncertainty. Developing a shared understanding of systemic risk can increase

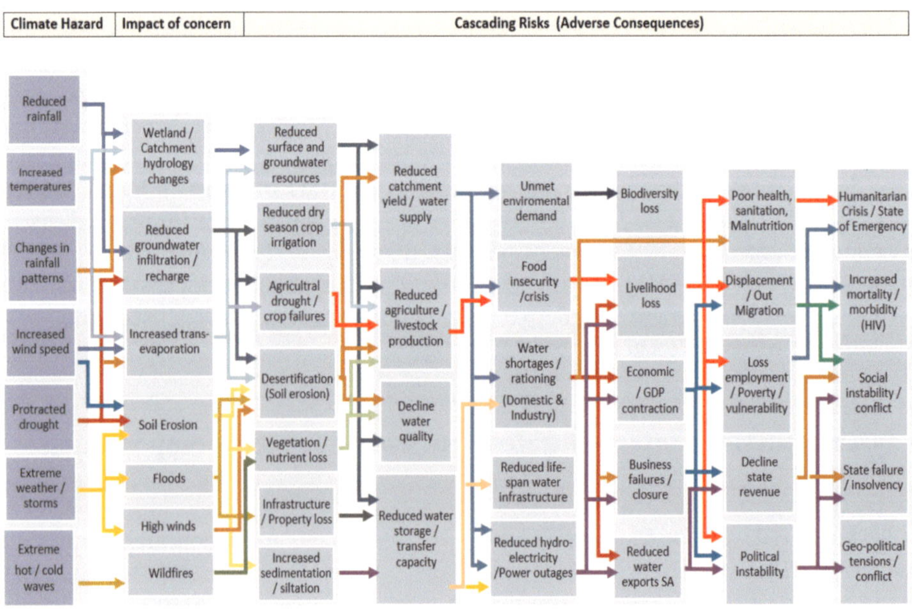

Figure 3.
SR with chain/cascading impacts (Source: GIZ/GIDRM & Resilience Solutions [12]).

stakeholder awareness of how the "whole" system behaves. This can support the identification of leveraged entry points and opportunities for joint actions that make optimum use of available resources across sectors and stakeholders.

In general, conventional risk assessment tools tend to focus on a single-risk approach, although new methodologies and enhanced tools are increasingly being developed to understand the interlinkages and identify the proximate and underlying causes of multidimensional risk. Understanding the multidimensional and systemic nature of risk is a prerequisite to developing holistic approaches that offer opportunities to both strengthening resilience and building sustainable development.

3. Principles, practice, and relevance of risk-informed development and its enabling environment

Increasingly, interconnected climate and disaster risks are on the rise, undermining progress in sustainable development. Reversing this trend requires enhanced risk-based decision processes to avoid creating risk through poor development choices and ensure that development is sustainable and resilient.

According to the UNDRR Global Assessment Report (GAR) 2022, the frequency, magnitude and impact of disaster and crisis is on the rise, with increasingly complex, interconnected risk cascading across geographies and sectors, pushing some countries toward existential limits (**Figure 4**).

The UNDRR GAR report [6] opined that by 2030, at the expiration of the Sendai Framework for Disaster Reduction (SFDRR), disasters will increase by 40%; droughts will double; extreme temperature events will triple; 340 million people will need humanitarian assistance, and ending poverty (a strong driver of disaster risk) will be elusive [3, 5, 6, 11, 12].

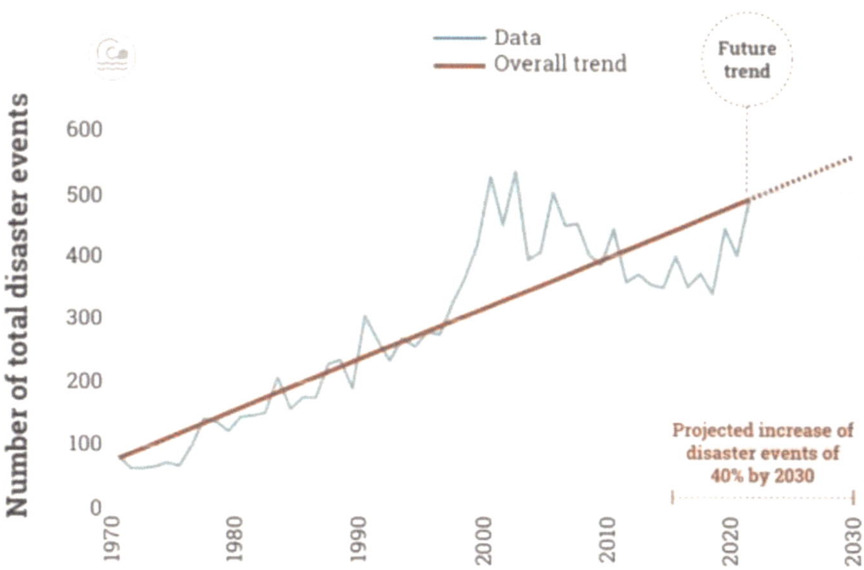

Figure 4.
UNDRR GAR 2022 Risk Trend (Source: UNDRR [6]).

Despite progress being made, the creation of risk is outstripping disaster risk reduction. The rapid accumulation of risk is not only a threat to development, but the development process itself can be a major driver of risk. Reversing this upward trend requires enhanced risk-informed approaches that are able to understand and manage both "*risks to*" and "*risk from*" development [12]. Risk-informed planning and decision-making is fundamental to preventing the creation of risk and a prerequisite for more resilient and sustainable development.

The core rationale for integrating risk reduction into development planning is based on the understanding that risk is a normal and inseparable part of development activity. Risk has both positive and negative connotations: negative risk is understood as the potential for loss and damage; while positive risk is the potential to realize benefits and opportunities caused by changes in the environment.

3.1 Enabling environment for risk-informed development (EE4RID)

The *enabling environment* for risk-informed development is defined as a complex web of policy, regulatory, organizational, procedural, and cultural conditions that can institutionalize risk within development policy, planning, and budgeting [1].

As a normal part of development activities, risk is determined by society's perceptions, needs, demands, decisions, behaviors, and practices, including the prevailing political economy, gender relations, and inequalities. In effect, the enabling environment forms the normative framework of an overall "risk governance" system required for the public sector to advance a risk-informed approach [12].

In general, complex societies comprise an intricate web of policy and planning instruments, institutional arrangements, stakeholders, and activities that are interconnected and can be mutually reinforcing, which make it relatively resistant to change. Accordingly, effecting change requires a whole-of-society, whole-of-government approach (from national to local level) whereby changes in institutional policy and legislation have to coevolve with changes in societal norms, values and individual behaviors (**Figure 5**) [12].

Figure 5.
Map of Lesotho showing the drainage system (Source: Adapted from DMA [14]; GoL [9]; LMS [10]; LVAC [15]).

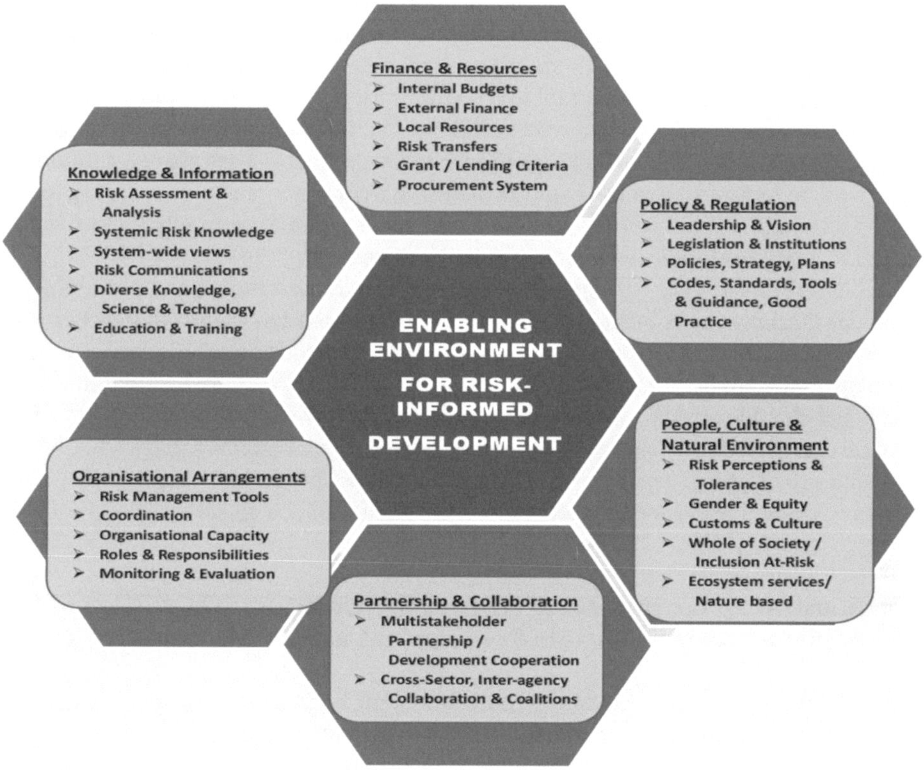

Figure 6.
EE for RID (Source: GIZ/GIDRM & RS [12]).

In the Republic of Colombia, the focus of the RID pilot project was to accompany the orientation process for the application of the "*Technical Guidance for Comprehensive Risk Assessment in the Context of Climate Change.*" With progress in sustainable development being undermined by increasing disasters, the guidance was developed by UNDRR and GIZ to provide a framework for the comprehensive assessment and management of multidimensional risk [16]. The developed EE4RID is presented in **Figure 6** and discussed in detail thereafter.

4. Understanding RID, DRR, and CCA nexus

Risk-informed development has its origins in the field of disaster risk reduction, although it remains a relatively new and still emerging concept. Within the UNDRR [17] definition of disaster risk reduction, RID aligns with "prospective" disaster risk reduction practices that seek to minimize the creation of new "negative" risk to a level the society considers acceptable. The primary reason the Sendai Framework will not achieve its expected outcome of a substantial reduction in disaster losses is because it is not effective in preventing or minimizing the creation of new risks. From a development perspective, RID involves understanding and capitalizing on "positive" risk opportunities that can strengthen resilience, improve development performance, and enhance competitive advantage.

DRR is about *prevent new risk, reducing existing* disaster risk, and *managing residual risk*, all of which contribute to strengthening resilience and therefore to the achievement of sustainable development [6]. DRR and CCA share common characteristics than differences, and both are precursors in building community resilience. DRR and CCA have common concerns in managing climate-related risks; they both share a common goal of reducing vulnerability and achieving sustainable development. They share a common conceptual understanding of the components of risk (product of exposure and vulnerability to hazards) and the processes of building resilience. DRR is often the first line of protection against weather- and climate-related disasters [5, 7, 14].

To build resilience, the RID approach also requires understanding and promoting socio-ecological processes including the use of nature-based solutions (NbS) as heightened in the Lesotho and Columbia case studies. Risk-informed development occurs within a specific landscape forming part of a socio-ecological system. Risk-based approaches should consider the dynamic interaction of development investments with the natural environment, particularly the functioning of ecosystems services that have important provisioning, regulating, and protection benefits that can influence risk factors of hazard, vulnerability, and exposure and build resilience [12, 15, 18].

5. Demonstrating the interconnectedness between SR, RID, DRR, and CCA for resilience and sustainable development using three case studies

A community that is not resilient to its recurrent hazards cannot achieve sustainable development. In the same token, any development projects within the community that are not resilient to recurrent hazards cannot be sustainable. Developing countries especially those in Africa generally have very low resilience to socio-natural disasters. In a study conducted by Khan et al. [19], using the IMF adopted index-based resilience that uses 62 indicators and focusing on the period between 1995 and 2019, the authors selected nine key indicators and examined 91 countries, 24 developed countries and 67 developing countries, to construct a disaster resilience index (DRI). The results showed that developed countries were more resilient to "natural disasters" with Switzerland at the top with a DRI of 0.787, followed by Germany and France with a DRI of 0.777 and 0.751, respectively. On the other hand, at the bottom of the ranking were Cameroon, Haiti, and Democratic Republic of Congo with DRI of 0.257, 0.249, and 0.227, respectively [19]. Though these results may not be conclusive, they however paint a general global picture on the distribution of risks.

The Global Initiative on Disaster Risk Reduction (GIDRM) commissioned Resilience Solutions (RS) to increase understanding and strengthen the enabling environment required for the successful application of risk-informed development (RID) planning and decision-making in different country contexts. The case studies that follow heighten the need to integrate SR, RID, DRR, and CCA in order to build community resilience and promote sustainable development.

5.1 Lesotho: "*Creating an Enabling environment for RID to build resilience in the water sector*"

5.1.1 Introduction

Lesotho like the rest of Africa is very vulnerable to the impacts of climate change. The hazard profile of Lesotho demonstrates many climate and environmental-related hazards, which then have cascading effect on the water, food, and energy nexus. Lesotho is heavily dependent on water resources, including rain-fed agriculture, and therefore very vulnerable to the impacts of climate change.

Notwithstanding the above, Lesotho is considered as the water tower of southern Africa where 40% of the Orange-Senqu basin water come from the mountain topography of Lesotho. The water, popularly known as the "white/liquid gold" of Lesotho, is then harnessed and sold to neighboring countries particularly South Africa through transboundary water transmission tunnels by the Lesotho Highland Water Project [9, 12, 20, 21]. The Orange-Senqu River basin also supplies water to neighboring countries like Namibia and Botswana.

Table 1 below presents important facts about Lesotho, while **Figure 1** shows the drainage system in Lesotho that provides the "white/liquid gold."

The Metolong Dam in the southern Puthasiana Catchment was used as a case study in this project, which was commissioned in 2020 by the GIZ-GIDRM to assess and identify any risk-informed development (RID) and the enabling environment (EE) related to river basin management and infrastructure systems to adapt to climate change and build local resilience [1, 12, 20].

Location	Completely surrounded by South Africa
Area	30,355 km^2
Administration	10 Administrative Districts (Butha-Buthe, Berea, Leribe, Mafeteng, Maseru, Mohale's Hoek, Mokhotlong, Qacha's Nek, Quthing, and Thaba Tseka) with Maseru as the capital city
Topography and ecological zones	Mountains (59%), foothills (15%), lowland (17%), and Senqu River Valley (9%)
Climate	Temperate with alpine influence (hot summers and very cold dry winters)
Annual rainfall	Varies (Senqu River valley with 500–1200 mm in north and eastern escarpment)
Annual evaporation rate	1400 mm in highlands to 1600 mm in lowlands
Drainage and water sector	Two main rivers, Senqu and Caledon (see **Figure 1**). LHDP and MDWSP main water projects
Population	2 million (2016)
Population growth rate	0.67% (2016). Retarded by the impacts of HIV/AIDS
Life expectancy	56 (2016)
Rural population	66% (1.3 million)
Adult literacy rate	86%
Unemployment rate	32.8%
Inflation rate	5.5% (2019)
Average GDP growth rate	1.6% (2018)
HIV/AIDS	23.6%
Major hazards	Drought, flood, extreme cold, snow, HIV/AIDS, COVID-19, climate change, environmental degradation
Main economic activities	Agriculture, textile industry, water sale (40% volume of water in the Orange Senqu Basin)

Table 1.
Lesotho overview facts.

5.1.2 Methodology

Based on existing normative frameworks, methodologies and practical resources developed by UN agencies, government, and nongovernmental organizations to support DRR and CCA mainstreaming, an initial conceptual framework for risk-informed development, were developed by GIDRM and RS [12]. The EE4RID framework was further elaborated and validated through a series of stakeholder consultations and direct experience gained from practical application in the RID pilot countries of Lesotho, Columbia, and Georgia. This initial Framework is flexible and can be expanded and modified (**Figure 6**).

In order to assess the enabling environment for RID to build resilience in the water sector in Lesotho, the following processes were followed:

- Identification of Disaster Risk Reduction (DRR) policies, climate change, strategies, and legal frameworks in Lesotho, which incorporate RID their relevance, existing gaps, and constrains in their implementation

- Undertook preliminary risk analysis to understand systemic and interconnected risks impacting Metolong Dam Water Supply Project (MDWSP) like catchment degradation/desertification, soil erosion (reduced surface water infiltration/groundwater recharge), dam reservoir sedimentation, reduced design life and storage, and loss of water supply resilience in extreme weather conditions induced by climate change.

- Conducted hazards and vulnerability assessment to identify key hazards and factors of vulnerability related to water sector in Lesotho.

- The EE baseline assessment involved a desk-based analysis of relevant country and sector policy and program documentation, together with virtual/face-to-face consultations interviews with relevant specialist-stakeholders.

- Identification of relevant stakeholders (Line Ministries, NGOs, and UN Agencies) relevant to water sector were mapped including their mandate, contact details, and identified focal person in that regard. This was done in order to have a database of relevant stakeholders on Water sector and the Metolong Dam [12].

Based on the components of the developed EE framework for RID, a baseline assessment on the status of RID in Lesotho was undertaken, with a particular focus on the water sector and river basin management, that is, MDWSP.

5.1.3 Results and findings

These findings are based on the survey that was conducted by the RID team in Lesotho involving identified stakeholders in the water sector; consultation with some key informants, review of documents including policies and plans, as well as field observation by the RID team.

5.1.3.1 Policy and regulation

- Lack of RID policy imperative and political ownership at strategic and operational levels.
- Strategic and operational plans do not take into account the impact of systemic risk and RID. The National Strategic Development plan II (NSDP II) while trying to map out a growth path for Lesotho and creating job and alleviating poverty, little or no considerations were placed on RID and the systemic/cascading effects of risks.
- Many water-related policies exist some with overlapping mandates. The implementation of these policies is weak, and there is little accountability mechanisms built into these policies.

5.1.3.2 Finance and resources

- Disaster management budgets primary focus on response and recovery, which are post-disaster activities.
- Investment criteria do not take systemic risk into account.
- Lack of coordination of external funding sometimes creates overlaps and gaps in executing RID.
- Lack of systemic risk analytical tools, standards, guidance, and best practices for RID approach.

5.1.3.3 Knowledge and information

- Understanding of systemic risk, RID, and EE was at different levels, and some stakeholders were still not able to link RID and EE in their daily organizational operations. This is an indication that the country needs trainings on systemic risk, RID, and EE as well as practical application of the concepts.
- RID remains a relatively new concept; many respondents were unsure of the differences between DRR activities and RID approaches. Some respondents had strong feelings that effective and efficient coordination of disaster-related activities is the main priority to address systemic risk.
- In general, stakeholders were able to identify specific threats and risks in Lesotho, although these were not seen from a systemic risk lens. Any future capacity-building initiative should fully explain the concepts of RID, EE, and systemic risk, with consideration as to how the RID approach can be incorporated into the NSDP III for a safer, more sustainable development in Lesotho [9, 15].
- Some respondents were able to identify the key actors and some limiting factors in the implementation of RID and how politics inhibits effective implementation of RID. This implies that the training of politicians should be considered for their

support and buy-in during the development and implementation of policies and legislations that support RID.

- Limited access to relevant risk information was observed. Information sitting at different departments are not easily disseminated to all the relevant stakeholders. Disaster management plans are not well communicated to all the users especially the most at risk communities.

5.1.3.4 Culture and people

- Lack of inclusion of at-risk people or whole of society approach within the development planning processes was observed. It was not also clear how the youth were engaged in the development planning processes.

- Gender issues still persist with limited empowerment of women in owning assets. Some legislations still consider women, even married women, as minors, and this may limit their ability to acquire assets, for example, obtaining bank loans without the approval of their husband. Also, only boys are given the cultural right to inheritance much to the disadvantage of the girl-child [15].

5.1.3.5 Partnership and collaboration

- Limited coordination and collaboration with other external development partners in Lesotho was identified. Most investments are tailored to address the current needs of the country without internalizing the possible negative impacts of any development that is not risk informed.

- GIZ has many good projects in Lesotho. It is still not clear whether these projects are integrating the principles and practice of RID in their activities. Another finding is that enough synergy is not built among these projects; there is still evidence of silo operations. For example, it was not clear how the RID team, the Public Infrastructure Engineering Vulnerability Committee (PIEVC) team, and Renoka project complement each other in the country [12].

5.1.3.6 Organizational arrangements

- There was lack of clarity on which government departments should lead the RID and EE approach. Though the Disaster management Authority (DMA) already has a legislative mandate without much authority to coordinate other departments during emergencies, other structures like the National Task Force for drought relief exist above the authority of the DMA. Besides, the DMA has limited capacity to monitor the implementation of RID in Lesotho.

- The need for coordination or working together beyond traditional boundaries was strongly emphasized for effective application of RID and EE in Lesotho. During PIEVC workshop, the stakeholders reiterated that collaboration and partnerships in risk assessments are key in combating the effects of climate change on the water sector. The ongoing activities supported by GIZ through Renoka and Integrated Catchment Management can be used as an entry point to advocate for the inclusion of RID and its enabling environment.

Based on the tasks and objectives for the RID project in Lesotho, which was executed by RS, the following were observed:

- RID project highlighted the relevance of an RID approach in Lesotho.

- Preliminary "national-level" risk assessment focused on water sector identified interconnected nature of risk, cascading across social, economic, and environmental capacities—informed by Lesotho disaster losses data—although Lesotho has no comprehensive disaster loss and damage database, but some relevant data is available

- Metolong Dam and Water Supply system is a Critical National Infrastructure (CNI) asset in Lesotho. However, the full risk and impact of climate change on the Metolong Dam and Water Supply Project (MDWSP) is not fully known.

- The understanding of systemic risks was used to undertake a risk analysis of Lesotho National Strategic Development Plan (NSDP) II (2018–2022). Although a new NSDP III plan is due for drafting, the current 2018–2022 plan has been extended for the next 5 years because the plan was not fully implemented due to COVID-19. The analysis emphasized the adverse impact of systemic risks manifested at operational/sectoral level as well as at the national/strategic level. Unless addressed, increasing systemic risk will pose an existential threat to the achievement of Lesotho's socioeconomic and development objectives, as currently defined within the NSDP II.

- Minimizing the creation of systemic risk will require coherent actions at both the strategic (NSDP) and operational (sectoral) levels where the ICM could provide a good entry point for risk-informed development [10, 12, 14, 20].

5.2 Colombia: "*Technical Guidance for Comprehensive Risk Assessment in the Context of Climate Change*"

5.2.1 Introduction

Although risk can be considered a normal and inseparable part of economic activities and development, the rapid accumulation of disaster and climate risk (and associated loss and damages) is largely rooted in inappropriate or "flawed" development. To reduce risk to an acceptable level, there is a growing need to better understand and address risk drivers within development planning, plans, and decision-making processes. This will require strengthening risk governance capabilities to address both "risks to" and "risks from" development. In the Republic of Colombia, the focus of the pilot project was to accompany the orientation process for the application of the "Technical Guidance for Comprehensive Risk Assessment in the Context of Climate Change." The framework below was used (**Figure 7**).

5.2.2 Methodology

The RID process methodology consisted of several core steps as indicated below (**Figure 8**):

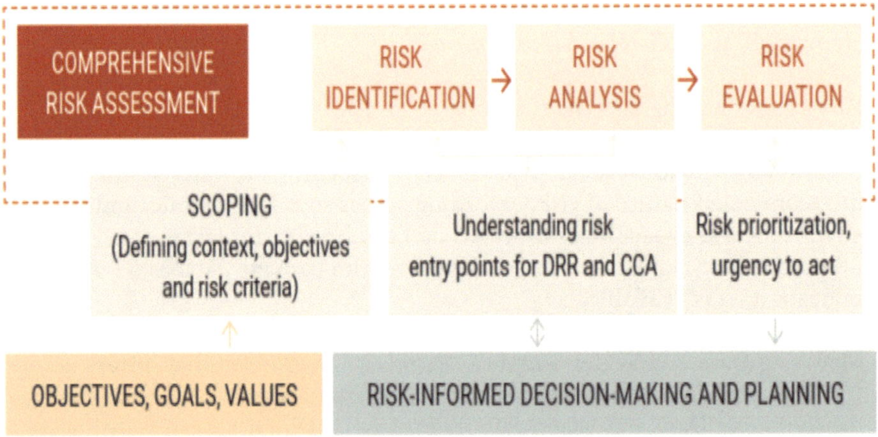

Figure 7.
Comprehensive risk assessment framework [12].

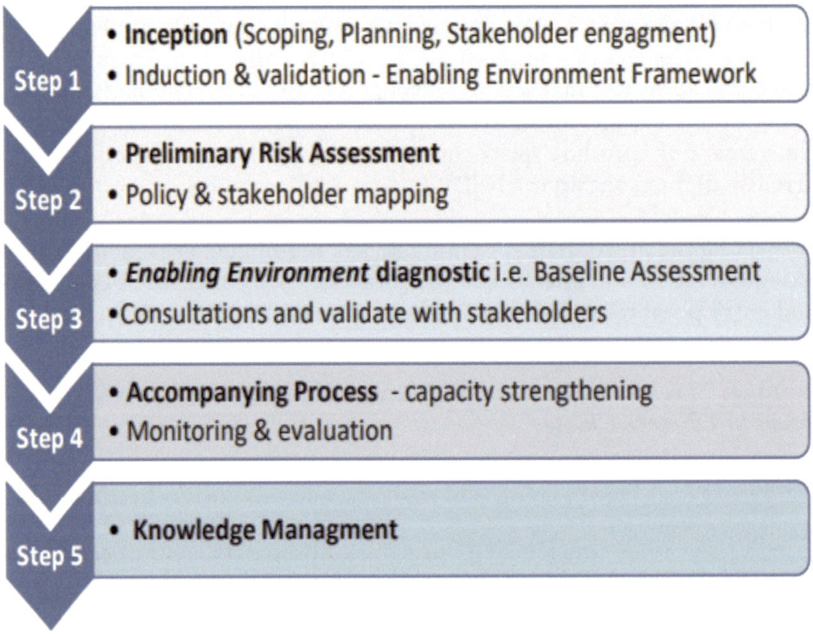

Figure 8.
The RID process methodology [12].

In the Colombia context, four main tools were selected for consideration for the RID:

i. National Development Plan 2018–2022;

ii. National Plan Climate Change Adaptation;

iii. National Plan Disaster Risk Management;

iv. A series of regional workshops organized by GIZ-Fondo Accion centered on "Tools and Challenges for Comprehensive Risk Management and Planning in a Changing Climate" [12, 22].

The main identified hazards in Columbia are indicated below (**Table 2**).

It is also worth noting that Colombia ranks among the top 10 countries with the highest number of forced displacement due to armed conflict.

5.2.3 Findings

The adoption of a risk-informed development approach has not been achieved to date in Columbia.

5.2.3.1 Policies and regulation

The strategic development paradigm in Columbia is still primarily driven by short-term need to promote rapid economic growth that places excessive strain on ecosystem protection and its regulating services. This exacerbates social inequalities, particularly for indigenous and agrarian-based people who are dependent on the declining natural resources.

5.2.3.2 Finance and resources

There were limited financial resources for disaster risk management, particularly at the local government level. Available funding is primarily focused on response and recovery triggered by high-severity events, although the cumulative impact of smaller

1. Hydro-meteorological • Floods • Storms • Strong winds	**5. Biological** • Coronavirus • Dengue/Zika/Chikunguna • Yellow fiver • Animal diseases and pest
2. Climatological • Wildfires • Droughts • Climate change—Extreme weather	**6. Environmental** • Deforestation/ecosystem degradation • Biodiversity loss • Desertification/soil erosion • Invasive species
3. Geohazard • Mass movements—Landslides/mud flows • Earthquake • Volcanic	**7. Chemical** • Industrial spillage • Pollution
4. Technological • Fire • Industrial accidents • Transport incidents • Cyberattack/data breach	**8. Conflict and instability** • Armed conflict/insecurity • Involuntary/forced displacement • Terrorist attacks • Organized crime—Illicit drug trafficking

Colombia Hazard/Threat Profile—Main Recurrent Hazards in Red.

Table 2.
Columbia risk profile.

scale events is bigger than that of larger, less frequent events, with greater relative impact on low-income households.

5.2.3.3 Organizational arrangements

There were fragmented and outdated sectoral and territorial policies and plans that are onerous to implement, siloed in disciplines and sectors, with a lack of coherent working relationship among disaster risk reduction, climate change, and development actors.

5.2.3.4 Knowledge and information

Another identified gap was that there was limited understanding outside of disaster management actors, of the relationship between development, disasters and risk, and the need for risk-informed approaches.

5.2.3.5 People, culture, and environment

There are underlying issues of poverty, gender and social inequalities, as well as exclusion, that marginalize at-risk people and result in grievances and conflict.

5.2.3.6 Partnership and collaboration

Though there are many role players identified for DRR and CCA, they work in silos. International donors and development agencies have also not fully embraced the concept and application of RID in their projects and activities.

Informed by discussions with relevant stakeholders, a broad range of entry points were identified across all six dimensions of the enabling environment that could support the integration of risk considerations into territorial and sectoral development planning processes. It was apparent the orientation process for the comprehensive risk assessment provided a strategic opportunity to increase understanding of the multidimensional risk and its relationship with development. The guidance applies a conventional risk assessment methodology with its main phases of risk identification, analysis, and evaluation. Results of the assessment are subsequently used for risk-informed decision-making and development planning, with a focus on integrating risk reduction measures.

Colombia is the second most biodiverse country in the world, hosting 10% of the planet's biodiversity. In line with principles of nature-based solutions and putting risk to human and ecological systems at the center, the strengthening of Colombia's rich and complex ecosystems could contribute to net biodiversity gains and support climate mitigation and adaptation, reduce the impacts of climate extreme events while contributing to the country's overall competitive advantage and transitioning to a greener more resilient economy [22].

Accordingly, a risk-informed development approach framed within socio-ecological systems could consider risk not only to avoid adverse consequences, but also as an potential to bring benefits and opportunities created by external changes.

An understanding of negative and positive risk can support multidisciplinary efforts to strengthen resilience and address both "risk to" and "risk from" development with multiple co-benefits.

The EE4RID tool can be locally contextualized to a range of development planning process and applications [12]. This includes mapping the current and future status of risk governance and identifying gaps and constraints and potential entry points to accelerate the adoption of a risk-informed approach. In doing so, the process highlighted aspects of the EE4RID dimensions "sub-categories" that needed to be strengthened, particularly in relation to the natural environment, local knowledge, and a whole-of-society approach [12].

The final stage of a comprehensive risk assessment is the integration of risk information drawn from the assessment into existing or new planning instruments.

However, RID is less about assessing risk to protect development but more about understanding how development choices can create risks that compound and cascade across geographies and sectors.

Given the high degree of complexity and uncertainty in assessing multidimensional risks, an alternative or complementary approach would be to develop an operational framework for strengthening resilience to sustain the functionality and/or critical services of a system or entity in the contexts of multiple (foreseen and unforeseen) risks and threats. The major variable in this approach is the critical functioning of the system.

The starting point would be to identify the existing capabilities, functionalities, operating parameters, and sources of resilience that are already present and influence the ability of the system to absorb, recover, and adapt to shocks and stresses.

Observing how systems perform when subjected to extreme conditions can reveal relative strengths and weaknesses, underlying risk drivers, critical interdependencies, and limits and boundaries that not readily apparent in "normal" operating conditions. Informed by these insights, the aim would be to build on existing capacities and improve the performance of the system and desired functions to minimize the creation of negative risk and capitalize on positive risk, including making optimal use of opportunities created by changes in the environment [12].

One of the key characteristics of a complex system is that the ability to perform one function is closely related to its ability to perform another.

As risk creation continues to outstrip progress in risk reduction, the development process itself can be a major driver of risk, whether through locating people and assets in exposed areas, rapid and unplanned urbanization, overexploitation of natural resources and ecosystems, social inequalities, poor governance, ignorance, greed, or misunderstanding.

From an RID perspective, the comprehensive risk assessment methodology as reviewed in the Colombian context puts development at the end rather than center stage of the risk process and underplays the importance of existing capacities and sources of resilience as the starting point to make development more sustainable. Transitioning to a more resilient risk-informed development pathway will require technical guidance for "next generation" assessment tools that considers risk as endogenous to development processes, requiring a stronger focus on analyzing and preventing "risk from" rather than "risk to" development [12].

5.3 Georgia: "*Development of landslide monitoring systems (piloted in selected communities), in partnership with the national environment agency and academia*"

5.3.1 Introduction

Despite an increasing understanding of the complexity of risks, these are not always adequately considered in development planning and programming, in regions and countries or by international donors. Against this background, the third phase of GIDRM was introduced toward strengthening RID, that is, an understanding of development that takes into account multifaceted, dynamic, interdependent, transnational, simultaneous, and systemic risks. GIZ Georgia program joined GIDRM Project at Stage III, with the technical support of Resilience Solutions. The aim was to better understand the enabling environment required for successful application of RID approaches. The specific project and sectoral focus for Georgia was the "*Development of landslide monitoring systems (piloted in selected communities), in partnership with the national environment agency and academia*" [12].

5.3.2 Methodology

An action research methodology using an innovative accompanying process evaluation approach was used. This involved Resilience Solutions consultants accompanying GIZ/GIDRM project staff and partners responsible for the planning and implementation of GIZ-supported projects. The project process included (**Table 3**):

Risk informed development context (landslide)-relevant policies	
• Concept for National Security	-Georgia Atlas on Natural Hazards and Risk
• National Strategy for disaster Risk Reduction 2017–2020	-National Climate Action Plan 2021–2030
• National Plan of Action for Capacity Development in DRR	-Low Emission Development Strategy
• Law on Protecting the Population and Territory from Natural and Man-made emergency Situations (2007)	-Sustainable Energy Action Plan
• Agriculture and Rural Development plan 2012–2027	-Forest Code
• Fourth National Communication on Climate Change	-Spatial Development and Construction Code
Regional Development plans/Program 2018–2021	-Socio-economic Development Strategy 2020
• National Environmental Action Programme	-UN Development Assistance Framework
• Building Codes and Construction Permits-Technical and Construction Supervision Agency • Ministry of Economy and Sustainable Development	-Civic Safety Law 2014 -National Biodiversity and Action Plan 2020

Table 3.
List of documents studied.

- Engagement with Stakeholders
- Conducting Desk Research/Documentation review
- Conducting multi-stakeholder workshops and trainings
- Raising Awareness/sensitization of national, regional, local stakeholders
- Participating at various events for knowledge sharing
- Producing/documenting assignment process for knowledge sharing

Furthermore, 20+ stakeholders were identified, and some of them consulted during the landslide RID baseline development (**Table 4**) [12].

5.3.3 Georgia country-specific findings

Below are the main type of hazards identified in Georgia (**Table 5**).

These findings are based on the desk research and stakeholder's consultation that was conducted by the RID team in Georgia involving consultation with some key informants; review of official documents which included policies and strategic plans. The assessment is based on the EE framework developed GIDRM & RS [12].

5.3.3.1 Policy and regulation

Policy architecture is fragmented; it is mostly focused on preparedness and response; compliance is week; mandate for RID approaches is weak.

DRR-related policies are missing the element of RID and still do not perceive risk through the lens of systemic risk. Several bylaws/enforcement mechanisms are still in

Risk-informed development context (landslides)—Key stakeholders	
1. Ministry of Environmental Protection and Agriculture (MEPA); Environment and Climate Department	11. Water Management Institute
2. Environmental Information and Education Centre	12. Ministry of Finance and Planning
3. National Environmental Agency	13. National Forestry Agency
4. Ministry of Regional Development & Infrastructure, including State Registry/Cadastre	14. National Statistics of Georgia (GEOSTAT)
5. State Security and Crisis Management Council	15. Disaster Prevention and Planning Division (EMA)
6. Emergency Management Agency (EMA)	16. Local/Municipal Government
7. Natural Disaster Prevention and Rapid Response Unit	17. Secretariat-Expert Advisory Council (EMA)
8. Ministry of Economy and Sustainable Development	18. Rural Development for Future Georgia (RDFG)
9. Department of Hydrometeorology	19. Geogia Red Cross Society
10. Department of Geology	20. Disaster Management Team—UN Inter agency

Table 4.
List of stakeholders.

Main types of hazards/threats in the region	
1. Weather/climate-related I. Flooding II. Droughts III. Strong winds IV. Rain/snow/hail storms and avalanches V. Extreme hot/cold temperatures	**2. Geophysical** I. Earthquakes (Seismic active region) II. Landslides III. Mudflows/Debris flows IV. Rockslides
3. Violent conflict I. Insecurity II. Forced displacements	**4. Environmental** I. Ecosystem collapse II. Desertification/deforestation
5. Technological hazards i. Industrial accidents ii. Dam/infrastructure failures iii. Cyberattacks	**6. Biological** i. Endemic diseases

Table 5.
Georgia hazard profile.

preparation; building codes, regulations, and standards were updated, but enforcement is weak. Strategic and operational plans do not consider systemic risks.

5.3.3.2 Finance and resources

Government budgets and incentives are limited and insufficient. The central budget is mostly spent on response and recovery. Technical/human resources and implementation mechanisms are inadequate. Public procurement and tenders are not risk informed.

Investment criteria do not take systemic risk into account, and there is lack of systemic risk analytical tools, standards, guidance, and best practices for RID approach.

5.3.3.3 Knowledge and information

There is limited understanding and training on RID concepts. RID remains a relatively new concept; many respondents were unsure of the differences between DRR, CCA activities, and RID approaches. Some respondents had strong feelings that effective and efficient coordination of disaster-related activities is the main priority to address systemic risk.

A National Risk atlas providing risk information has been produced although the atlas underutilized.

In general, there is lack of risk analysis tools and lack of good practice examples.

5.3.3.4 Organizational arrangements

According to National Response Plan of Georgia [23], EMS is responsible for DRR. Though the EMS already has a legislative mandate, it has limited authority and capacity to coordinate other ministries during emergencies. Other structures like the National Security Council exist, which are above the authority of the EMS.

Generally, there is limited local coordination, implementation, capacities and competencies, accountability, and enforcement on risk management within the different organizations; a lack of RID standards/guidance and tools for systemic risk assessments is common.

The institution/agency/department to lead/coordinate and advance on RID is unclear.

5.3.3.5 People and culture

There is little public awareness and little inclusion of risk bearers (at-risk people) in development planning. Participation of the local population/community members and inclusion in the RID planning process does not happen or is very weak with no public demand for more risk-informed approaches. There is no risk awareness campaign for the population living in the risk zones or on inappropriate livelihood/agricultural practices. Disasters magnify existing social inequalities and further disadvantage those who are already more vulnerable.

5.3.3.6 Partnership and collaboration

A lack of coordination of external development cooperation funding creates overlaps and gaps and missing opportunities for collaboration in executing RID approaches.

Fragmentation across disciplines and sectors needs interdisciplinary cross-sector collaboration.

5.4 Recommendations

Most of the recommendation discussed here also applies to Lesotho and Columbia. They are discussed under the Georgia case study to avoid duplication.

5.4.1 Policy and regulation

Institutional and policy architecture in Georgia is complex, even when focused on a particular typology of hazards (i.e., landslides). Therefore, the solutions must be holistic, collaborative across traditional boundaries.

Relevant policy and regulatory frameworks should be strengthened to explicitly support a risk-informed approach to development, including defining the roles and responsibilities of lead organizations at national/local levels. Finally, strategies should be based on long-term visions with realistic timeframes for meaningful change in development planning.

5.4.2 Finance and resources

National government needs to provide incentive structures to national and local actors to integrate consideration of systemic risks within investment decision-making/development planning processes.

Development cooperation and international financial mechanisms have an important role of model risk-informed approaches in high-risk sectors/locations, that is, exemplar of RID good practices.

More effort should be placed into engagement with private sector actors supported by greater risk analysis of development and investment decisions.

5.4.3 Knowledge and information

Key entry point for advancing RID is raising awareness of RID and EE concepts and understanding of systemic risk among relevant stakeholders. Training/technical guidance is needed on risk assessments, including the application of the existing risk maps/atlases at regional/local levels.

The role of higher education/research institutions could be better utilized to develop RID tool, standards, and knowledge products and support education and knowledge management to share risk information and understanding of RID approaches. There is a need to engage young people in order to build knowledge and skills.

5.4.4 Organizational arrangement

There is a need to strengthen the risk governance and RID implementation capabilities within Georgia. Greater clarity of lead agency roles and responsibilities is required, including cross-sector planning, coordination, and the application of procedures and tools in support of more integrated risk-informed development program. Going forward, mechanisms for monitoring and accountability for risk-informed development will need to be developed.

5.4.5 People and culture

A lack of inclusion of at-risk people or whole-of-society approach within the development planning processes was heightened. In general, an enabling environment requires a combination of top-down actions that provide political commitment and guidance for a risk-informed approach, which goes together with bottom-up actions that harness local knowledge and increase risk-understanding and risk governance capabilities of relevant actors and risk bearers at the national/regional/local levels, that is, closest to the point of implementation of territorial and/or sector-based development interventions [11, 12]. This will require participatory planning processes that are inclusive of at-risk people, particularly the most vulnerable groups to ensure no one is left behind [12, 13].

5.4.6 Partnership and collaboration

GIZ and other international donors/investors can model risk-informed projects in - high-risk areas that will help to build the evidence base to inform policy changes and strengthening RID mandate. Working collaboratively with other RID allies and development partners can be an effective way to share knowledge and good practices.

There is a need to look more at systems, not individual hazards, and to work across disciplines and sectors not in silos [12, 13].

6. General conclusions on the chapter

Disasters and development are two faces of the same coin. Development can either increase community vulnerability and exposure to disasters or reduce them. On the other hand, disasters can set back hard-earned development gains or provide windows of opportunity for development, hence the emphasis to build back better in disaster recovery. We need to reduce risk both to development and from development through risk-informed development (RID). For RID to hold water, it needs a well-oiled enabling environment. Six of those broad entry points were discussed in this chapter. While DRR is acknowledged as a good practice, this chapter lays more emphasis on building community resilience to multiple hazards, which then will support sustainable development. This chapter also brings to the fore that risks are complex, interconnected with compounding and cascading impacts. It then recommends a holistic risk assessment that not only looks at multi-hazards but assesses risk from a system-thinking lenses in DRM and RID strategic planning processes. The chapter presents three case studies to elucidate the importance of RID and its enabling environment supported by SR thinking to build resilience and support sustainable development. The approach sounds complex, but it is achievable and the way forward in building community resilience and ensuring sustainable development. The unique contribution of this chapter is the fact that it integrated SR, RID, DRR, CCA into a holistic framework to build resilience and promote sustainable development.

Acknowledgements

The authors wish to acknowledge the GIZ-GIDRM who commissioned the project on RID and developing an enabling environment for RID in three different countries to the Resilience Solution team and for granting permission to the authors to publish the results of the project. In the same vein, the authors wish to acknowledge the contributions of the local experts in the three countries of Lesotho, Columbia, and Georgia.

Author details

Johanes A. Belle[1*], Carolina Velásquez[2], Marcus Oxley[3] and Ketevan Getiashvili[4]

1 University of the Free State, Disaster Management Training and Education Centre for Africa (DiMTEC), Bloemfontein, South Africa

2 Department of Geography, Florida State University, USA

3 Resilience Solutions Associates, UK

4 National RID Expert, Georgia, USA

*Address all correspondence to: belleja@ufs.ac.za

References

[1] UNDP (United Nations Development Programme). RISK-INFORMED DEVELOPMENT: A Strategy Tool for Integrating Disaster Risk Reduction and Climate Change Adaptation into Development. New York: UNDP; 2020

[2] UNDP (United Nations Development Programme). Risk-Informed Development: From Crisis to Resilience. New York: UNDP; 2019

[3] UNDP (United Nations Development Programme). The UNDP Approach to Risk-Informed Development. Berlin: UNDP; 2022

[4] FSD & DCF (Financing for Sustainable Development and Development Cooperation Forum). Risk Informed Development Cooperation and Its Implication for Allocation and Use of Official Development Assistance (ODA): Lessons for the Decade of Action to Deliver SDGs-Full Study. 2021. Available from: https://www.un.org/development/desa/financing/sites/www.un.org.development.desa.financing/files/2021-03/2021ODA%20Full%20Study_final.pdf [Accessed: 09 October 2021]

[5] IPCC (Intergovernmental Pannel on Climate Change). Mitigation of Climate Change. Working Group III Contribution to the Sixth Assessment Report of the Intergovernmental Panel on Climate Change: Summary for Policymakers. Geneva: IPCC; 2022

[6] UNDRR (United Nations Office for Disaster Reduction). GAR 2022-Everything that You Can Find on Africa. 2022. Available from: https://www.undrr.org/publication/africa-regional-assessment-report-disaster-risk-reduction [Accessed: 07 December 2022]

[7] WRR (World Risk Report). The World Risk Report 2023. Berlin: Bündnis Entwicklung Hilft; 2023

[8] CRED-EMDAT (Centre for Research on the Epidemiology of Disasters). 2022 Disasters in Numbers. Brussels: CRED; 2022

[9] GoL (Government of Lesotho). Lesotho Economy. 2021. Available from: https://www.gov.ls/lesotho-economy/ [Accessed 20 December 2021]

[10] LMS (Lesotho Meteorological Services). Climate of Lesotho. 2023. Available from: https://www.lesmet.org.ls/home/open/Climate-of-Lesotho#:~:text=On%20the%20other%20hand%2C%20mean,%C2%B0C%20in%20the%20highlands [Accessed: 16 June 2022]

[11] UNDRR (United Nations Office for Disaster Reduction). Midterm Review of the Implementation of the Sendai Framework for Disaster Risk Reduction 2015-2030 in Sub-Saharan Africa. Geneva: UNDRR; 2023

[12] GIZ/GIDRM & RS (Deutsche Gesellschaft fur Internationale Zusammenarbeit/Global Initiative for Disaster Risk Management and Resilience Solution). Risk Informed Development Process Evaluation and Capacity Building Project Reports. Bonn: GIZ; 2023

[13] UNISDR (United Nations International Strategy for Disaster Reduction). Strategic Approach to Capacity Development for Implementation of the Sendai Framework for Disaster Risk Reduction. A Vision of Risk-Informed Sustainable Development by 2030. Geneva: UNDRR; 2018

[14] DMA (Disaster Management Authority). Lesotho Multi-Hazard Contingency Plan 2020–2023. Maseru: WB and GFDRR; 2020

[15] LVAC (Lesotho Vulnerability Assessment Committee). Annual Vulnerability Assessment Report. Maseru: DMA; 2021

[16] UNDRR & GIZ (United Nations Office for Disaster Reduction and Deutsche Gesellschaft fur Internationale Zusammenarbeit). Technical Guidance on Comprehensive Risk Assessment and Planning in the Context of Climate Change. Bonn: GIZ; 2021

[17] UNDRR. Sendai Framework for Disaster Risk Reduction 2015-2030. Sendai: United Nations Office for Disaster Risk Reduction; 2015

[18] IUCN (International Union for Conservation of Nature). IUCN Global Standard for Nature-Based Solutions. Gland: IUCN; 2020

[19] Khan MTI, Anwar S, Sarkodie SA, Yaseen MR, Nadeem AM, Ali Q. Comprehensive disaster resilience index: Pathway towards risk-informed sustainable development. Journal of Cleaner Production. 2022;**366**:132937

[20] CBL (Central Bank of Lesotho). Lesotho Economic Outlook 2020–2021. 2021. Available from: https://www.centralbank.org.ls/images/Publications/Research/Reports/Economic%20Outlook/LEO_UPDATES/LEO_Update_-_March_2021.pdf [Accessed: 20 December 2021]

[21] WB (World Bank). Water for Lesotho's Lowlands: Metolong Dam and Water Supply Program. 2021. Available from: https://www.worldbank.org/en/results/2020/10/21/water-for-lesothos-lowlands-metolong-dam-and-water-supply-program [Accessed: 21 November 2021]

[22] GIDRM (Global Initiative on Disaster Risk Management). Disaster Risks, Migration, and Forced Displacement: Understanding the Link Between Disaster Risk, Migration and Forced Displacement in Colombia. 2023. Available from: https://www.preventionweb.net/publication/disaster-risks-migration-and-forced-displacement-understanding-link-between-disaster?utm_source=PreventionWeb&utm_campaign=a9713ff245-PreventionWeb+daily&utm_medium=email&utm_term=0_b73053c1c6-a9713ff245-466466861 [Accessed: 27 November 2023]

[23] National Response Plan of Georgia. Preventionweb [Online]. 2008. Available from: https://www.preventionweb.net/files/globalplatform/ENGPridonSadunishvili.pdf [Accessed: 23 April 2023]